CANCER BIOLOGY. IV
Differentiation and Carcinogenesis

Senior Contributors

John R. Cann, Ph.D., Department of Biophysics and Genetics, University of Colorado Medical Center, Denver, Colo.

Albert Dorfman, M.D., Ph.D., Department of Pediatrics, University of Chicago, Chicago, Ill.

Gerald M. Edelman, M.D., Ph.D., Rockefeller University, New York, N.Y.

Alan Garen, Ph.D., Department of Molecular Biophysics and Biochemistry, Yale University, School of Medicine, New Haven, Conn.

Walter B. Goad, Ph.D., Theoretical Division, University of California, Los Alamos Scientific Laboratory, Los Alamos, N.M.

Leroy E. Hood, M.D., Ph.D., Division of Biology, California Institute of Technology, Pasadena, Calif.

Abraham Hsie, Ph.D., Biology Division, Oak Ridge National Laboratory, Oak Ridge, Tenn.

Edwin S. Lennox, Ph.D., The Salk Institute, San Diego, Calif.

Vincent T. Marchesi, M.D., Ph.D., Department of Pathology, Brady Memorial Laboratory, Yale University School of Medicine, New Haven, Conn.

Philip I. Marcus, Ph.D., Department of Microbiology, University of Connecticut, Storrs, Conn.

Beatrice Mintz, Ph.D., Institute for Cancer Research, Philadelphia, Pa.

Aaron A. Moscona, Ph.D., Department of Biology, University of Chicago School of Medicine, Chicago, Ill.

Bert W. O'Malley, M.D., Department of Cell Biology, Baylor College of Medicine, Houston, Texas

G. Barry Pierce, M.D., Department of Pathology, University of Colorado Medical Center, Denver, Colo.

Theodore T. Puck, Ph.D., Department of Biophysics and Genetics, University of Colorado Medical Center, Denver, Colo.

Arthur Robinson, M.D., Department of Biophysics and Genetics, University of Colorado Medical Center, Denver, Colo.

Leo Sachs, Ph.D., Department of Genetics, Weizmann Institute of Science, Rehovot, Israel

Gordon H. Sato, Ph.D., Department of Biology, University of California, San Diego, La Jolla, Calif.

For a listing of Workshop contributors, see page 234

ADVANCES IN PATHOBIOLOGY 6

CANCER BIOLOGY. IV
Differentiation and Carcinogenesis

Edited by

Carmia Borek, Ph.D.
Cecilia M. Fenoglio, M.D.
Donald West King, M.D.

College of Physicians and Surgeons of Columbia University
New York

Based on a series of lectures and laboratory workshops presented at the Given Institute of Pathobiology of the University of Colorado in Aspen, Colorado, August 1976

Stratton Intercontinental Medical Book Corp. / New York

Courses Sponsored by the Given Institute, 1977

July 5–9	Cell Control Mechanisms
July 5–August 5	Pathology Course
July 11–15	Nuclear Transplantation of Mammalian Cells
July 18–22	Education I—Genetics and Chromosomes
July 18–22	Toxicology Forum: The Effect of New Legislation
July 25–29	Immunoperoxidase and Immunofluorescence
August 1–5	Education II—Receptors
August 1–5	Advances in Cancer Biology I—Hepatic and Biliary Lipids
August 8–12	Association of HLA with Disease Susceptibility
August 15–19	Advances in Cancer Biology II—Cancer Virology
August 22–26	Molecular Cloning
August 29–Sept. 2	Education III—Immunology

PUBLISHED IN THIS SERIES TO DATE

1. Fenoglio CM, Borek C, King DW (Eds): Cell Membranes—Structure, Receptors, and Transport (1975)
2. Pascal RR, Silva F, King DW (Eds): Cancer Biology, I—Induction, Regulation, Immunology and Therapy (1976)
3. Fenoglio CM, Goodman R, King DW (Eds): Developmental Genetics (1976)
4. Fenoglio CM, King DW (Eds): Cancer Biology, II—Etiology and Therapy (1976)
5. Borek C, King DW (Eds): Cancer Biology, III—Herpes Virus (1976)
6. Borek C, Fenoglio CM, King DW (Eds): Cancer Biology, IV—Differentiation and Carcinogenesis (1977)

ADVANCES IN PATHOBIOLOGY is published
under the general Series Editorship of
Dr. Donald West King.

Stratton Intercontinental Medical Book Corp.
381 Park Avenue South
New York, N.Y. 10016

LC 77-3067. ISBN 0-913258-48-2
Printed in U.S.A.

Contents

LABORATORY WORKSHOPS

Foreword

This symposium is in honor of Theodore Puck, commemorating his 60th birthday. Dr. Alan Garen, one of his former colleagues, has agreed to describe his national scientific accomplishments, including his well known cloning of mammalian cells, the Denver classification of chromosomes and, most recently, his work on the auxotrophic mutants. Dr. Harry Ward, Dean of the University of Colorado Medical School will cite his accomplishments as Chairman of one of the first Departments of Biophysics in this country and later as Director of the Eleanor Roosevelt Institute for Cancer Research. His contributions to the scientific stature of the school, his teaching of medical students, residents and graduate students, and the development, with Arthur Robinson, of a genetics counseling program are well recognized.

I would like to comment on his contributions to the Given Institute of Pathobiology. Since the inception of these conferences in 1964, Ted Puck has been one of the most loyal supporters of this program. In the early 1970's, during the development of the physical facility, Dr. Puck was highly instrumental in persuading Mrs. Elizabeth Paepcke to contribute part of her beautiful garden. Along with Vice President John Conger, President Smiley, later Dean Harry Ward, Vice President Robert Aldrich and President Thieme, Aspenites Mayor Eve Homeyer, Attorney Robert Delaney, and Architect Harry Weese, he contributed greatly and continues to do so to the development of this thriving institute.

The goals of the Institute include the development of an educational program in Biology and Medicine. Initially, this focused on a course in pathology, which was later expanded to a Basic Medical Sciences Review Course for U.S. students in foreign medical schools. The latter three-year program has been completed this year with great success and encouraged hundreds of students in foreign countries to transfer into U.S. medical schools during their third year. Although it is directed by Drs. Leland Stoddard, Robert Anderson, and Rolla Hill, Ted Puck participated in and strongly supported this program. He also encouraged John Bowers of the Macy Foundation to have two educational conferences at the Institute on "Methods of Education" and "Women in Medicine."

Four years ago the Institute decided to develop wet laboratory courses, and this year six were given. Each consists of a single technic, lasting one week in length, for the purpose of introducing new research methods to the scientific community at large. Directed by Dr. Oscar Ris, Dr. Puck's laboratory, including Drs. Waldren, Kao, and Seeds, conducted many of these courses and continues to support them with supplies, equipment and expertise on an instant basis when needed.

In the past 12 years, the scientific program has expanded from one conference to some seven seminars each summer, alternating each week with courses sponsored by the Continuing Postgraduate Medical Education Department of the University of Colorado Medical Center (under the direction of Dr. Frank Cozzetto).

The Executive Program Committee of the Institute, including David Talmage (Chairman), C. Henry Kempe, and Ted Puck, as well as the course directors selected for a particular course, have imbued the Institute with a quality of excellence that allows it to attract speakers and participants from throughout the world. The programs have concentrated in the areas of cell and molecular biology, with applications to health sciences. A monograph series, *Advances in Pathobiology*, is published annually. The number of people wishing to attend, including medical and graduate students, resident, postdoctoral fellows, and junior and senior investigators, has increased markedly each year.

Along with his many other accomplishments, the Given Institute of Pathobiology stands strong today as a result of Dr. Puck's interest and support.

Donald West King

Theodore T. Puck, Ph.D.

The publication of the 1976 conference on "Differentiation in Cell Biology" held at the Given Institute of Pathobiology in Aspen is a fitting Festschrift volume to celebrate Dr. Theodore T. Puck. His many contributions over the past 30 years represent foundation blocks of modern cellular biology without which this conference could not have been held. His love and support for the "Aspen idea" helped give rise to the fine facility of the Given Institute.

The University of Colorado School of Medicine owes much to this man. He established our national role in basic sciences when he joined our faculty in 1948 as Professor and Chairman of the Department of Biophysics. Not only did he develop a strong department, but he assisted the school in recruiting faculty in virtually every department. In 1962, he helped develop the Eleanor Roosevelt Institute for Cancer Research and became its first director. This program continues under his leadership, making not only fundamental contributions to biologic science, but more recently, serving as a model for the bridging of basic and clinical sciences.

He and his associates are among the "fathers" of modern somatic cell research. Selected achievements in over 200 contributions include the development of *in vitro* technics for clonal growth of mammalian cells, fundamental studies on the effect of radiation, chromosomal identification that led to the "Denver Classification," life cycle analysis of mammalian cells, and concepts of genetic markers.

Ted Puck is "The Professor" in the State of Colorado. He speaks with equal facility to students in elementary schools as to colleagues in the National Academy of Science. Consultation with State legislators and the governor allowed us to establish one of the first genetic counseling and cytogenetic centers in the country. In fact, he has the ability to make the Regents of the University of Colorado speechless, no mean feat in itself.

This is a remarkable man. Our lives have been enriched by him.

Harry P. Ward, M.D.
Dean, University of Colorado
School of Medicine

Theodore T. Puck, Ph.D.

The accomplishments in molecular biology stand as one of the major intellectual landmarks of the past thirty years. The field was shaped by the entry into biology of scientists originally trained in mathematics, physics, and chemistry, who shared the belief that biologic processes could be understood in terms of the same laws that governed physical and chemical processes and that the awesome complexities of biologic systems could be unraveled by using appropriate organisms, methodologies, and ingenuity.

Ted Puck was ideally qualified to participate in this approach. He had been a student of the distinguished physical chemist, James Franck, at the University of Chicago and a youthful associate of such major figures in the emerging field of molecular biology as Leo Szilard and Max Delbruck. His own career has been closely interrelated with the history of the Department of Biophysics at the University of Colorado Medical Center in Denver, which he organized in 1948, and the Eleanor Roosevelt Institute for Cancer Research, which he organized in 1962 and still serves as director. Publications by Ted and his associates at these two institutions have formed a continuous record of original and fundamental contributions. Each of us could single out favorites from the bibliographical list on the following pages.

I will not attempt to discuss the over-all impact of these broad achievements in a brief introduction, but cannot resist commenting on a few. The elegant technic for preparing pure clones of individual mammalian cells by plating on agar containing a feeder layer of irradiated cells, reported by Puck and Marcus in 1955, revolutionized the study of mammalian cells in culture by providing a quantitative basis for biochemical, genetic, radiologic, immunologic, and physiologic experiments. Shortly afterward they used this new technic to obtain the first reliable data on the radiosensitivity of mammalian cells, which substantiated Ted's earlier warnings that the "permissible" levels of radiation for industry and medicine were dangerously high. All of us are the beneficiaries of his courageous efforts to bring about a reduction of these levels and impose more rigorous safeguards for radioactive materials. Concomitant with the radiosensitivity studies were those in which Puck and Fisher used the cloning technic to obtain stable mutants of the human tumor line *HeLa*. Subsequently, the Chinese hamster ovary line was shown by Kao and Puck to have particularly favorable properties for genetics. In a definitive series of papers they proceeded to demonstrate the feasibility of manipulating mammalian cells as microorganisms, successfully adapting various powerful methods from prokaryotes such as selection of auxotrophic mutants and complementation analyses. With the advent of cell fusion methods it became possible to map human genes, and the current work in Ted's laboratory, involving the use of human-hamster hybrids to study the genetics and immunology of surface antigens on human cells, is an exciting new chapter of his research. These pioneering efforts have helped to establish mammalian cells as a major branch of molecular biology.

Those of us who have been associated with Ted Puck had the good fortune to experience first-hand an atmosphere in which the highest scientific standards merged with a profound concern for human values. In the closing words of his book, *The Mammalian Cell as a Microorganism,* Ted himself gives eloquent expression to this credo: "It would appear that the only healthy biological solution to the dilemma that confronts modern man is a new intellectual and moral synthesis, giving expression to human values in a way that is consistent with the needs of man as a biological and social organism, and making available to all mankind the almost limitless potential for development which the new biological science appears to offer."

Alan Garen, Ph.D.

THEODORE T. PUCK

Born: 24 September 1916, Chicago, Illinois

Ph.D. University of Chicago, 1940

1948–1967: Professor and Chairman, Department of Biophysics, University of Colorado Medical Center, Denver.

1962–Present: Director, Eleanor Roosevelt Institute for Cancer Research, Denver, Colorado.

1967–Present: Research Professor, Department of Biophysics and Genetics, University of Colorado Medical Center, Denver, Colorado.

Honors

Lasker Award, 1958; Borden Award, 1959; Stearns Award, University of Colorado, 1959; General Rose Memorial Hospital Award, 1960; Member of the National Academy of Science, 1960; Heidelberger Lecturer, College of Physicians and Surgeons, 1958; Harvey Society Lecturer, 1959; Squibb Lecturer, 1959; Annual Research Lecturer, Council on Research and Creative Work, 1962; Karl F. Muenzinger Lecturer, 1963; First Distiguished Lecturer Series, University of Tennessee, 1967; Lifetime Research Professorship, American Cancer Society, 1966; Distinguished Service Award, University of Chicago Medical Alumni Association, 1969; Voice of America Lecturer, 1972; Fellow, American Academy of Arts and Sciences, 1972; Louisa Gross Horwitz Prize, Columbia University, 1973; Member, National Academy Institute of Medicine, 1974; National Lecturer, Sigma Xi, 1975, 1976; member, Commission on Physicians for the Future.

Married to a fellow scientist and lovely companion, Mary, and has three beautiful physician daughters—Sterling, Jennifer, and Laurel.

Selected Publications

Franck J, French CS, Puck TT: The fluorescence of chlorophyll and photosynthesis. J Phys Chem 45: 1268, 1941.

Robertson OH, Bigg E., Puck TT, Miller BF: The bactericidal action of propylene glycol vapor on microorganisms suspended in air. I. J Exp Med 75: 593, 1942.

Puck TT, Robertson OH, Lemon HM: The bactericidal action of propylene glycol vapor on microorganisms suspended in air. II. The influence of various factors on the activity of vapor. J Exp Med 78: 387, 1943.

Puck TT, Wise H: Studies in vapor-liquid equilibria. I. A new dynamic method for the determination of vapor pressures of liquids. J Phys Chem 50: 4, 1946.

Puck TT: The mechanism of aerial disinfection by glycols and other chemical agents. I. Demonstration that the germicidal action occurs through the agency of the vapor phases. II. An analysis of the factors governing the efficiency of chemical disinfection of the air. J Exp Med 85: 729–757, 1947.

Puck TT: An automatic dewpoint meter for the determination of condensible vapors. Rev Sci Instrum 19: 16, 1948.

Puck TT: A reversible transformation of T1 bacteriophage. J Bact 57: 647, 1949.

Wise H, Puck TT, Failey CR: Studies in vapor-liquid equilibria. II. The binary system triethylene glycol-water. J Phys Coll Chem 54: 734, 1950.

Puck TT, Garen A, Cline J: The mechanism of virus attachment to host cells. I. The role of ions in the primary reaction. J Exp Med 93: 65 (1951).

Puck TT, Wasserman K, Fishman AP: Some effects of inorganic ions on the active transport of phenol red by isolated kidney tubules of the flounder. J Cell Comp Physiol 49: 73, 1952.

Puck TT: The first steps of virus invasion. Cold Spring Harbor Symp Quant Biol 18: 149, 1953.

Puck TT, Tolmach LJ: Mechanism of virus attachment to host cells. IV. Physicochemical studies of virus and cell surface groups. Arch Biochem Biophys 51: 229, 1954.

Puck TT, Lee HH: Mechanism of cell-wall penetration by viruses. I. An increase in host-cell permeability induced by bacteriophage infection. J Exp Med 99: 481, 1954.

Puck TT, Marcus PI: A rapid method for viable cell titration and clone production with HeLa cells in tissue culture. (The use of X-irradiated cells to supply conditioning factors.) Proc Natl Acad Sci USA 41: 432, 1955.

Puck TT, Marcus PI: Action of X-rays on mammalian cells. J Exp Med 103: 653, 1956.

Puck TT, Marcus PI: Cieciura SJ: Clonal growth of mammalian cells *in vitro*. Growth characteristics of colonies from single HeLa cells with and without a "feeder" layer. J Exp Med 103: 273, 1956.

Puck, TT, Fisher HW: Genetics of somatic mammalian cells. I. Demonstration of the existence of mutants with different growth requirements in a human cancer cell strain (HeLa). J Exp Med 104: 427, 1956.

Tjio JH, Puck TT: Genetics of somatic mammalian cells. II. Chromosomal constitution of cells in tissue culture. J Exp Med 108: 259, 1958.

Puck TT: Action of radiation on mammalian cells. III. Relationship between reproductive death and induction of chromosome anomalies by X-irradiation of euploid human cells *in vitro*. Proc Natl Acad Sci USA 44: 772, 1958.

Puck TT, Cieciura SJ, Robinson A: Genetics of somatic mammalian cells. III. Long-term cultivation of euploid cells from human and animal subjects. J Exp Med 108: 945, 1958.

Tjio JH, Puck TT: The somatic chromosomes of man. Proc Natl Acad Sci USA 44: 1229, 1958.

Puck TT: Quantitative studies on mammalian cells *in vitro*. Rev Mod Physics 31: 433, 1959.

Robinson A et al: A proposed standard system of nomenclature of human mitotic chromosomes. (Denver Conference, April, 1960) Lancet 1: 1063, 1960; JAMA 174: 159, 1960; J Hered 51: 214, 1960; Am J Hum Genet 12: 384, 1960; and Pediatrics 27: 485, 1961.

Oda M, Puck TT: The interaction of mammalian cells with antibodies. J Exp Med 113: 3, 599, 1961.

Puck TT, Steffen J: Life cycle analysis of mammalian cells. I. A method for localizing metabolic events within the life cycle, and its application to the action of colcemide and sublethal doses of X-irradiation. Biophys J 3: 379, 1963.

Puck TT, Robinson A: Some perspectives in human cytogenetics. *In:* Cooke RE (Ed): Biologic Basis of Pediatric Practice—Infancy, Childhood and Adolescence. New York, McGraw-Hill, 1968, vol. II, p 1407.

Puck TT: Cellular aspects of the mammalian radiation syndrome. I nucleated cell depletion in the bone marrow. Proc Natl Acad Sci USA 52: 152, 1964.

Puck TT: Studies of the life cycle of mammalian cells. Cold Spring Harbor Symp Quant Biol 29: 167, 1964.

Robinson A, Puck TT: Sex chromatin studies in newborns: Presumptive evidence for external factors in human nondisjunction. Science 148: 3666, 1965.

Puck TT: Cell turnover in mammalian tissues. I. Use of cell depletion measurements to estimate X-ray reproductive survival curves. Proc Natl Acad Sci USA 54: 1797, 1965.

Puck TT: Cellular aspects of the mammalian radiation syndrome. II. Cell depletion in bone marrow, spleen and thymus of young mice. Rad Res 27: 272, 1966.

Puck TT: The mammalian cell. *In:* Cairns J, Stent GS, Watson JD (Eds): Phage and the Origins of Molecular Biology. Cold Spring Harbor Symp Quant Biol 31: 275, 1966.

Puck TT, Hill HZ: Enzyme kinetics in mammalian cells. I. Rate constants for metabolism in erythrocytes of normal galactosemic and heterozygous subjects. Proc Natl Acad Sci USA 57: 1676, 1967.

Puck TT, Kao F-T: Genetics of somatic mammalian cells. V. Treatment with 5-bromodeoxyuridine and visible light for isolation of nutritionally deficient mutants. Proc Natl Acad Sci USA 58: 1227, 1967.

Puck TT, Waldren CA, Jones C: Mammalian cell growth stimulation by fetuin. Proc Natl Acad Sci USA 59: 192, 1968.

Puck TT, Kao F-T: Genetics of somatic mammalian cells. VI. Use of an antimetabolite in analysis of gene multiplicity. Proc Natl Acad Sci USA 60: 561, 1968.

Kao F-T, Puck TT: Genetics of somatic mammalian cells. VII. Induction and isolation of nutritional mutants in Chinese hamster cells. Proc Natl Acad Sci USA 60: 1275, 1968.

Cox DM, Puck TT: Chromosomal non-disjunction: The action of colcemid on Chinese hamster cells *in vitro*. Cytogenetics 8: 158, 1969.

Kao F-T, Johnson RT, Puck TT: Genetics of somatic mammalian cells. VIII: Complementation analysis on virus-fused Chinese hamster cells with nutritional markers. Science 164: 312, 1969.

Kao F-T, Puck TT: Genetics of somatic mammalian cells. IX. Quantitation of mutagenesis by physical and chemical agents. J Cell Physiol 74: 245, 1969.

Kao F-T, Chasin L, Puck TT: Genetics of somatic mammalian cells. X. Complementation analysis of glycine-requiring mutants. Proc Natl Acad Sci USA 64: 1284, 1969.

Hsie AW, Puck TT: Mammalian cell transformations in vitro. I. A morphological transformation of Chinese hamster cells produced by dibytyryl cyclic adenosine monophosphate and testosterone. Proc Natl Acad Sci USA 68: 358, 1971.

Hsie AW, Jones C, Puck TT: Mammalian cell transformations *in vitro*. II. Further changes in differentiation state accompanying the conversion of Chinese hamster cells to fibroblast form by dibutyryl adenosine cyclic 3′:5′-monophosphate and testosterone. Proc Natl Acad Sci USA 68: 1648, 1971.

Puck TT, Wuthier P, Jones C, Kao F-T: Genetics of somatic mammalian cells. XIII: Lethal antigens as genetic markers for study of human linkage groups. Proc Natl Acad Sci USA 68: 3102, 1971.

Kao F-T, Puck TT: Genetics of somatic mammalian cells. XIV. Genetic analysis *in vitro* of auxotrophic mutants. J Cell Physiol 80: 41, 1972.

Puck TT, Waldren CA, Hsie AW: Membrane dynamics in the action of dibutyryl adenosine cyclic 3′:5′-monophosphate and testosterone on mammalian cells. Proc Natl Acad Sci USA 69: 1943, 1972.

Kao F-T, Puck TT: Genetics of somatic mammalian cells. XVI. Demonstration of a human esterase activator gene linkaged to the AdeB gene. Proc Natl Acad Sci USA 69: 3273, 1972.

Puck TT: The Mammalian Cell as a Microorganism: Genetic and Biochemical Studies *in vitro*. San Francisco, Holden-Day, Inc., 1972.

Wuthier P, Jones C, Puck TT: Surface antigens of mammalian cells as genetic markers. II. J Exp Med 183: 229, 1973.

Porter K, Puck TT, Hsie AW, Kelley D: An electron microscope study of the effects of cyclic AMP on Chinese hamster ovary cells. Cell 2: 145, 1974.

Patterson D, Kao F-T, Puck TT: Genetics of somatic mammalian cells: Biochemical genetics of chinese hamster cell mutants with deviant purine metabolism. Proc Natl Acad Sci USA 71: 2057, 1974.

Jones C, Wuthier P, Puck TT: Genetics of somatic cell surface antigens. III. Further analysis of the A_L marker. Somatic Cell Genetics 1: 235, 1975.

Puck TT: Mammalian cell regulation. *In:* Fenoglio CM, Goodman R, King DW (Eds): Advances in Pathobiology, 3, Developmental Genetics. New York, Stratton Intercontinental, 1976.

Kao F-T, Jones C, Puck TT: Genetics of somatic mammalian cells: Genetic, immunologic and biochemical analysis with Chinese hamster cell hybrids containing selected human chromosomes. Proc Natl Acad Sci USA 73: 193, 1976.

Introduction

Developmental biology—with its central issue, cellular differentiation—has emerged within recent years as a field upon which many disciplines have converged. It has profited greatly from new knowledge and technics in rapidly advancing areas such as cellular and molecular biology and genetics. The study of differentiation consists now of a collection of related problems approached from different angles in a variety of biologic systems. Closely linked to the exploration of controls in normal growth is the pursuit of the mechanisms underlying neoplastic conversion.

The present symposium has been organized with an attempt to focus on some of the key problems and approaches in the areas of cellular differentiation and neoplasia by bringing together investigators from diverse but closely allied areas. In a series of workshops, some of the advanced technics used in cellular and molecular biology and genetics have been explored. These methods, which are invaluable tools in studies related to differentiation and cancer, are also presented in this volume.

The problem of cellular differentiation is one of the regulation and stable restriction of gene expression during embryonic development. It involves processes which select specific segments of the genome which will be expressed and processes which play a role in the actual expression of a selected segment.

Development in multicellular organisms is associated with cell movement, cell division and cell-cell interation. Cells transmit information about their presence, identity and function by contact through cell surface materials ultimately giving rise to differentiated tissues and organs (see Moscona, Garen). These events are most evident in the embryogenesis of higher organisms and are also found in teratocarcinomas (see Pierce, Mintz). Patterns of recognition prevail in the organism in complex immunologic systems where the categorization of self and non-self are of utmost importance (see Edelman, Hood, Lennox). While regulation at a genetic level has long been recognized, there exist phenotypic patterns of behavior in cells closely affected by the cellular environment which cannot be explained satisfactorily by gene programming alone. Such phenotypic patterns found in normal cells (see Edelman, Moscona, Dorfman), neoplastic cells (see Edelman, Lennox, Sachs, Pierce, Mintz) and in some established cell lines (see Puck, Hsie) strongly suggest that there exists an additional level of regulation which is epigenetic. One such mechanism may operate largely at the level of the cell surface (see Edelman, Sachs). This control system may function by the coordination of interacting macromolecules (see Edelman, Goad, Cann), both surface receptors and submembranous fibrillar elements. It appears to be transmembranous in nature and inextricably related to the fluid structure of the cell membrane and to the molecular features of its surface receptors. These are mainly glycoproteins which traverse the matrix of the bilayer lipid membrane as integral membrane proteins. Their glycosylated portion projects extracellularly where it can bind to other molecules (see Marchesi, Edelman). The inner portion of these polypeptides projects into the cytoplasm, where a number of interactions with cytoplasmic components such as microtubules can take place (see Edelman, Marchesi, Sachs) and can be affected by a variety of antimitotic drugs and hormones (see Edelman, Puck, Hsie, Sachs).

The most extensively characterized surface glycoproteins, at present, are glycophorin, the major integral membrane protein in the human red blood cell (see Marchesi), and the major histocompatibility antigens, HL-A in man and H-2 in the mouse. They are integral membrane proteins present on cells in all tissues and high on lympocytes (see Edelman). The latter antigens are specified and controlled by genes in a complex multigene locus involved in immune recognition (see Hood, Edelman, Lennox). Their complex and intriguing structure has been determined (see Edelman). The histocompatibility antigens are responsible for the recognition and destruction by cytotoxic lymphocytes of allogeneic cells bearing foreign antigens and of syngeneic cells which have acquired new surface antigens such as virally infected cells or tumor cells (see Edelman, Lennox). The histocompatibility antigens possess a wide range of genetic polymorphism (see Hood, Edelman, Lennox). A similar degree of diversity is found in tumor-specific transplantation antigens (TSTA's) present on tumors induced by chemical carcinogens (see Lennox). In spite of the large body of experimentation, it is still unclear whether the diversity of the TSTA's has its origin in the polymorphism of the histocompatibility antigens. The diversity of the TSTA's apparently bears no relationship to the embryonic antigens (see Lennox).

With the development of chromosomal banding technics which enable the identification of individual chromosomes, it has become possible to analyze the genetic basis of normal and abnormal development and the effects of environmental factors on these developments (see Puck, Robinson, Sachs). Procedures of isolating mammalian cell mutants and of somatic cell hybridization have further enabled the ongoing detailed genetic analysis of the human cell surface and the linkage between specific human chromosomes and defined human cell surface antigens (see Puck).

Molecular genetics has had an immediate impact on the study of differentiation of eukaryotic cells at the level of gene transcription and translation (see O'Malley, Garen, Dorfman), and in studies on the evolution and organization of multigene families (see Hood). Temporal genetic determination has been studied in developing chick limbs and suggested to begin by an amplification of specific DNA sequences already present in multiple copies (see Dorfman). The expression of new genes coding for the synthesis of differentiated products is inducible by hormones, as exemplified by estrogen and progesterone induction of chick oviduct differentiation (see O'Malley) and by ecdysone induction of development in *Drosophila* (see Garen). The induction of oviduct differentiation is a positive form of regulation whereby the steroid binds in the cytoplasm of target cells to an identified dimer receptor. The complex transverses the nucleus, where it acts directly on gene transcription by binding to tissue-specific nonhistone proteins and to DNA (see O'Malley). Other types of inducers active in the differentiation of mammalian stem cells into macrophages and granulocytes have been characterized (see Sachs).

If gene expression in undifferentiated normal cells can be activated following interactions with appropriate inducers, would some neoplastic cells which are at the stage of undifferentiated stem cells respond in a similar manner? In other words, given the appropriate conditions and environment which have been proven to permit normal development and differentiation, will certain neoplastic cells respond and differentiate to the same degree as the normal cells and will they cease to appear malignant? The answer is yes (see Sachs, Pierce, Mintz), as seen in various systems such as leukemic cells (Sachs), some carcinomas (Pierce), and teratocarcinoma cells (Pierce, Mintz). Ideas that have long dominated, suggesting that neoplastic conversion is attributable to a mutation in DNA leading to a loss of differentiation and to a malignant phenotype, must be expanded to allow for these new experimental findings. These data suggest that the stability and heritability of some neoplasms may be due to mechanisms similar to those in normal development. Carcinogenesis could result from an activation of gene loci in suitable target cells (see Pierce). A carcinogen could act by causing a mutational change or by altering the milieu, thereby eliciting a modification in cellular gene expression (see Mintz). The resulting neoplasia would be "a failure of the normal progression of differentiation" (see Mintz) or an aberrant differentiation (see Pierce) which then responds to different controls of growth. By modifying the immediate environment of certain neoplastic cells to allow for optimal conditions (e.g., by the presence of specific inducers), normal differentiation may proceed in part (see Sachs) or completely (see Sachs, Pierce, Mintz). These processes can be demonstrated *in vivo* and *in vitro*. In certain systems (see Sachs), neoplastic cell maturation, accompanied by a normal phenotype and loss of malignancy, has been linked to the presence of specific chromosomes and can proceed even though the cell karyotype may be abnormal. A most dramatic case demonstrating the differentiation of tumor cells into normal cells has been the differentiation of mouse teratocarcinoma cells which had been injected into normal blastocysts and implanted in a foster mother. These blastocysts developed into normal mosaic mice in which all tissues had contributions from the injected tumor cells (see Mintz).

The Editors

Somatic Cell Genetics in Problems Related to Differentiation

Theodore T. Puck, Ph.D

The purpose of this paper is to illustrate by means of some experiments carried out in the Eleanor Roosevelt Institute during the past year how the methods of somatic cell genetics and genetic biochemistry can contribute to understanding in the area of mammalian cell regulation.

The approaches have involved the combination of a variety of basic technics: production of additional auxotrophic mutants; biochemical analysis to establish the site of the metabolic block in each such mutant; use of these mutants to prepare specific hybrids; use of these hybrids for gene mapping; use of these hybrids for genetic and biochemical analysis of cell surface antigens and exploration of their relationship to developmental and pathological problems; study of the mutagenic process; and analysis of the action of regulatory molecules like cyclic AMP.

PREPARATION OF NEW MUTANTS

Auxotrophic Mutants of Chinese Hamster Cells

Mutants are required in order to provide markers which make possible a large variety of genetic and biochemical experiments. Auxotrophic mutants are especially valuable: One can readily secure clean, well defined mutants which grow with 100% plating efficiency in some media and display no growth whatever in others; single gene mutants are readily isolated by means of the BUdR-visible light selection technic; and determination of the biochemical block is relatively simple since the nature of the auxotrophy affords a clue to the actual point of block, which can be confirmed by feeding and radioactive tracer experiments.

Dr. O. Hankinson has isolated new mutants requiring glutamate and

From the Eleanor Roosevelt Institute for Cancer Research, and the Department of Biophysics and Genetics, University of Colorado Medical Center, Denver, Colo.

Acknowledgments: This investigation was supported by a grant from the Max C. Fleischmann Foundation and by Grant No. HD02080 from the National Institute of Child Health and Human Development, DHEW. (Contribution No. 224)

alanine [1]. Mutants were also isolated which lack branched chain amino acid transaminase. These experiments, carried out by Drs. C. Jones and E. Moore, resulted in the production of mutants unable to convert alpha keto isovaleric acid to valine, alpha keto isocaproic acid to leucine, and alpha keto beta methylvaleric acid to isoleucine. Each mutant examined was found to lack all three amino acid transaminase activities. Moreover, each could be mutated in a single step back to wild-type behavior in which all three activities are simultaneously restored. Thus, these experiments provide genetic evidence which complements previously available biochemical evidence that the same enzyme is involved in all three of these transaminase activities [2].

Dr. D. Patterson and his collaborators have used the genetic biochemical approach to analyze the biosynthetic pathways involved in purine and pyrimidine synthesis in the CHO-Kl cell. By appropriate adjustment of the "minimal" and "enriched" media of the BUdR-visible light technic [5a], large numbers of mutants in a single biochemical pathway or even in a particular region of the pathway can be isolated. Dr. Patterson and his coworkers have now isolated nine different auxotrophic mutants in the pathway leading from ribose-5-phosphate to adenine monophosphate. The two most recent members of this series are those designated adenine H, which cannot convert inosinic acid to adenyl succinate, and adenine I, which cannot convert the latter compound to AMP [3]. All of these behave like single-gene, recessive, constitutive mutants. The availability of this family of mutants makes it highly convenient to examine feedback relationships in this biosynthetic chain and to search for regulatory mutants which will exhibit different behavior from the presumed constitutive mutants. For example, mutants have now been obtained which require adenine for growth only in the presence of other specific metabolites. High hopes are entertained that the system may lend itself to the characterization of regulatory mechanisms in this pathway since the adenine biosynthetic pathway is required for synthesis of DNA, RNA, ATP, cyclic AMP, and a variety of critical co-enzymes. It appears reasonable to expect that a variety of important cellular control mechanisms should become amenable to study.

One interesting result is worth noting: $HGPRT^-$ mutants have been isolated by a number of investigators by means of the resistance they exhibit to killing by thioguanine. Since these mutants presumably lack the enzyme HGPRTase, they are unable to utilize hypoxanthine or its antagonists and are, therefore, resistant to the killing action of the latter. It should be possible to isolate from mutagenized populations of CHO cells identical $HGPRT^-$ mutants by the use of the BUdR-visible light technic applied to the isolation of cells unable to utilize hypoxanthine for growth. Such experiments were carried out [4] and the theoretical expectation was completely confirmed.

Mutants with Respect to Biosynthetic Steps in Alternative Metabolic Pathways

Proline can be synthesized in mammalian cells either from glutamate or from ornithine [5b,5c]. The function served by the existence of these two alternative pathways and the mode of their control are not understood. If mutants blocked at different points in this multiple pathway system were produced, important experimentation analyzing the nature of the control mechanisms involved should become possible. For example, mutants in one, the other, or both pathways for proline synthesis could be used for hybridization experiments with tissue cells which exhibit one or another of the two pathways for proline synthesis. In this way, dominance and recessive relationships could be determined and complementation analysis could be carried out. Such experiments have been undertaken by Dr. Annette Baich in our laboratory, who has succeeded in producing the four different expected forms of Chinese hamster cells corresponding to the phenotypes in which one, the other, both, and neither pathway for proline biosynthesis are active.

Finally, it has been possible to add new classes of mutants for genetic studies in mammalian cells by means of the technic of replica plating designed by Drs. T. Stamato and L. Hohmann in our laboratory [6]. The use of replica plating permits isolation of mutants for which selective media do not exist, including situations in which the test procedure needed to reveal the mutant status of the clone requires killing of the cells. The replica plating procedure is simple and readily permits the examination of 10^4 or 10^5 colonies. Mutant clones deficient for G6PD and LDH-A have readily been isolated after mutagenesis of a parental population. In addition, by means of special modifications of this technic Dr. Stamato has been able to isolate an ultraviolet-sensitive clone which promises to be of particular value in analysis of radiation-repair mechanisms.

PREPARATION AND USES OF HUMAN-CHINESE HAMSTER HYBRIDS

Preparation of Hybrids Containing Specific Single Human Chromosomes

The hybrids which are particularly useful in genetic and genetic biochemical experiments are those which contain only one or a few specific human chromosomes along with the complete or almost complete set of the CHO-K1 chromosomes. A variety of these have now been prepared and their cytogenetic analysis has been carried out by Dr. Fa-Ten Kao and his co-workers. The hybrids currently available include those containing human chromosomes numbered 8, 11, 12, 14, and 21. Some hybrids with single and some with specific combinations of these human chromosomes have been prepared. Identification of the human chromosomes contained by the mutant is carried out by karyotype analysis with banding and by isozyme analysis.

Chromosomal Localization of Human Genes with the Aid of Human-CHO Hybrids

Given a stable hybrid which contains only a single human chromosome, the genetic locus of all the human gene products produced by this hybrid can be assigned to its contained human chromosome. By this means, the chromosomal identity of the genetic determinants for seven human markers have been identified, as demonstrated in Table 1. Gene mapping is essential to provide markers for particular chromosomes and chromosomal regions and to determine the role of geometrical factors in control mechanisms which regulate gene expression in somatic cells.

TABLE 1. Chromosomal Localizations Carried Out with Human-CHO Hybrids

Marker Name	Chemical Identity	Marker Nature	Human Chromosome Number
a_1	Glycophorin antigen	cell surface antigen	11 short arm
a_2	Unidentified	cell surface antigen	11 short arm
a_3	Unidentified	cell surface antigen	11 short arm
ade C	phosphoribosyl glycineamide synthetase (EC 6.3.1.3)	enzyme (purine synthesis)	21
gly A	serine hydroxymethyl transferase (EC 2.2.2.1)	enzyme (glycine synthesis)	12
ade B	formylglycineamide ribonucleotide amidotransferase (EC 6.3.5.3)	enzyme (purine synthesis)	4 or 5
pro	glutamic γ-semialdehyde synthetase	enzyme (proline synthesis)	10

Cell Surface Antigens on Human-CHO Hybrids

Cell surface structures provide the sensing elements for cell-cell recognition and interaction [7] which provide the basis for tissue formation and for sensing the nutritional, hormonal and other molecular constituents of the cellular environment. These structures appear to hold the key to many of the fundamental aspects of differentiation and may well be involved in the expression of malignancy and other kinds of cell pathology in the mammal. Immunologic methods offer particularly sensitive and powerful methods for studying many of these cell structures. Application of these methods to the human-Chinese hamster hybrids developed in this laboratory appears to make possible a variety of important studies. These investigations were initiated several years ago by Paul Wuthier, C. Jones and the author. The studies have been expanded to include E. Moore, F. T. Kao, S. Nielson, and A. Robinson [8,9,10,14a].

When cell survival is measured by means of the single cell plating procedure, it is found that antisera to human cell surface antigens kill 99% or more of the human cell population in the presence of standard complement at antibody concentrations in the neighborhood of 0.02%, but produce no killing whatever of Chinese hamster cells under identical conditions even when 10 or 100 times as much antiserum is employed. A similar high degree of specificity for the homologous cell is found with antisera to Chinese hamster cells. Therefore, specific hybrids containing particular human chromosomes can be tested with antisera made against particular human cells taken directly from biopsies or grown in tissue culture. In this way it is possible to determine whether a given human chromosome causes the hybrid cell to produce surface antigens which are common to those of the human cell which elicited the test antiserum. We were able to identify mutually exclusive families of antigens characteristic of particular human-CHO hybrids, which were designated A_L, B_L, C_L, etc. The work with the A_L hybrid is illustrative. Cytogenetic and isozyme analysis revealed the A_L hybrid to contain one single human chromosome: #11. Therefore, the A_L marker includes all of the human cell surface antigens contributed by chromosome #11 to the hybrid. This hybrid was killed by antisera to human fibroblasts, human lymphoblasts, and the HeLa cell. Antiserum against human lymphoblasts was exhaustively adsorbed by cells of the A_L hybrid so that all of the specific killing activity against this hybrid was removed. The resulting antiserum was then tested with other hybrids. When another susceptible hybrid was found, its cell surface antigenic character was named B_L. The process was repeated in this way, defining the antigenic families corresponding to C_L, D_L, etc.

The advantage of using cell survival curves for measurement of interaction between cell and antiserum in the presence of complement, as compared with other immunologic procedures, is three-fold: (a) The curves obtained are quantitative, free of subjective influence, and highly reproducible. (b) The difference between susceptible and resistant cell reactions with the antiserum is highly definitive. For example, a given serum which provides an end point by an agglutination reaction of eight two-fold dilutions may yield a titer by the single cell plating technic of 20,000 or more killing units where one killing unit is defined as the amount of serum needed to reduce the plating efficiency to 37%, as measured in the region where the survival curve declines exponentially. (c) The single cell survival curve method provides a means for identification, quantitation, and isolation of mutants and other variant cells that are resistant to an antiserum action which is lethal to the parental cell. These mutants make possible the kinds of genetic analysis described in the following paragraphs.

Having established the presence of human antigenic markers connected with the presence of a particular human chromosome, the next step is to resolve these into individual genetic markers. This has been done in case of the A_L system by means of the following considerations: if the immunologic markers here under study are characteristics associated with specific tissues, one might then expect one of two different situations. On the one hand, one might find a single cell surface antigenic structure characteristic of every different tissue in the mammalian body. This system, however, would be far less efficient than one in which a certain basic array of such cell surface antigens was available from which particular combinations would be utilized on the cell membranes of the different kinds of tissue cells. Such a combinatorial arrangement would greatly increase the flexibility in the operation of cellular recognition and regulatory systems.

If the latter scheme is indeed operative in the body, it should be possible to find human tissues which exhibit different combinations of the individual A_L antigens. Such tissue cells would make possible a resolution of the A_L complex into component parts, since antibodies to some antigens could be selectively adsorbed out from a complex antiserum, leaving behind a more limited set of antibodies for easier immunogenetic analysis.

There are two necessary and sufficient conditions which permit identification of tissue cells which exhibit some but not all of the activities of the A_L antigenic complex. Such cells, when injected into a rabbit, must produce antisera highly lethal to A_L hybrid cells and, in addition, the cells of such a tissue must be unable to remove *all* of the killing activity for A_L hybrid cells from a more complete anti-A_L serum. Thus, cells of various tissues were tested by these two criteria. By such analysis the human red cell was found to contain some but not all of the antigens of the A_L complex; i.e., the red cell was demonstrated to produce antisera which killed A_L hybrid cells with great effectiveness, yet exhaustive adsorption of an antiserum like that produced by the HeLa cell, with human red cells, failed to remove more than a tiny part of its killing activity against the A_L hybrid.

In such fashion it has been possible so far to isolate three different A_L

antigenic activities present on the surface of the hybrid containing human chromosome #11. These have been named a_1, a_2, and a_3. Antisera prepared against various human cells and against the A_L hybrid itself behave in the expected fashion when titrated for killing action against the various test cells, both before and after exhaustive adsorption with cells assigned particular phenotypes. However a more definitive test of these antigenic assignments arises from genetic analysis of these markers. When the A_L hybrid cells are treated with mutagenic agents, and then plated in the presence of specific antisera, mutants lacking the corresponding antigens are readily isolated. Thus, all of the four different stable phenotypes with respect to the behavior of a_1^+ and a_2^+ antigens have been obtained—i.e., $a_1^+a_2^+$, $a_1^-a_2^+$, $a_1^+a_2^-$, and $a_1^-a_2^-$. Of the eight different phenotypes expected with the various possible combinations of a_1, a_2, and a_3, six have so far been prepared as clonal stocks.

These various mutant hybrids can also be used as direct antigens. If these are injected into Chinese hamsters, the antiserum collected is active only against the human constituents in a fashion which has previously been demonstrated by other investigators for human-mouse hybrids. (This procedure affords only a small volume of antiserum. Therefore, we prefer to inject specific hybrid cells into a rabbit or sheep and free the antisera so obtained from Chinese hamster cell antibodies by exhaustive adsorption with the parental CHO-Kl cell.)

By means of analyses of this kind, the human red cell was shown to contain the a_1 and a_3 surface antigens, but not a_2. Thus, some but not all of the loci of human chromosome #11 which are involved in the cell surface antigen production are expressed in the formation of human RBC. Analysis of the distribution of these and other cell surface antigens on the cells of other human tissues is proceeding. It is obvious that the surface antigens studied by these technics differ from those of the human HLA system which are carried on the cells of all or almost all tissues. The antigens which are the subject of this study exhibit tissue specificities of a kind which make it seem reasonable to associate them with the process of tissue differentiation.

The next step in such an experimental program is chemical identification of the antigens already established by immunogenetic means. Where chemical isolation of different cell membrane components has already been accomplished, extremely convenient approaches are possible. Thus, glycophorin, the principal glycoprotein system of human red cells, has been isolated in other laboratories [12,13]. In collaboration with Dr. Marchesi's group, we have carried out experiments demonstrating that glycophorin exhibits a high degree of immunologic identity with the a_1 and a_3 antigens, but displays no cross-reaction whatever with a_2. The presumption, then, is that a_1 and a_3 contain unique chemical sequences, each of which resembles that of specific components of the glycophorin system. These relationships should facilitate resolution of the glycophorin system into its individual components and chemical identification of the differences in their structures.

Since these cell surface antigens form excellent genetic markers, they can be studied by complementation analysis. In experiments on the a_1 marker carried out by C. Jones and E. Moore, mutant A_L hybrids were prepared which were a_1^- and which also possess auxotrophic markers such as Gly^- and Ade^-. Pairs of such mutant hybrids were then hybridized again in medium, lacking both of the required nutrilites of the two parental forms. In this way, only complementing hybrids will form colonies. The result is a hybrid cell to which four parental cells have contributed genetic material. Such tetraparental hybrids were examined for the activity of the a_1 marker. By this kind of complementation analysis, it has been found that three different complementing a_1^- mutants exist and preliminary evidence for a fourth is already available. These experiments demonstrate that mutation in several genes can lead to loss of the a_1 marker. Jones and Moore have provisionally interpreted these results in terms of a model in which successive biosynthetic steps are needed to produce the fully active a_1 antigen, a model which would lend itself to placing at least part of the immunologic specificity in the carbohydrate fraction of the glycophorin molecule.

GENE MAPPING

Regional mapping of the genes controlling these cell surface antigens can be achieved by subjecting A_L hybrid cells to mutagenesis by agents producing recognizable deletions. Cytogenetic analysis then permits correlation between specific chromosomal deletions and a specific marker loss. Such analysis has been carried out for the a_1, a_2, a_3, and LDH-A markers on human chromosome #11. It demonstrated that a_1 and a_3 and the lactic dehydrogenase A marker are present on the short arm of this chromosome, while a_2 is on the long arm. It is hoped that measurement of the frequency with which particular pairs of markers are lost after mutagenesis may provide true gene mapping of these markers on each arm of the chromosome.

APPLICATION TO DIFFERENTIATION PROCESSES AND TO HUMAN CELLULAR DISEASES

The cell surface antigens which have been uncovered by this program offer an interesting contrast to the human HLA antigens. The latter structures appear to be present on all or most tissues of any individual. The antigens considered here exhibit tissue specificity and, therefore, should form useful tools in studying normal developmental processes and the diseases which involve changes in these cell surface structures.

The rationale of this approach involves preparation of antisera specific for particular cell surface antigens. Such antisera can be applied to cells from particular human tissues and their uptake measured in different cell types. Uptake of such specific antibody can be demonstrated either by immunofluorescence or by the horseradish peroxidase methodology. In every case, careful controls must of course be carried out. Preliminary experiments demonstrate that cells from a human adult brain exhibited a_1^+ activity. This approach may well have application in the study of normal and pathologic developmental processes.

MUTAGENESIS AND REPAIR PROCESSES

A principal concern of modern cellular biology is the nature of mutagenesis and carcinogenesis, their interrelationships, and factors which affect repair of genetic damage. Agents which inhibit aspects of DNA repair in mammalian cells have been studied in a variety of laboratories. One of the principal difficulties involved in such studies lies in the fact that agents which affect such repair systems may also exercise a more general cellular toxicity which prevents careful study of the repair system itself.

Dr. C. Waldren and his co-workers have devised means by which the action of agents like caffeine can be studied without interference by their intrinsic cell-killing action [15]. When cells are treated with mutagenic agents like x-rays or ultraviolet irradiation, caffeine is added only for the 16 hour period after the exposure to the mutagenic agent. Under these circumstances, the effect of caffeine on repair mechanisms is still exerted maximally, but its general toxicity for the cells can be eliminated. Such experiments have demonstrated that contrary to previous claims, caffeine inhibits intrinsic cellular mechanisms of repair for x-rays as well as for ultraviolet light damage. This experiment appears to furnish the explanation for the initial shoulder in the x-ray survival curve as being due to the operation of repair processes. It also makes possible studies of the actions of various agents on mammalian repair systems, a study of profound importance for both its fundamental and practical implications.

Recent estimates have placed the amount of cancer which is caused by environmental agents to lie in the neighborhood of 60–90% of all human malignancy. Since cancer accounts for approximately one-third of all human deaths, and since the rate of cancer incidence has been increasing even beyond the informed projections of experts, the magnitude of the human problem caused by environmental carcinogens is enormous.

Similarly, much of human genetic disease, both single gene defects and chromosomal aberrations, has been attributed to the presence of physical, chemical, and biological mutagens in the environment. Human chromosomal disease alone affects 0.5 – 1% of all live babies born in this country and is responsible for much spontaneous abortion. Obviously, then, if it were possible to institute rapid, effective and economically feasible screening procedures which would identify and so permit elimination of environmental mutagens and carcinogens, an enormous health contribution could result. The demonstration that many, if not all, carcinogens are also mutagens (either directly or after metabolic transformation in the body) makes it imperative to devise a program for the screening of mutagens which threaten human populations.

Several procedures have already been proposed for this purpose. Perhaps the best known of these is the Ames test [16], which utilizes specific bacterial auxotropic mutants and scores those cells which have been mutagenized so as to revert to prototrophy. It is an excellent test by virtue of its reliability and ease of execution. However, several features not found in this test would recommend themselves for inclusion in a procedure designed to detect mutagens and carcinogens.

It would be most desirable to have a test which could detect deletions, which are nonreversible lesions. Ideally, it should be possible to detect both small- and large-scale deletions. Such mutations are not readily observed in a unichromosomal organism, where most deletions are lethal and even nonlethal ones are not readily reversible. Further, the use of a forward mutation rather than a reversion should permit a higher yield of detectable mutants and so increase the sensitivity of the test, especially if the reversion can occur in a chromosome unnecessary for cell multiplication. It would also seem important to detect processes like loss of an entire chromosome, which is a concomitant of nondisjunction, the process responsible for the majority of new genetic insults in man. Finally, it would be advantageous if one could use human genes carried on human chromosomes for such a test.

A system which appears to provide all these advantages is the A_L human-Chinese hamster cell hybrid which displays a_1, a_2, and a_3 surface antigen activity. These cells are chromosomally stable and have been maintained in continuous cultivation for several years. Forward mutations in mutagenized populations are readily scored by plating the cells in the presence of the appropriate antiserum and complement. Moreover, by scoring for the simultaneous loss of genes on both sides of the centromere, loss of the entire chromosome can readily be detected. Cytogenetic confirmation can easily be carried out. Since one is measuring mutation in human genes carried on human chromosomes, a reasonable resemblance to situations in man may be anticipated. Study is now being directed to the incorporation of these operations into a routine screening test which might make possible simple large-scale applications.

The approach described here for study of cell-surface antigens appears to offer several important features: it simultaneously defines families of specific cell surface antigens and their chromosomal loci; it makes possible genetic analysis by mutation, dominance-recessiveness investigation, complementation analysis, and at least regional mapping of the genes on the chromosomes; it permits resolution of a chromosomal set of antigens into individual antigens with specific genetic loci; it appears to furnish information about cell surface antigens which are tissue-specific and, therefore, presumably involved in differentiation and development; it gives promise of being applicable to a wide variety of tissues; and it promises to permit studies on regulatory control of cell surface antigens, since agents which modulate such activities can be tested to determine which cell surface antigens are regulated coordinately or independently, and whether the mapping positions of the genetic loci bear any relationship to their participation in coordinate control mechanisms.

ACTION OF CYCLIC AMP AND ORGANIZATION OF THE CELLULAR MICROTUBULAR SYSTEM

In 1970, Dr. A. W. Hsie in our laboratory first observed a morphologic change induced by dibutyryl cyclic AMP (DBcAMP) in our standard CHO-Kl cell. Simultaneously, I independently observed a similar change to be produced by addition of testosterone to the medium. Both phenomena required relatively high concentrations of their respective agents, which frequently involved toxic or at least growth inhibitory actions. We combined forces and reagents, and found a synergistic effect to be produced in reliable fashion, which did not involve toxic actions to the cells, but which altered their behavior in a manner which was definitive, reversible, and reproducible.

It was established that treatment with DBcAMP and testosterone of cells which *in vitro* display the stigmata associated with malignant transformation, cause them to lose the morphologic and other characteristics of the transformed condition. The reaction of DBcAMP is synergized by agents like testosterone, prostaglandins, or testololactone. The compact, pleomorphic, heavily knobbed and randomly multiplying cells are caused by the specified treatment to stretch out into spindle-shaped forms, lose their surface knobs, adopting a smooth membrane which is considerably more tranquil in its surface movements, and form colonies in which the cells associate closely in parallel with their long dimension. Simultaneously, the cells lose their ability to grow in suspension, even though their growth on surfaces is unaffected. Morphologic findings similar to ours but on a different cell type were simultaneously described by Pastan and his associates

[17,18]. A number of other investigators have since demonstrated similar effects in a variety of malignantly transformed cells [19]. We have given the name "Reverse Transformation" to this phenomenon in which transformed cells take on the habitus of a normal fibroblast.

In further experiments carried out in collaboration with Dr. K. Porter, it was demonstrated that Reverse Transformation is accompanied by an organizing process of the cellular microtubules in which these structures change from a randomly oriented, disorganized pattern into a highly ordered system of parallel microtubules which traverse the entire length of the cell [21]. It seems highly likely that the conversion of this random disorganized system into a patterned structure is the phenomenon which underlies the change in cell morphology attending reverse transformation. Experiments also demonstrated that organization of the microfibrils is a concomitant action of the reverse transformation process and that both the cell surface and the inner cell structures are involved in this profound change in habitus.

Protein synthesis is not required for the morphologic change in either direction [22]. Reverse Transformation can be effected by a variety of cyclic AMP derivatives, and even by cAMP itself, provided that the hydrolytic action of the cellular phosphodiesterases is inhibited by agents like theophylline or 3-isobutyl-methylxanthine. Finally, experiments still in progress indicate that DBcAMP can modulate the activity of cell surface antigens. Dr. A. W. Hsie, in his paper appearing in this volume, presents further recent experiments carried out by him and his co-workers in this system.

The foregoing experiments make it obvious that cyclic AMP is critically involved in conversion of the cell from a loose, disorganized structure with a random array of microtubules into a highly ordered pattern. Studies from several laboratories have now become available demonstrating that cyclic AMP affects a large variety of mammalian cells in a fashion basically similar to that of the CHO-Kl cell. Obviously, the highly ordered pattern would appear likely to promote coordinated cell function as required by normal cells, as opposed to the regulation-failure which is characteristic of cancer. The question then arises: Does the microtubular system communicate with the genetic structure which is the ultimate repository of information which controls protein synthesis and, therefore, ultimately all biosynthetic activities?

A possible working hypothesis which we are currently attempting to test visualizes the microtubules when properly organized as making attachment to particular sites in or around the genetic appartus, so as to cover up or expose different regions of the genome. This suggestion may not seem so radical when it is recalled that in mitosis, microtubular structures extend from the centrioles to the centromeric regions of each chromosome.

Perhaps, then, similar direct or indirect attachment sites affecting other chromosomal regions may exist in interphase, so that the microtubules help regulate the spectrum of the genome available for specific genetic biochemical activities. The hypothesis postulates a direct connection between the morphologic state of the cell and regulation of the activity of its genome.

The various papers presented in this conference illustrate the power and excitement which now characterize mammalian somatic cell biology. My work in science has been blessed far beyond my desserts in that it has been possible to see a few original ideas develop to fruition and to have had a small effect on human health problems. But treasured most of all has been the opportunity to be associated with friends, colleagues, students, and teachers who have demonstrated how scientific creativity can be combined with affection for co-workers and concern for the problems of mankind.

REFERENCES

1. Hankinson I: Mutants of the Chinese hamster ovary cell line requiring alanine and glutamate. Somatic Cell Genet 2: 497–507, 1976.

2. Jones C, Moore EE: Isolation of mutants lacking branched-chain amino acid transaminase. Somatic Cell Genet 2: 235–243, 1976.

3a. Patterson D: Biochemical genetics of Chinese hamster cell mutants with deviant purine metabolism. III. Isolation and characterization of a mutant unable to convert IMP to AMP. Somatic Cell Genet 2: 41–53, 1976.

3b. Patterson D: Biochemical genetics of Chinese hamster cell mutants with deviant purine metabolism. IV. Isolation of a mutant which accumulates adenylsuccinic acid and succinylaminoimidazole carboxamide ribotide. Somatic Cell Genet 2: 189–203, 1976.

3c. Tu AS, Patterson D: Biochemical genetics of Chinese hamster cell mutants with deviant purine metabolism. Enzymatic studies of two adenine-requiring mutants. Biochem Genet, in press, 1976.

4. Patterson D, Jones C: Biochemical genetics of Chinese hamster cell mutants with deviant purine metabolism. Isolation, selection, and characterization of a mutant lacking hypoxanthine-guanine phosphoribosyl transferase activity by nutritional means. Somatic Cell Genet 2: 429–439, 1976.

5a. Puck, TT, Kao FT: Genetics of somatic mammalian cells. V. Treatment with 5 bromodeoxyuridine and visible light for isolation of nutritionally deficient mutants. Proc Nat Acad Sci USA 58: 1227–1234, 1967.

5b. Kao FT, Puck TT: Genetics of somatic mammalian cells. IV. Properties of Chinese hamster cell mutants with respect to the requirement for proline. Genetics 55: 513–529, 1967.

5c. Valle D, Downing SJ, Harris SC, Phang JM: Proline biosynthesis: multiple defects in Chinese hamster ovary cells. Biochem Biophys Res Comm 53: 1130–1136, 1973.

6. Stomato TD, Hohmann LK: A replica plating method for CHO cells using nylon cloth. Cytogenet Cell Genet 15: 372–379, 1975.

7. Moscona AA: Surface specification of embryonic cells: lectin receptors, cell recognition, and specific cell ligands. *In:* Moscona AA: The Cell Surface in Development. New York, John Wiley & Sons, pp. 1974, 67–100.

8. Puck TT, Wuthier P, Jones C, Kao FT; Genetics of somatic mammalian cells XIII. Lethal antigens as genetic markers for study of human linkage groups. Proc Natl Acad Sci USA 68: 3102–3106, 1971.
9. Wuthier P, Jones C, Puck TT: Surface antigens of mammalian cells as genetic markers II. J Exp Med 138: 229–244, 1973.
10. Jones C, Wuthier P, Puck TT: Genetics of somatic cell surface antigens. III. Further analysis of the A_L marker. Somatic Cell Genet 1: 235–246, 1975.
11. Moore EE, Jones C, Puck TT: Cell surface antigens. IV. Immunological correspondence between glycophorin and the a_1 cell surface antigen. Cytogenet Cell Genet 17: 89–97, 1976.
12. Marchesi VT, Tillack TW, Jackson RL et al: Chemical characterization and surface orientation of the major glycoprotein of the human erythrocyte membrane. Proc Natl Acad Sci USA 69: 1445–1449, 1972.
13. Jackson RL, Segrest JP, Kahane I, Marchesi VT: Studies on the major sialoglycoprotein of the human red cell membrane. Isolation and characterization of tryptic glycopeptides. Biochemistry 12: 3131, 1973.
14a. Kao FT, Jones C, Puck TT: Genetics of somatic mammalian cells: genetic, immunologic, and biochemical analysis with Chinese hamster cell hybrids containing selected human chromosomes. Proc Natl Acad Sci USA 73: 193–197, 1976.
14b. Kao FT, Jones C, Puck TT: Paper in preparation, 1976.
15. Waldren CA, Rasko I: The action of caffeine on the survival of x- and uv-irradiated mammalian cells. Radiat Res, in press, 1976.
16. McCann J, Choi E, Yamasaki E, Ames BN: Detection of carcinogens as mutagens in the *Salmonella* microsome test: assay of 300 chemicals. Proc Natl Acad Sci USA 72: 5135–5139, 1975.
17. Hsie AW, Puck TT: Morphological transformation of Chinese hamster cells by dibutyryl adenosine cyclic 3′:5′-monophosphate and testosterone. Proc Natl Acad Sci USA 68: 358–361, 1971.
18. Johnson GS, Friedman RM, Pastan I: Restoration of several morphological characteristics of normal fibroblasts in sarcoma cells treated with adenosine 3′:5′-monophosphate and its derivatives. Proc Natl Acad Sci USA 68: 425–429, 1971.
19a. Prasad KN, Hsie AW: Morphological differentiation of mouse neuroblastoma cells *in vitro* by dibutyryl adenosine cyclic 3′:5′-monophosphate. Nature [New Biol] 233:141–142, 1971.
19b. Furmanski P, Silverman DJ, Lubin M: Expression of differentiated functions in mouse neuroblastoma mediated by dibutyryl 3′:5′-monophosphate. Nature [New Biol] 233: 413–415, 1971.
20. Willingham MC, Pastan I: Cyclic AMP and cell morphology in cultured fibroblasts: effects on cell shape, microfilament and microtubule distribution, and orientation to substratum. J Cell Biol 67: 146–159, 1975.
21. Porter KR, Puck TT, Hsie AW, Kelley D: An electron microscope study of the effects of dibutyryl cyclic AMP on Chinese hamster ovary cells. Cell 2: 145–162, 1974.
22. Patterson D, Waldren CA: The effect of inhibitors of RNA and protein synthesis on dibutryl cyclic AMP mediated morphological transformations of Chinese hamster ovary cells *in vitro*. Biochem Biophys Res Comm 50: 566–573, 1973.

Cell Surface Modulation

Gerald M. Edelman, M.D., Ph.D.

A number of lines of investigation are now converging on the goal of understanding the organization, specificity, and control properties of the cell surface. The picture that has emerged so far differs in many ways from that of a decade ago. It appears that: (1) the lipid bilayer is a fluid structure in which embedded surface proteins can diffuse; (2) although this diffusion appears to be random and independent for different surface proteins, perturbation of the surface by cross-linkage of glycoproteins can lead to restriction of receptor mobility; (3) this global modulation effect (anchorage modulation) is mediated by cytoplasmic structures, particularly microtubules; and (4) local surface modulation may also occur, in which the interaction of a cell surface receptor with another protein can change the pattern presented to other cells or to interacting molecules in solution.

Both global and local surface modulation phenomena depend upon the architecture of the cell membrane and specifically upon the lateral mobility of the receptors. A variety of changes in the structural pattern or the dynamics of cell surface receptors can be seen in different animal cells, providing examples of surface modulation (Table 1). Although a general and completely logical classification of surface modulation phenomena is not presently feasible, a provisional classification such as that in Table I serves to emphasize the widespread occurrence of these phenomena.

Antigenic modulation [4,5] in which cellular interaction with antibodies leads to a decrease in a particular surface antigen has been studied in various systems. A similar phenomenon has also been observed [17,27] in paramecia: interaction of antibodies with a specific serologic type of surface antigen leads to disappearance at the surface of that antigen and appearance of antigens of a new type. Viral interactions may lead to alterations in the cell surface, including the appearance of new antigens [25], a decrease in the amount of surface histocompatibility antigens [24], and viral budding at particular sites [7]. Transformation by oncogenic viruses has also been observed to produce alterations in cell surface glycoproteins—for example, disappearance of the LETS (large external transformation sensitive) protein [21]—as well as gross changes in surface glycolipids [20].

From the Rockefeller University, New York, N.Y.

Acknowledgments: This work was supported in part by PHS grants AI 11378, AI 09273, AM 04256, and HD 09635, and by grant RF 70095 from the Rockefeller Foundation.

TABLE 1. Phenomena Connected with Surface Modulation Events in Various Systems

Noncovalent interactions with local alteration of specific receptors
Antibody and antigen binding (antigenic modulation [4,5])
Alteration of histocompatibility antigens [11,19,32,33,41]
Viral attachment and budding [7]
Insulin binding [10,18] and other hormonal interactions
Noncovalent interactions with global alteration of the cell surface
Cross-linkage of surface glycoproteins by lectins with anchorage modulation [14,16,36,38,39]
Capping [34]
Covalent alteration resulting in surface modulation
Proteolytic cleavage of surface glycoproteins [21,30]
Action of glycosyl transferases [29]
Complex cellular interactions
Sperm-egg interaction [28]

Hormonal interactions, such as the binding of insulin, can result in alterations of the behavior of insulin receptors on cells as well as in their disappearance [10,18]. Fertilization in certain species results in a highly specialized surface alteration of eggs that prevents polyspermy [28].

The molecular mechanisms of most of the phenomena listed in Table 1 are presently unknown. Only in a small number of cases can we relate the phenomenology to structure, and then only in a preliminary way. Most of these are related to the immune system, but even in those that are not, specific antibodies have proven to be key tools for elucidation of molecular events in surface modulation phenomena.

In this paper, I wish briefly to consider some examples of global and local surface modulation and their possible relation to cell function. These examples include global anchorage modulation induced by cross-linkage of cell surface glycoproteins, local modulation of histocompatibility antigens and its role in immune recognition of cells, and evidence for the existence of local modulation in cell interactions of the embryonic chick neural retina.

GLOBAL SURFACE MODULATION AND ITS RELATIONSHIP TO CELL FUNCTION

Anchorage Modulation

Taylor et al. [34] have described the results of perturbing the cell surface by divalent antibodies to a particular receptor, a process which results in so-called "patch formation" and subsequently in "cap formation." Cross-linkage by the specific divalent antibody results in a diffusion-controlled nucleation into patches of the particular receptors binding that antibody. If the cell is metabolizing actively, these cross-linked patches of receptors are then systematically gathered within minutes to one pole of the cell to become caps and to undergo endocytosis or be cast off. Patching, which is the primary molecular event in this entire process, results in surface modulation, as does subsequent capping. To date, however, no physiologic function has been identified for these modulation events.

Although the existence of capping implies that the cell possesses structures capable of inducing active movements of aggregates of receptors, it does not reveal much about the anchorage of individual receptors. The key observation relating to the reversible anchorage of individual receptors was made when nonsaturating amounts of the plant lectin concanavalin A (Con A), which binds to the carbohydrate portion of various cell surface proteins, were added to lymphocytes [14,36,38–40]. Under these conditions, subsequent addition and binding of specific antibodies against various receptors failed to induce patches and therefore also did not induce caps. This restriction (or anchorage modulation) was reversed upon removing the lectin, and thus it did not result from permanent interference with metabolism or from cell death. Moreover, even local application of Con A to one portion of the cell surface resulted in the restriction of the movement of the receptors on the rest of the cell surface [14]. This indicates that only a small fraction of the receptors need be bound for all cell surface receptors to be restricted in their movements. Anchorage modulation is therefore a propagated phenomenon [14], resulting from amplication of the effects of the initial binding signal and its extension to the whole cell surface. Anchorage modulation of a variety of receptors has been studied mainly in lymphocytes, but it also has been observed on a variety of different cells [14], with the H-2 antigen used as a marker, and therefore it is a general phenomenon.

A number of experiments [14,16,36,38–40] indicate that anchorage modulation is triggered by cross-linkage of certain surface glycoproteins. Furthermore, it appears to be mediated by a collection of submembranous structures rather than by changes at the outside of the bilayer or changes in lipid fluidity. The conclusion [12,16] that this global surface change is mediated by cytoplasmic structures is supported by experiments with drugs that disrupt microtubular structures or interfere with their assembly. If colchicine or various *Vinca* alkaloids are added to lymphocytes in concentrations ranging from 10^{-6} to 10^{-4} M, anchorage modulation is reversed in many of the cells, and this effect is itself reversed by removal of the drugs. Inactive derivatives of colchicine such as lumicolchicine, which has no effect on microtubules, do not reverse anchorage modulation [36,39]. In experiments with colchicine-resistant cell lines, it has recently been shown [2] that colchicine must enter the cytoplasm to affect anchorage modulation.

It is not known how the surface receptors reversibly interact with microtubules, but inasmuch as microtubules are not present at the inner lamella of the lipid bilayer, an additional linkage molecule appears to be required. If microfilaments are involved, they must be in a cytochalasin-resistant form, for this drug inhibits capping, but has no effect on anchorage modulation. Besides cytochalasin-resistant microfilaments, the linkage might consist of various assembly states of tubulin subunits [37] or of an additional protein, possibly α-actinin.

We have proposed a minimal model [12] to account for anchorage modulation. Obviously, the details of this model will have to change as knowledge of the molecules involved accumulates. In this model (Fig. 1) the appropriate surface modulating assembly (SMA) has a tripartite structure: (1) a subset of glycoprotein receptors that penetrate the membrane and confer specificity on the system; (2) various actin-like microfilaments and their associated proteins, such as myosin, conferring the properties of coordinated movement necessary for capping; and (3) dynamically assembling microtubules, both to provide anchorage of the receptors and to allow propagation of signals to and from the cell surface.

It is assumed that the receptors can exist in two states, anchored and free. Cross-linkage of certain glycoprotein receptors alters the various equilibria between the microfilaments, microtubules, and their subunits and induces a propagated assembly (see [14]) of microtubules as well as fixation of microfilaments. As a result, there is a shift to a larger proportion of an-

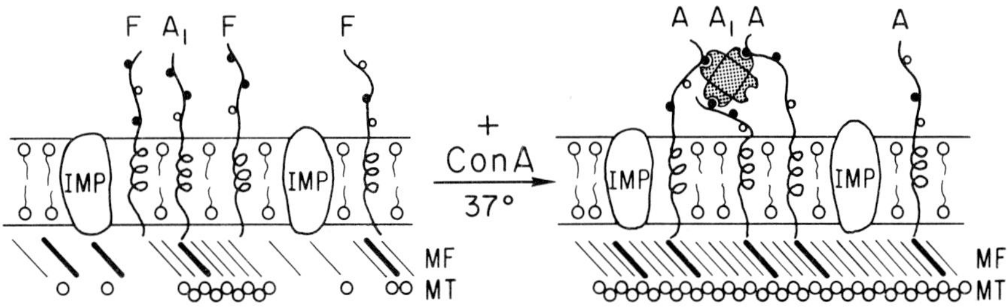

FIG. 1. Anchorage modulation. This process is proposed to be mediated by a surface modulating assembly (SMA) consisting of receptors, microfilaments (MF), and microtubules (MT). A_1, preexistent anchored state; A, induced anchored state shown after binding of tetravalent Con A (stippled); F, free state of surface receptors; IMP, intramembraneous particles. Heavy bars in MF region represent either cytochalasin-resistant microfilaments or an unidentified anchoring molecule. Cross-linkage of certain glycoproteins leads to propagated anchorage and microtubular assembly. (Figures 1, 2A-B and 3 are reproduced by permission from Edelman GM: Surface modulation in cell recognition and cell growth. Science 192: 218–226, 1976.)

chored receptors. Conversely, changes in the state of the cytoplasmic microfilaments and microtubules can alter the mobility and distribution of surface receptors. Disruption of microtubules by drugs would still leave the microfilaments and their associated proteins free to induce capping, either directly or by inducing lipid flow [6]. Obviously, alternative models of the SMA are possible. The main purpose in formulating the present model is to emphasize the need to explain some rather extraordinary phenomena that arise after perturbing the cell surface. Whatever its detailed structure turns out to be, the evidence for action of the SMA is now persuasive. A direct structural and chemical demonstration of the postulated interactions is necessary to confirm or reject this particular model of the SMA.

Although the major function of the SMA is unknown, there is now evidence to suggest that anchorage modulation can inhibit the process of mitogenic induction of lymphocytes [13] and the continued division of transformed cells [35], as well as affect the alteration of cell shape [31]. This raises the intriguing possibility that components of the SMA may be involved in regulating and coordinating signals for induction of cell movement and growth as well as cell interaction. If so, its action would be as a negative control element; there is so far no evidence that it is involved directly in the signals themselves.

This notion that the SMA may be a negative regulator of cell activities has been described elsewhere [12] and will not be pursued at length here. Two sets of observations are worth noting briefly, however. First, chicken fibroblasts infected with *ts* mutants of Rous sarcoma virus showed, at the permissive temperature (37°), a correlation between unrestrained growth, altered morphology, and disordered microfilament networks and microtubular arrays as detected by fluorescent antibodies against actin and tubulin [15]. At the same temperature, these cells also showed greatly reduced anchorage modulation as tested with Con A. At nonpermissive temperatures [41°], however, the morphology, microfilament and microtubule structure, and anchorage modulation all resembled those of normal fibroblasts. These observations, which were made in the absence of drugs, are in accord with the suggested implication of microtubules in surface modulating events. In addition, however, they also imply that components of the SMA are direct or indirect targets for the action of the product of the *src* gene that is responsible for cellular transformation by the virus. This raises the possibility that the deregulation of growth and the disturbance of the SMA may be related [15].

Another set of biochemical observations indicates that anchorage modulation may not prevent mitogenic signaling from occurring but may affect the transfer of a cytoplasmic signal to the cell nucleus. Lymphocytes in which modulating doses of Con A are present do not undergo mitogenesis [35]. If, however, Con A is released from the cell surface after 48 hours by addition of α-methyl mannoside, a competing sugar, the cells rapidly undergo mitogenic transformation and DNA synthesis. The use of a new cell-free assay for detecting the onset of DNA synthesis (i.e., initiation of replication) indicates that Con A-stimulated lymphocytes have a factor in their cytoplasm that is not present in unstimulated lymphocytes [22]. This factor (or group of factors) was observed in a great variety of proliferating cells but was not found in resting cells. Curiously enough, it is present in cells subject to anchorage modulation in the presence of high doses of Con A, yet these cells do not undergo mitogenic alteration until Con A release, as already mentioned. This raises the possibility that in the totally modulated cell, the SMA may act to prevent transport of such essential factors to the cell nucleus.

Clearly, the relation of cell surface modulation to cell function, particular growth control and the control of movement requires much more study. Although the data are scanty and the models provisional, enough information exists to indicate that this is a particularly rewarding line of inquiry. At the very least, it points up an outstanding unsolved problem: to determine the nature of those structures responsible for coordinating the commitment of the cell to division, motion, and interaction with other cells.

LOCAL MODULATION AND CELL RECOGNITION

Local Modulation Involving H-2 Antigens in the Immune Recognition of Cells

The major histocompatibility antigens, termed HLA in man and H-2 in the mouse, are a group of surface glycoproteins possessing extensive genetic polymorphism. Many lines of evidence indicate that they are responsible for allograft rejection by cytotoxic T lymphocytes [23]. This is probably not the major function of these surface antigens, for the same types of cytotoxic lymphocytes are also capable of recognizing and destroying virally-infected syngeneic cells [11] and syngeneic tumor cells [32,33]. Indeed, recent studies [32,33,41] suggest that an interaction between H-2 antigens and viral antigens may be necessary for the action of such cytotoxic T cells. The complete understanding of this interaction requires a detailed structural analysis of these glycoproteins. Recent sequence work (reviewed in [9]) has provided a model of these molecules on the cell surfaces and a structural basis for their extreme polymorphism.

A clue to the function of histocompatibility antigens was provided by studies [11] prompting the hypothesis that these molecules may be modified by viral infection. This opens up the possibility that these antigens are involved in local surface modulation. More recently, it has been shown

[19,32,33] that the recognition and lysis of tumor cells by H-2 compatible lymphocytes or syngeneic lymphocytes require the participation of the H-2 antigens on the surface of the tumor cells. In functional studies designed to test this hypothesis [32,33], mouse lymphoid tumor cells (P388, histocompatibility type H-2^d) were used to immunize mice of the H-2 compatible strain BALB/c. The interesting finding is that specific antiserum to the H-2^d antigen blocked cytotoxic lysis of P388 cells (Fig. 2A). In contrast, irrelevant antiserums or those against other H-2 specificities did not block this lysis, or blocked it very poorly. Similar results were obtained both with another tumor line, EL4 (H-2^b) and by other workers [19] in a different system. An extensive analysis [33] has indicated that it is the H-2 antigens

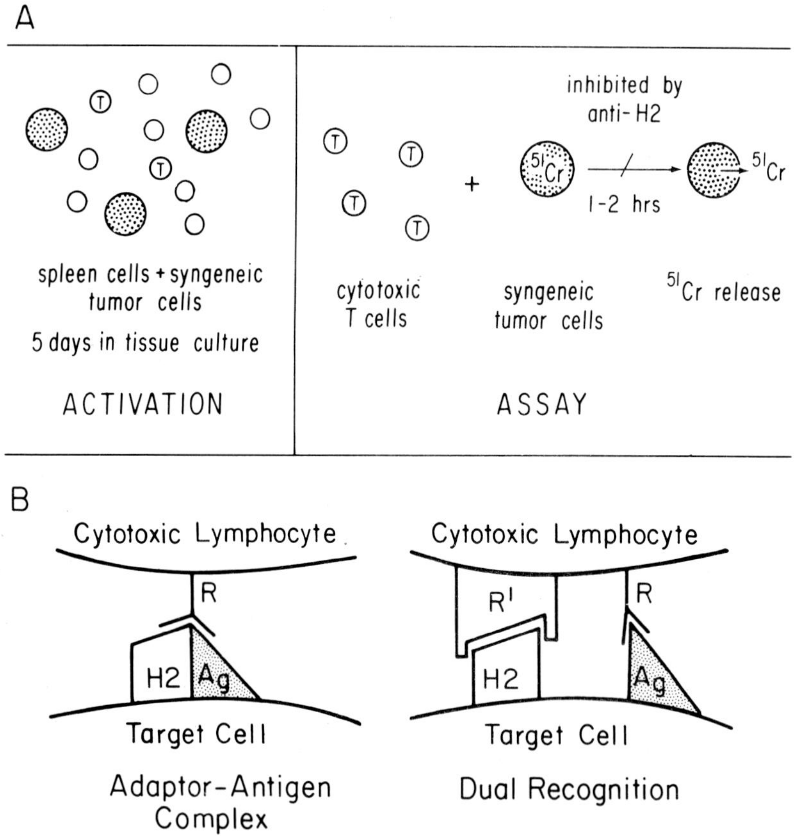

FIG. 2. (A) Assay for demonstrating the participation of H-2 antigenes in killing (^{51}Cr release) of syngeneic tumor cells by T lymphocytes [32,33]. (B) Alternative models for the immune recognition of tumor cells [32,41]. Abbreviations: R, T cell receptor; Ag, tumor viral antigen; R′, second receptor recognizing H-2 antigen.

on the target cell that are involved in the recognition or killing events. It is particularly significant that H-2 antigens on the target tumor cells can participate in the interactions with the cytotoxic lymphocytes even though these lymphocytes are of the same H-2 type as the target.

Two models involving surface modulation [32,41] appear to provide reasonable explanations of these findings (Fig. 2B). Either there is a dual recognition of H-2 and tumor-associated antigens on the cell surface by two separate receptors on the cytotoxic T lymphocytes, or a physical complex is formed between H-2 antigens and tumor-associated antigens on the target, and this modulated cell surface complex is recognized by the T lymphocytes. Although the dual recognition hypothesis cannot so far be excluded, there is some preliminary evidence [32,33] that is in accord with the occurrence of an adaptor-antigen complex containing H-2 molecules (Fig. 2B). This evidence was obtained by using separate antiserums directed against either H-2 antigens or against viral antigens associated with the tumors in order to patch and cap both antigens together on the tumor cell surface. For example, capping and patching of H-2 molecules by appropriately absorbed antiserums that were specific for the H-2 antigen resulted in co-patching and co-capping of the viral antigens. Reciprocal experiments in which Rauscher leukemia virus antigens were capped by antiserums directed against the viral antigens resulted in partial redistribution of H-2 antigens into caps.

These experiments have been extended and confirmed, particularly to rule out the presence of any unexpected cross-reactivity between the various antiserums. The results are in accord with the hypothesis [32] that H-2 molecules may serve as adaptors that combine with viral antigens in the cell surface to form hybrid antigens containing elements of self (H-2) and non-self (virus). An adaptor-antigen complex may be recognized by a subclass of T lymphocytes having a repertoire of receptors directed against such hybrid antigens. In the adaptor hypothesis, the lateral mobility of receptors provides grounds for specific interactions between glycoproteins on the cell surface and foreign molecules such as those of viruses. This hypothesis provides an example of local noncovalent modification of the cell surface followed by recognition of the modulation by a specific set of antibody receptors on T lymphocytes. This implies an asymmetric interaction between syngeneic cells in which definite, but not absolute, restrictions are imposed by the specificity of the H-2 gene product.

Possible Roles of Local Modulation in Cell Interactions of Chick Embryonic Neural Tissues

It is tempting to suppose [26] that specific cell-cell interactions in developing tissues are mediated by tissue-specific ligands analogous to the antibodies involved in the recognition of the difference between self and non-self. Indeed, the H-2 antigens are specified by genes at a complex genetic locus [23] that is known to be involved in immune recognition and that has been proposed to function in embryonic cell recognition [1,3].

Recent experiments [30] on developing chick brain and retina suggest, however, that initial cell interactions measured by a two-cell assay are mediated by a surface protein or cell association molecule (CAM) that is shared by both tissues. This is perhaps not surprising, in view of the close relationship of brain and neural retina. Indeed, CAM was found on practically all cells of the developing retina and brain.

An antibody preparation reactive with CAM was found to block cell interactions and was able to bind each tissue to nylon fibers to which it was coupled (Fig. 3). Of particular interest was the finding [30] that the age-dependent increases in cell numbers during development are correlated with the adhesiveness of retinal and brain cells to fibers coated with antibodies that bind to CAM. A similar correlation was seen in the two-cell assay for cell-cell interactions among brain and retina cells of different ages. This finding suggests that the surface of developing cells is altered with develop-

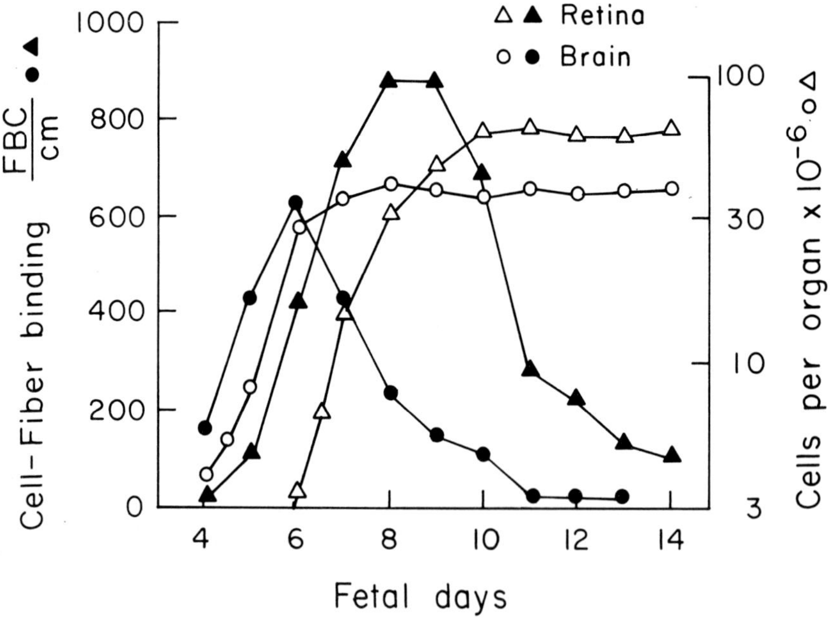

FIG. 3. Correlation between increase in cell numbers of chick brain and retina and the ability of these cells to bind specifically to nylon fibers coated with antibodies reactive with CAM [30]. The decline in cell-fiber interactions occurs at the time the increase in cell numbers reaches a plateau. Abbreviation: FBC/cm, fiber binding cells per centimeter of fiber.

mental age and that this alteration is correlated with the net increase in cell numbers (Fig. 3). The decrease in the adhesiveness of late stage cells in tissue culture might reflect the increase with time in the size of a subpopulation of cells in one stage of the cell cycle. At the molecular level, it might also be the result of surface events related to anchorage modulation.

Detailed understanding of these events will depend on counting the number of CAM molecules per cell as a function of age. The obvious possibility that these molecules are removed from the surface has not yet been excluded, for example. To account for the change with age, the main possibilities to be distinguished are whether the *synthesis* of CAM is a function of age or whether it is shed or cleaved from the cell membrane at a certain period of development. In any case, there is a suggestion that local modulation of a cell surface molecule contributes to changes in the adhesiveness of developing tissues. The challenge is now to define the role of such molecules *in vivo*.

CONCLUSION

This article has been concerned with some evidence for global and local surface modulation and for transmembrane control *of* cell receptors as well as *by* cell surface receptors. Investigation of this set of phenomena raises the possibility that, together with specific surface receptors, the submembranous microtubules and microfilaments may play a role in regulatory events connected with cell growth, movement, and interactions. In view of the generality of global modulation phenomena, such a surface modulating assembly is probably found in a variety of cells. On the other hand, local modulating events are probably highly specific and diverse as seen in the examples of immune recognition and embryonic cell adhesion discussed here.

All of these phenomena indicate the importance of certain properties of the cell surface membrane, such as fluidity, the transmembranous nature of most complex receptors, and their interaction with cytoplasmic proteins. In addition, they imply that during evolution there were stringent requirements on the cell surface for the development of coordinating mechanisms to control cell functions as well as specific cell-cell interactions.

REFERENCES

1. Artzt K, Bennett D: Analogies between embryonic (T/t) antigens and adult major histocompatibility (H-2) antigens. Nature 256: 545–547, 1975.
2. Aubin JE, Carlsen SA, Ling V: Colchicine permeation is required for inhibition of con-

canavalin A capping in Chinese hamster ovary cells. Proc Natl Acad Sci USA 72: 4516–4520, 1975.
3. Bodmer WF: Evolutionary significance of the HL-A system. Nature 237: 139–145, 1972.
4. Boyse EA, Stockert E, Old LJ: Modification of the antigenic structure of the cell membrane by thymus-leukemia (TL) antibody. Proc Natl Acad Sci USA 58: 954–961, 1967.
5. Boyse EA, Old LJ: Some aspects of normal and abnormal cell surface genetics. Ann Rev Genet 3: 269–290, 1969.
6. Bretscher MS: Directed lipid flow in cell membranes. Nature 260: 21–23, 1976.
7. Choppin PW, Compans RW: Reproduction of paramyxoviruses. *In* Fraenkel-Conrat H, Wanger RR (Eds): Comprehensive Virology. New York, Plenum, 1975, Vol. 4, pp 95–178.
8. Craig SW, Cuatrecasas P: Mobility of cholera toxin receptors on rat lymphocyte membranes. Proc Natl Acad Sci USA 72: 3844–3848, 1975.
9. Cunningham BA, Henning R, Milner RJ et al: Structure of murine histocompatibility antigens. *In* Proceedings of the XLI Cold Spring Harbor Symposium on Quantitative Biology: Origins of Lymphocyte Diversity, 1976, in press.
10. DeMeyts P, Roth J, Neville DM Jr et al: Insulin interactions with its receptors: experimental evidence for negative cooperativity. Biochem Biophys Res Commun 55: 154–161, 1973.
11. Doherty PC, Zinkernagel RF: T cell mediated immunopathology in viral infections. Transplant Rev 19: 89–120, 1974.
12. Edelman GM: Surface modulation in cell recognition and cell growth. Science 192: 218–226, 1976.
13. Edelman GM: Surface alterations and mitogenesis in lymphocytes. *In* Clarkson B, Baserga R (Eds): Control of Proliferation in Animal Cells: Cold Spring Harbor Conferences on Cell Proliferation. New York, Academic Press, 1974, Vol. 1, pp 357–379.
14. Edelman GM, Wang JL, Yahara I: Surface modulating assemblies in mammalian cells. *In* Goldman R, Pollard T, Rosenbaum J (Eds): Cell Motility. Cold Spring Harbor Laboratory, 1976, Vol. 3, pp 305–321.
15. Edelman GM, Yahara I: Temperature-sensitive changes in surface modulating assemblies of fibroblasts transformed by mutants of Rous sarcoma virus. Proc Natl Acad Sci USA 73: 2047–2051, 1976.
16. Edelman GM, Yahara I, Wang JL: Receptor mobility and receptor-cytoplasmic interactions in lymphocytes. Proc Natl Acad Sci USA 70: 1442–1446, 1973.
17. Finger I: Surface antigens of *P. aurelia. In* van Wagtendonk WJ (Ed): Paramecium, A Current Survey Amsterdam, Elsevier, 1974, pp 131–164.
18. Gavin JR III, Roth J, Neville DM et al: Insulin-dependent regulation of insulin receptor concentrations: a direct demonstration in cell culture. Proc Natl Acad Sci USA 71: 84–88, 1974.
19. Germian RN, Dorf ME, Benacerraf B: Inhibition of T lymphocyte-mediated tumor-specific lysis by alloantisera directed against H-2 serological specificities of the tumor. J Exp Med 142: 1023–1028, 1975.
20. Hakomori S: Cell density-dependent changes of glycolipid concentrations in fibroblasts, and loss of this response in virus-transformed cells. Proc Natl Acad Sci USA 67: 1741-1747, 1970.
21. Hynes RO: Role of surface alterations in cell transformation: the importance of proteases and surface proteins. Cell 1: 147–155, 1974.
22. Jazwinski SM, Wang, JL, Edelman GH: Initiation of replication in chromosomal DNA induced by extracts from proliferation cells. Proc Natl Acad Sci USA 73: 2231–2235, 1976.

23. Klein J: Biology of the Mouse Histocompatibility-2 Complex. New York, Springer, 1975.
24. Lilly F, Pincus T: Genetic control of murine viral leukemogenesis. Adv Cancer Res 17: 231–277, 1973.
25. Möller G (Ed): Immunological surveillance against neoplasia. Transplant Rev 7, 1971.
26. Moscona AA: Cell aggregation. *In* Bittar EE (Ed): Cell Biology in Medicine New York, Wiley, 1973, pp. 571–591.
27. Nanney DL: Ciliate genetics: patterns and programs of gene action. Ann Rev Genet 2: 121–140, 1968.
28. Piko L: Gene structure and sperm entry in mammals. *In* Metz CB, Monroy A (Eds): Fertilization. New York, Academic Press, 1969, Vol. 2, pp. 325–403.
29. Roth S, White D: Intercellular contact and cell-surface galactosyl transferase activity. Proc Natl Acad Sci USA 69: 485–489, 1972.
30. Rutishauser U, Thiery J-P, Brackenbury R et al: Mechanism of adhesion among cells from neural tissues of the chick embryo. Proc Natl Acad Sci USA 73: 577–581, 1976.
31. Rutishauser U, Yahara I, Edelman GM: Morphology, motility, and surface bahavior of lymphocytes bound to nylon fibers. Proc Natl Acad Sci USA 71: 1149–1153, 1974.
32. Schrader JW, Cunningham BA, Edelman GM: Functional interactions of viral and histocompatibility antigens at tumor cell surfaces. Proc Natl Acad Sci USA 72: 5066–5077, 1975.
33. Schrader JW, Edelman GM: Participation of the H-2 antigen of tumor cells in their lysis by syngeneic T cells. J Exp Med 143:601–614, 1976.
34. Taylor RB, Duffus WPH, Raff MC, dePetris S: Redistribution and pinocytosis of lymphocyte surface immunoglobulin molecules induced by anti-immunoglobulin antibody. Nature 233: 225–229, 1971.
35. Wang JL, McClain DA, Edelman GM: Modulation of lymphocyte mitogenesis. Proc Natl Acad Sci USA 72: 1917–1921, 1975.
36. Yahara I, Edelman GM: Effects of concanavalin A on the mobility of lymphocyte surface receptors. Exp Cell Res 81: 143–155, 1973.
37. Yahara I, Edelman GM: Electron microscopic analysis of the modulation of lymphocyte receptor mobility. Exp Cell Res 91: 125–142, 1975.
38. Yahara I, Edelman GM: Modulation of lymphocyte receptor mobility by concanavalin A and colchicine. Ann NY Acad Sci 253: 455–469, 1975.
39. Yahara I, Edelman GM: Modulation of lymphocyte receptor redistribution by concanavalin A, anti-mitotic agents and alterations of pH. Nature 246: 152–155, 1973.
40. Yahara I, Edelman GM: Restriction of the mobility of lymphocyte immunoglobulin receptors by concanavalin A. Proc Natl Acad Sci USA 69: 608–612, 1972.
41. Zinkernagel RF, Doherty PC: Immunological surveillance against altered-self components by sensitized T lymphocytes in lymphocytic choriomeningitis. Nature 251: 547–548, 1974.

Molecular Features of Integral Membrane Proteins of the Human Red Cell Membrane

V. T. Marchesi, M.D., Ph.D.

Integral membrane proteins have been operationally defined as those polypeptide chains which resist solubilization from membranes by simple buffer washings. Such procedures ordinarily release approximately 40–50% of the total protein components of normal human red cell membranes, but the so-called integral membrane proteins require detergents or harsh denaturing agents before they can be associated from the lipid elements of the membrane. On this basis it has been considered likely that these proteins form an integral part of the membrane structure. Although this idea was largely speculative when it was first suggested by Singer, it is now clear that the structural features of some of the integral membrane proteins are remarkably consistent with this idea. This review will attempt to summarize what is now known about the molecular features of glycophorin A, the major sialoglycoprotein of the human red cell membrane, and the polypeptide components called band 3 which represent the most abundant integral membrane protein of this membrane.

GLYCOPHORIN A

Glycophorin A has been isolated by a number of different laboratories using a wide variety of protein solvents (see reviews [4,14,27] for references), and as a result of an intensive amount of study it is now possible to construct a provisional model of its primary structure and offer some ideas as to how and where it is oriented in the intact red cell membrane.

Recently the complete amino acid sequence and the sites of oligosaccharide attachment have been determined [33] and this is shown in Figure 1.

The polypeptide portion of glycophorin A, which comprises approximately 40% of the total dry weight, is made up of 131 amino acids. The distribution of amino acids is rather interesting in that there is a very high concentration of threonine and serine residues located near the N-terminal end of the polypeptide chain. There is also a striking concentration of nonpolar amino acids located approximately midway between the N-ter-

From the Department of Pathology, Yale University, New Haven, Conn.

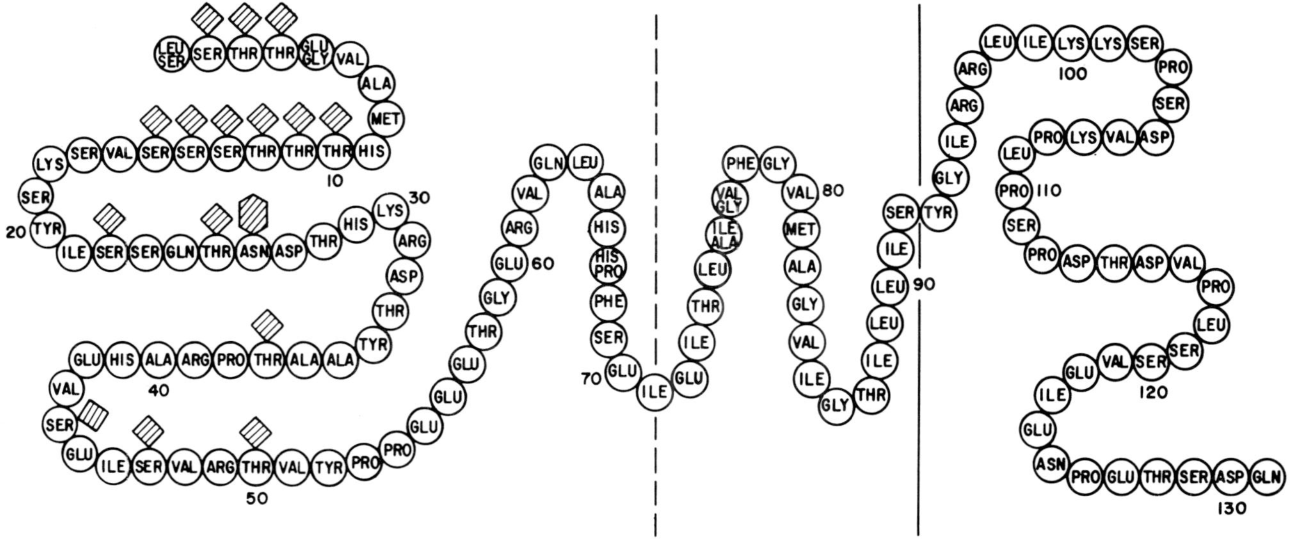

FIG. 1. The amino acids of glycophorin A are arranged in this diagram to simulate, in a very general way, the positions they might have if the glycophorin molecule were to be arranged perpendicular to the lipid bilayer of the membrane. The limits of the bilayer are defined by the two vertical lines. The solid vertical line, which passes between residues 92 and 93, should be the approximate location of the polar groups of the inner half of the phospholipid bilayer. This assignment is based on the results of enzymatic iodination of tyrosine 93 and the distribution of ferritin-antibody conjugates directed against antigenic determinants defined by residues 102–118. Since we do not have comparable data with regard to the amount of the N-terminal end of glycophorin A which is buried with the lipid bilayer, we can only guess at the location of this outer lamella of the bilayer relative to the glycophorin molecule; hence the outer edge of the bilayer is defined by the dashed vertical line.

minal third of the polypeptide and the C-terminal third. In the diagram shown in Figure 1, the segment starting from amino acid 71 and extending through amino acid 90 is depicted between two vertical lines, and this is thought to be the segment of polypeptide which is buried within the lipid bilayer. The reasons for assigning this orientation will be described below.

It is also evident from this diagram that there are 16 different oligosaccharide chains attached to the polypeptide chain. It is significant that the glycosylated residues start from the N-terminal end of the molecule and extend to residue 50, and no additional sugar residues have been detected in amino acids beyond residue 50. Since all the glycosylated residues are concentrated at one end of the molecule and some of these sugars probably represent receptor sites for lectins and perhaps for some blood group antigens, this end of the molecule has been provisionally designated as the *receptor domain*.

The amino acid sequence of the C-terminal 40 amino acids is striking in that this portion of the polypeptide chain contains a large number of charged amino acids with a peculiar clustering of acidic residues at the very C-terminal segment of the chain. This portion of the polypeptide chain also contains a substantial number of prolines which may play a major role in determining the conformation of this part of the molecule.

The 22 amino acid stretch of peptide which connects the glycosylated receptor region with the C-terminal end (Fig. 1) is composed solely of nonpolar residues, and it is logical to suggest that this segment is the part of glycophorin that interacts with the lipids of the membrane.

Glycophorin A Spans the Lipid Bilayer

There is a considerable amount of experimental data in support of the idea that glycophorin A and probably other integral membrane proteins of the red cell are in a transmembrane configuration. This idea was originally based on the attempts to differentially label portions of the polypeptide chain in intact cell membranes versus membrane preparations which had been prepared by osmotic lysis. Bretscher originally showed that a segment of the polypeptide chain of glycophorin A was not accessible to labeling by a radioactive isotope introduced outside intact cells; however, this inaccessible segment could be labeled when the cell membranes were damaged or rendered permeable to the reagent [3]. One interpretation of this experiment was that part of the polypeptide chain of the molecule extends outside the intact red cell and another part is either buried within the membrane or situated in the cytoplasmic compartment of the cell. This experiment was repeated in several laboratories, using a more gentle labeling procedure, lactoperoxidase iodination, with comparable results [18,24]. Under ideal conditions, lactoperoxidase should selectively iodinate tyrosine residues.

In theory, lactoperoxidase iodination should be an effective probe for analyzing the orientation of glycophorin A, since this molecule has four tyrosines, three of which are located in the glycosylated portion of the polypeptide chain, while the fourth is close to the C-terminal end. Thus, three tyrosines should be labeled when intact cells are incubated with lactoperoxidase, while the fourth, located at position 93, should not be labeled unless the permeability of the membrane is broken. Several investigators have reported the expected results, although other investigators were not able to reproduce these findings [22,25].

Some investigators also consider the positive findings to be ambiguous and suggest that they could be the result of any one of a number of technical artifacts. For example, it has been suggested that rearrangements of membrane proteins might take place as a result of osmotic lysis. It has also been suggested that chemical modifications of membrane proteins which result from the labeling procedures themselves could also lead to rearrangement of the proteins or changes in membrane permeability or both. In either case any one of these changes could serve to block or modify the reactivity of certain exposed proteins or, alternatively, expose other proteins which are normally not exterior to the lipid barrier. Others have also pointed out the need to use mild labeling reagents at saturating levels to lessen the likelihood of selectively labeling either the most accessible or the most reactive groups of proteins rather than sampling the complete spectrum of externally arranged molecules. In addition to these many theoretical caveats, some investigators have failed to differentially label the C-terminal segment of the glycophorin A molecule when labeling reagents were introduced inside the cell. Thus they argue that some of the earlier studies may have been artifactual. For all these reasons, it must be admitted that the evidence for the transmembrane orientation of glycophorin A produced by labeling studies is not definitive.

In the light of the controversy surrounding the results of radiolabeling, we recently reinvestigated this problem and attempted to study the orientation of glycophorin A in intact red cells, using ferritin conjugated antibodies directed against specific polypeptide segments of this molecule [7]. We have immunized rabbits with the cyanogen bromide fragment containing residues 82 through 131 and have found that this segment is antigenic. A radioimmunoassay was developed in order to assess the specificity of the antisera and to determine what segments of the polypeptide chain bore the determinants. By a series of purification steps, we have determined that the antibodies which were used to prepare ferritin conjugates were directed specifically against a 17 amino acid peptide fragment, residues 102 to 118, which is just adjacent to the C-terminal end of the polypeptide chain. Antibodies were further purified by immunoadsorption with a small tryptic peptide also derived from the C-terminal end of the molecule. These antibodies were found to be unreactive with intact cells and were unreactive to peptides derived from the N-terminal glycosylated end of the polypeptide chain. These antibodies also showed remarkable specificity in that they did not react with other sialoglycopeptides isolated from human red cell membranes nor with comparable glycoproteins isolated from a variety of different animal red cell membranes.

Since there was good reason to suspect that the antigenic determinants assigned to residues 102 to 118 were probably on the cytoplasmic side of the red cell membrane, the only available technic for applying labeled antibodies to the inner surfaces of intact cell membranes is to expose conjugated sera to thin sections of red cells which are either fixed and embedded in albumin or rapidly frozen in sucrose media. Such thin sections permit complete access of antibody-conjugates to subcellular sites without having to resort to osmotic lysis or other tricks to break down membrane barriers. Appropriately prepared frozen thin sections are also suitable for high resolution transmission electron microscopy.

Ferritin-antibody conjugates of such sera were found to localize exclusively to sites which were distributed uniformly along the inner surfaces of the intact red cell membranes. No staining was seen on sections prepared from red blood cells from other species, consistent with the immunochemical findings, nor on sections of human red blood cells which were pretreated with unconjugated blocking antisera. Thus, these results provide the first direct evidence that glycophorin A has a transmembrane orientation in intact human erythrocytes.

Glycophorin A May Exist as Multimeric Complexes in the Membrane

Recent studies from this and other laboratories suggest that glycophorin A has a remarkable capacity to form multimeric complexes in the presence of SDS [13,17]. Almost all glycophorin preparations which are analyzed by SDS acrylamide gels contain multiple PAS-staining bands, usually designated PAS-1, 2, and 3. These bands correspond to molecular weights of 83,000, 45,000, and 25,000 when they are compared with the mobilities of standard proteins. However, we are not sure that these values represent the true molecular weights of the glycoproteins since they are known to migrate anomalously on SDS gels. PAS-1 is the predominant form and represents approximately 75% of the total. When this pattern was first obtained it was not clear whether the multiple bands represented a series of chemically distinct polypeptide chains or whether the smaller PAS staining bands represented degradation products of the larger component. In a series of recent studies we have shown that two of the PAS staining bands (PAS-1 and PAS-2) are interconvertible and their relative amounts depend on the conditions used to prepare the glycoproteins for SDS electrophoresis. Glycophorin A appears to exist as a dimeric form, which corresponds to PAS-1, and this dimeric form seems to be stabilized by noncovalent associations between hydrophobic segments of its polypeptid chains [13]. This association can be disrupted by heating the glycoproteins in SDS or by modifying the methionine located in position 81 by specific alkylation in the presence of denaturing agents [26]. The conditions for the selective alkylation are extremely stringent, in that high concentrations of alkylating agents must be used in the presence of either urea or guanidine, and the glycoprotein must have been previously delipidated with organic solvents. Since the presence of either lipid or SDS interfere with the alkylation of methionine 81, and in so doing render the molecular aggregate less susceptible to dissociation, it is conceivable that when the molecule is oriented in the intact cell membrane the surrounding membrane lipids serve a similar function. On the basis of this we might speculate that glycophorin A actually exists as dimeric or possibly even multimeric forms *in situ*. If this proves to be true then the possibility exists for monomer-multimer association-dissociation reactions to take place in membranes which could play a role in regulating the functions of these macromolecules. This hypothetical idea is consistent with recent studies on the apparent mobility of membrane components and the important role the lipid matrix may play in regulating this process.

THE BAND 3 INTEGRAL PROTEIN

Band 3 appears as a diffuse band when RBC membranes are analyzed by SDS-gels and migrates as if it were approximately 90,000 daltons molecular weight. Parts of this polypeptide are labeled when radioactive reagents are added to intact cells [2], and a portion of the polypeptide chain is removed when intact cells are incubated with chymotrypsin or pronase [29,31]. It has also been shown that the polypeptide segments of the band 3 can be labeled by radioactive reagents when they are introduced inside the red cell ghost membranes, or when they are applied to inside-out vesicles. The conclusion of these experiments is that band 3 has parts of its polypeptide chain exposed to the external surface of the cell and parts exposed to the cytosol.

Band 3 may exist in a dimeric form in the membrane since it can be cross-linked by bifunctional reagents and oxidized by appropriate oxidizing conditions to what appears to be a disulfide linked dimeric form [28,35,36].

Whether it exists as such without prior cross-linking or oxidation remains to be seen.

Band 3 has been isolated and purified by a number of laboratories [10,12,31] and found to contain approximately 5–8% carbohydrate on a dry weight basis. This carbohydrate is composed of mannose, galactose, and N-acetylglucosamine in the approximate ratios of 1:2::2 and, in addition, traces of fucose and glucose are frequently found. This composition suggests that the oligosaccharides are of the complex type which are linked to asparagine residues by N-glycosidic bonds.

Since the band 3 polypeptide binds to concanavalin A and ricinus communis lectins [1,21,34], the nonreducing ends of the oligosaccharide units are likely to be mannose and galactose and all the N-acetylglucosamine residues should be in the core components. The oligosaccharide units of the band 3 should have an aggregate molecular weight of 5000 to 8000 if the band 3 polypeptide is really 90,000; thus, it is likely that there will be several oligosaccharide units per individual polypeptide chain. Since pronase digestion results in the production of at least two large fragments of band 3 polypeptide, but only one, the smaller, contains all the sugar, it is conceivable that most of the oligosaccharides are clustered in one particular segment of the polypeptide chain [30]. However, Jenkins and Tanner [16] have recently proposed an S-shaped variant of the model for band 3 on the basis of results obtained from thermolysin peptide maps of iodinated peptides generated by proteolytic digestion of different forms of the band 3 component. It seems likely that the precise molecular arrangement of the carbohydrate units in this molecule will not be settled until the peptide portion containing these sugars is isolated and sequenced.

Band 3 May be Part of an Anion Channel

Cabantchik and Rothstein combined specific inhibitor studies with covalent-labeling reagents and seem to have succeeded in identifying polypeptides of the membrane which mediate anion movements [5,6]. These investigators produced a stilbene derivative, [^{3}H]DIDS, which has the capacity to inhibit anion permeability of red cells and also is able to form covalent bonds with the polypeptides of the membrane to which it is bound, since it contains an isothiocyanate group capable of reacting with free amino groups. Incubation of intact red blood cells with this reagent resulted in the binding of [^{3}H]DIDS to approximately 300,000 sites per red cell, and this was accompanied by complete inhibition of anion exchange. When appropriately treated red cells were lysed and the membranes prepared and analyzed by SDS-gels, essentially all the [^{3}H]DIDS was bound to the band 3 polypeptides. If the cells are incubated with pronase under conditions in which the band 3 polypeptides are degraded to a 65,000 molecular weight peptide, most of the [^{3}H]DIDS is still associated with the latter fragment. Thus, [^{3}H]DIDS seems to bind to the exposed portions of the band 3 polypeptides, but these binding sites are still sufficiently internal, either within the polypeptide itself or the membrane, so that the bulk of the label is inaccessible to digestion by this proteolytic agent.

Ho and Guidotti [15] have found that phosphate transport is inhibited reversibly by sulfanilate anions and irreversibly by the isothiocyanate derivative of sulfanilate. This reagent also binds to approximately 300,000 sites on each red cell, and the radioactive derivatives is attached to polypeptides which migrate in the band 3 regions when the membranes are analyzed by SDS-gels.

Thus, both sets of observations show clearly that the movement of anions is inhibited if the polypeptides which migrate in the band 3 region are modified by the attachment of certain covalent reagents which can otherwise compete with the anion sites. Although a small percentage of the [^{3}H]DIDS label was also associated with the major sialoglycoprotein, the authors feel that this result is unrelated to the inhibition of anion movement, and probably results from the proximity of the sialoglycopeptides to the band 3 peptides. Cabantchik and Rothstein concede that it is not clear whether the polypeptides which migrate in the band 3 regions are chemically homogeneous. Thus it is still a possibility that the anion carrier is represented by one of several different polypeptides which have the same electrophoretic mobilities. However, both sets of investigators argue that the amount of anion flux in red cells is such that a substantial fraction of the protein would have to be involved even if the carrier mechanisms have turnover rates comparable to those of rapid enzymes.

Some preliminary reconstitution studies also support the idea that the integral polypeptide chains of the red cell membrane mediate anion transport [23]. Triton X-100 extracts of red blood cell ghosts (composed largely of band 3, the sialoglycopeptides and several minor glycopeptides) increase (3 to 10-fold) the flux of sulfate anions out of synthetic lecithin vesicles. When the same extracts are prepared from red cells which were pretreated with [^{3}H]DIDS, the anion transport inhibitor, they do not share this sulfate flux enhancing property. DIDS also inhibited sulfate fluxes in vesicles prepared with intact membrane proteins although the inhibition was only partial and was found to be more variable, ostensibly because of scrambled arrangement of the different protein-lipid complexes in such vesicles. These results seem extremely promising, but clearly need further investigation.

INTEGRAL PROTEINS MAY FORM MOBILE COMPLEXES IN THE MEMBRANE

One of the most intriguing and still largely unanswered questions in membrane biology concerns the three dimensional arrangement of proteins at the cell surface. Do membrane proteins occupy fixed positions relative to some internal cytoplasmic structure, or are they in continuous motion, constantly changing their positions within the bilayer? One of the earliest suggestions concerning the arrangement of membrane proteins was the idea that monomolecular sheets of polypeptide coated the phospholipid polar groups of the lipid bilayer. This idea now seems unlikely for many reasons. Most of the evidence suggests that the bulk of the membrane proteins of the red cell membrane exist as discrete macromolecular complexes which are attached to the lipid bilayer either by insertion of hydrophobic segments into the hydrocarbon interior (like glycophorin A) or bound to polar groups of lipids electrostatically.

Our conceptions about the arrangement of proteins in membranes have also been heavily influenced by the spectacular experiments demonstrating membrane fluidity and phospholipid mobility. Since the lipid matrix in which the protein molecules must be inserted is in a constant state of flux, many have reasoned that it is likely that protein molecules are also in a mobile state, although it is not clear whether they are simply floating passively in the lipid sea or are directed by some other agents independent of the surrounding lipids.

A vivid demonstration of the fact that some membrane proteins are mobilizable by external forces has been provided by the results of applying multivalent ligands to lymphocytes and other cells in suspension. Appropriately labeled ligands form patches and caps on such cells within minutes after their application. Other evidence for the mobility of surface components has been provided by cell fusion studies in which it has been shown that surface antigens can freely intermix over the combined surfaces of hybrid cells at a remarkably rapid rate [11]. Both sets of observations demonstrate rather unequivocally that certain types of surface macromolecules are capable of moving within the plane of the membrane. Unfortunately, the dramatic nature of these experiments may be overemphasizing the potential importance of the mobility of membrane proteins. It is likely that only certain types of membrane proteins are completely mobile and others may be more or less in a fixed position at the cell surface relative to some cytoplasmic structures.

It might also be useful to point out that we still do not know whether any proteins in the human red cell membrane are potentially mobile, or whether all of the major integral membrane proteins are firmly fixed at their respective sites.

Band 3 and glycophorin A are probably associated with the intramembranous particles seen by freeze-etching [20,32]. Since these particles do not appear to move within the plane of the intact membrane (based on their inability to patch or cap by the usual technics) it is reasonable to conclude that these proteins are not mobile. The intramembranous particles are movable, however, when osmotically-lysed membranes are incubated with proteases [32] or pH 5.5 buffers [8].

Some investigators have suggested that the major integral proteins of the red cell membrane may not be freely mobile because of the well developed submembranous protein network which is thought to exist on the inner surface of this membrane.

INTERACTIONS BETWEEN INTEGRAL MEMBRANE PROTEINS AND OTHER MEMBRANE COMPONENTS

There have been several recent provocative studies purporting to show that glycophorin A may have specific interactions with other membrane proteins. In particular, it has been suggested that the buried or cytoplasmic segment of glycophorin A may have the capacity to interact with the spectrin polymers and together this complex may have some important function in regulating the stability of the membrane or in determining the relative positions of different membrane proteins. This exciting idea is based on the capacity of antispectrin antibodies to modify the topographic distribution of glycophorin A molecules [19] and attempts to modify the distribution of intramembranous complexes by manipulating the polymeric state of spectrin [9]. Another piece of circumstantial evidence in support of this idea is the recent demonstration that anti-spectrin antibodies and anti-glycophorin A antibodies (directed against the C-terminal segment) both localize to identical sites along the inner surfaces of intact red cell membranes.

However, it must be pointed out that all of these observations are extremely circumstantial, in that they simply show that spectrin and glycophorin A peptide segments are in close proximity to each other along the inner surface of the membrane. A perturbation in any one of these components might be expected to modify the arrangement or the behavior of the neighboring components. Thus it is our feeling that although the idea that spectrin and glycophorin A may represent a functional complex is extremely provocative, there is not sufficient data to really support this generalization. It will be important to determine whether in fact peptide segments of glycophorin A do have the capacity to bind to portions of spectrin polymers. If this can be established, perhaps more meaningful experiments can be designed to determine what the relationship between these two molecules is in the intact membrane and what the functional implications might be.

REFERENCES

1. Adair WL, Kornfeld S: Isolation of the receptors for wheat germ agglutinin and the rincinus communis lectins from human erythrocytes using affinity chromatography. J Biol Chem 249: 4696, 1974.
2. Bretscher MS: A major protein which spans the human erythrocyte membrane. J Mol Biol 59: 351, 1971.
3. Bretscher MS: Major human erythrocyte glycoprotein spans the cell membrane. Nature [New Biol.] 231: 229, 1971.
4. Bretscher MS: Membrane structure: some general principles, Science 181: 622, 1973.
5. Cabantchik ZI, Rothstein A: Membrane proteins related to anion permability of human red blood cells. J Membrane Biol 15: 207, 1974.
6. Cabantchik ZI, Rothstein A: Membrane proteins related to anion permeability of human red blood cells, II effects of proteolytic enzymes on disulfonic stilbene sites of surface proteins. J Membrane Biol 15: 227, 1974.
7. Cotmore S, Furthmayr H, Marchesi VT: J Mol Biol, in press, 1977.
8. Elgsaeter A, Branton D: Intramembrane particle aggregation in erythrocyte ghosts. J Cell Biol. 63: 1018, 1974.
9. Elgsaeter A, Shotton DM, Branton D: Intramembrane particle aggregation in erythrocyte. Biochim Biophys Acta 426: 101, 1976.
10. Findlay JBC: The receptor proteins for concanavalin A and lens culinaris phytohemagglutinin in the membrane of the human erythrocyte. J Biol Chem 249: 4398, 1974.
11. Frye LD, Edidin M: The rapid intermixing of cell surface antigens after formation of mouse human heterakaryons. J Cell Science 7: 319, 1970.
12. Furthmayr H, Kahane I, Marchesi VT: Isolation of the major intrinsic transmembrane protein of the human erythrocyte membrane. J Membrane Biol 26: 173, 1976.
13. Furthmayr H, Marchesi VT: Subunit structure of human erythrocyte glycophorin A. Biochemistry 15: 1137, 1976.
14. Guidotti G: Membrane proteins. Ann Rev Biochem 41: 731, 1972.
15. Ho MK, Guidotti G: A membrane protein from human erythrocyte involved in anion exchange. J Biol Chem 250: 675, 1975.
16. Jenkins RE, Tanner, MJA: Separation and some properties of the major proteins of the human erythrocyte membrane. Biochem J 147: 393, 1975.
17. Marton LSG, Garvin JE: Subunit structure of the major human erythrocyte glycoproteins depolymerization by heating ghosts with sodium dodecyl sulfate. Biochem Biophys Res Commun 52: 1457, 1973.
18. Morrison M, Mueller TJ, Huber CT: Transmembrane orientation of the glycoproteins in normal human erythrocytes, J Biol Chem 249: 2658, 1974.
19. Nicolson GL, Painter RG: Anionic sites of human erythrocyte membranes. J Cell Biol. 59: 395, 1973.
20. Nicolson GL, Singer SJ: The distribution and asymmetry of mammalian cell surface saccharides utilizing ferritin-conjugated plant agglutinins as specific saccharide stains. J Cell Biol 60: 236, 1974.
21. Pinto da Silva P, Nicolson GL: Freeze-etch localization of concanavalin a receptors to the membrane intercalated particles of human erythrocyte ghost membranes. Biochim Biophys Acta 363: 311, 1974.

22. Reichstein E, Blostein R: Arrangement of human erythrocyte membrane proteins. J Biol Chem 250: 6256, 1975.
23. Rothstein A, Cabantchik ZI, Balshin M, Juliano R: Enhancement of avion permeability in lecithin vesicles by hydrophobic proteins extracted from red blood cell membranes. Biochem Biophys Res Commun 64: 144, 1975.
24. Segrest JP, Kahane I, Jackson RL, Marchesi VT: Major glycoprotein of the human erythrocyte membrane: evidence for an amphipathic molecular structure. Arch Biochem Biophys 155: 167, 1973.
25. Shin BC, Carraway KL: Lactoperoxidase labeling of erythrocyte membrane proteins. Biochem Biophys Acta 345: 141, 1974.
26. Silverberg M, Furthmayr H, Marchesi VT: The effect of carboxymethylating a single methionine residue on the subunit interaction of glycophorin A. Biochemistry 15: 1448, 1976.
27. Singer SJ, Nicolson GL: Fluid mosaic model of the structure of cell membranes. Science 175: 720, 1972.
28. Steck TL, Dawson G: Topographical distribution of complex carbohydrates in the erythrocyte membrane. J Biol Chem 249: 2135, 1974.
29. Steck TL: The organization of proteins in human erythrocyte membranes. *In:* Fox CF (Ed): Membrane Research. New York, Academic Press, 1972, p. 71.
30. Steck TL: Cross-linking the major proteins of the isolated erythrocyte membrane. J Mol Biol 66: 295, 1972.
31. Tanner MJA, Boxer DH: The major human erythrocyte membrane protein: evidence for an s-shaped structure which transverses the membrane twice and contains a duplicated set of sites. Biochem J 129: 333, 1972.
32. Tillack TW, Scott RE, Marchesi VT: The structure of erythrocyte membranes studied by freeze-etching, J Exp Med 135: 1209, 1972.
33. Tomita M, Marchesi VT: Amino-acid sequence and oligosaccharide attachment sites of human erythrocyte glycophorin. Proc Natl Acad Sci USA 72: 2964, 1975.
34. Triche TJ, Tillack TW, Kornfeld S: Localization of the binding sites for the ricinus communis agaricus bisporus and wheat germ lectins on human erythrocyte membranes. Biochim Biophys Acta 394:540, 1975.
35. Wang K, Richards FM: An approach to meanest neighbor analysis of membrane proteins. J Biol Chem 249: 8005, 1974.
36. Wang K, Richards FM: Reaction of Dimethyl 1-3,3-diathiobis-propionimidate with intact human erythrocytes. J Biol Chem 250: 6622, 1975.

Studies on the Mechanism of Embryonic Cell Recognition

A. A. Moscona, Ph.D., M. Moscona, Ph.D. and R. E. Hausman, Ph.D.

The organization of cells in the embryo and their assembly into multicellular patterns—tissues and organs—is clearly the outstanding feature of embryogenesis; it represents one of the central problems in biomedical sciences, and its elucidation is important also for the understanding of various developmental malformations and of aberrant cell behavior in cancer. Exploration of this problem has focused attention on the mechanisms which enable cells in the embryo to identify one another and to associate preferentially into tissue-forming complexes (for further discussion see [12,14]).

It is generally known that in order to construct the various tissues in the embryo, the different cells must first become reshuffled, sorted out and segregated into distinct groupings and associated in histogenetic configurations. The cells are able to accomplish this because they possess on their surfaces mechanisms which enable them to *recognize* one another and to *adhere selectively* into tissue-forming patterns. These mechanisms of embryonic cell recognition evolve in the embryo progressively and coordinately with cell differentiation; their ontogeny and modulation are directed—like other aspects of cell differentiation—by genetic information and by influences from the cell's microenvironment (including cell-cell interactions). Thus, as the various cell classes and cell types arise and develop in the embryo, the cell surfaces become differentially specified, or encoded with molecular "labels"; these labels project outwardly the changing phenotypic identities of the cells, and they determine the cells' recognition properties, i.e., their morphogenetic affinities and adhesive selectivities. So that, as cells in the embryo make contracts, they can "tell" by their sur-

From the Departments of Biology and Pathology, University of Chicago, Chicago, Ill.

Acknowledgments: The studies from the authors' laboratory referred to in this essay have been supported by research grants from the National Institute of Child Health and Human Development (HD-01253) and, in part, the University of Chicago Cancer Center Institutional grant from the National Cancer Institute (CA-14599).

This essay is dedicated in friendship and respect to Theodore Puck, a pioneer in modern cell biology.

faces if they are, or are not, of the "right" kind; accordingly, they either adhere and aggregate into tissue-forming configurations, or they separate and move on.

Reaggregation of Embryonic Cells *in vitro*

The detailed mechanisms which determine how and why cells make the choice between union and separation are difficult to explore in the intact embryo, for obvious technical reasons. Therefore, when we first became interested in these problems [15], we started to devise and employ "simplified" experimental model systems, based on reaggregation *in vitro* of cell suspensions from embryonic chick and mouse tissues. In these experiments the dispersed cells reassemble and reform their typical multicellular patterns; this makes it possible to analyze under controlled conditions the mechanisms involved in the construction of tissues from cellular building blocks.

The key to this experimental approach has been the development of methods for separating embryonic tissues into suspensions of viable cells; this was made possible by the use of trypsin [9,10]. When tissues isolated from mouse or chick embryos (e.g., retina, heart, kidney, etc.) are gently treated with trypsin, the enzyme cleaves intercellular bonds, degrades proteins at the cell surface and between cells, and increases the internal fluidity of the cell membrane; as a result, the cells detach from one another and can be dispersed live into a suspension. When such a cell suspension is gently swirled in a flask on a gyratory shaker, the cells reaggregate into progressively larger clusters; within these clusters the different cells become reshuffled, sort out and reconstruct their characteristic tissue pattern (see below).

This process of morphogenetic cell reaggregation requires protein synthesis [17], in order to enable cells to regenerate cell-surface (cell-linking) materials that had been removed by trypsinization of the cells. It is of interest that the synthesis of some of these proteins is mediated by gene products with a half-life of about 2 hours; this suggests that the properties of the embryonic cell surface which are important for morphogenetic cell reassociation are subject to continued genomic control.

In addition to the requirement of biosynthetic processes, morphogenetic cell reaggregation is dependent on certain structural features of the cell periphery. These structural features deserve emphasis because a full appreciation of the morphologic dynamics of the cell periphery is, in our opinion, essential for a meaningful analysis and a coherent interpretation of the biochemical mechanisms involved in cell interactions. When freshly dissociated neural retina cells (from chick embryos) were examined by a scanning electron microscope, they were found to have formed numerous elongated projections [2]. By means of these long probe-like processes the discrete cells make their initial contacts. The projections bridge cells across considerable distances and pull them together. As the cells come closer, the processes between them are retracted, and the bodies of the cells make contact. The emerging aggregates continue to increase in size by adding on single cells and small cell clusters, so that within 12 to 24 hours, large cell aggregates are present in the culture (see [2] for details).

One must emphasize that such elongated processes are not a peculiarity of experimentally dissociated cells, but that they are also a characteristic and normal surface-feature of cells in the embryo, as they undergo morphogenetic assembly and organization [19]; these processes tend to disappear as the cells complete tissue morphogenesis and cell contracts become stabilized. The fact that disaggregation of embryonic tissue causes the reappearance of these structures suggests that when the cells are disconnected, the cell surface reverts to a more "immature" condition conducive to morphogenetic reassembly of the cells. In fact, if the appearance of these projections is experimentally prevented, or if the cells are no longer able to form this particular kind of processes, morphogenetic reaggregation does not occur [2].

Within the aggregates, the cells become gradually rearranged, sorted out, and positioned so as to reestablish their characteristic tisssue-pattern. For example, in reaggregates of embryonic chick neural retina, the different retinocytes become organized in layers and reform the characteristically stratified neuroretinal tissue. Subsequently, the reconstructed retina tissue continues to differentiate structurally and biochemically, including formation of synapses and the appearance of retina-characteristic enzymes and other products [21]. Reaggregated heart cells reconstruct into pulsating cardiac tissue [22]; kidney cells reform into nephric structures, and embryonic mouse or chick brain cells restitute to form brain tissue (see [12] for review and further details).

When cells from two different tissues of the embryo are combined in the same suspension (for example, retina cells and liver cells), the cells from the different tissues segregate and reform into their characteristic tissue-patterns separately. (In this case, the formation would result in separate masses of retina and liver.) This *tissue-specific* sorting out of embryonic cells in cell aggregates reproduces *in vitro* a process that is of central importance in normal embryogenesis: for, in order to give rise to the various tissues, the different cells in the embryo must become spatially segregated into separate and distinct groupings. This suggests the existence of mechanisms for *tissue-specific cell recognition,* of histotypic cell affinities which cause cells that belong to the same tissue to adhere preferentially to one another, and to segregate from others.

Studies on the Molecular Basis of Embryonic Cell-Affinities

Our work on this problem has suggested the following working hypothesis: embryonic cell-cell recognition and selective cell adhesion represent two facets of the same mechanism, a mechanism mediated by specific cell-surface constituents which function as intercellular ligands or specific cell-cell receptors (sometimes referred to as "recognition-sites"). The hypothesis proposes that the specificities of embryonic cell recognition and cell adhesion are determined by the chemical characteristics of these cell-ligands as well as by their topographic organization on the cell surface—i.e., by a combinational "code" which (like other biologic codes) depends on both the quality and the configuration of the units of which it consists. Furthermore, the possibility was suggested that cells of different tissues might be specified by different cell ligands, and that such disparities could be responsible for tissue-specific cell recognition; on the other hand, the organization and positioning of cells within a single tissue could depend primarily on the topographic-temporal patterns of the ligands, i.e., on their organization on cell surfaces.

The gist of the hypothesis was that cells with complementary surface-displays of ligands would be capable of mutual affinities, namely, would show *positive* recognition and associate morphogenetically. In contrast, noncomplementarity or absence of the appropriate ligands would result in *negative* recognition, reflected in transient, nonspecific cell adhesion, or nonadhesion. In essence, this hypothesis has been conceptually inspired by the notion of "chemo-specification" of cell surfaces, originally proposed for neurons by Sperry [23]; it aims to conform with Pauling's suggestion that complementariness of molecular structures is responsible for biologic specificity in general [18], and with the views expressed by Weiss [24] and Monod [9] that the shape-recognizing and stereospecific binding properties of proteins provide the key to understanding the phenomena of specific cell interactions. While the cell-ligand hypothesis was considered by many to be rather iconoclastic when it was first proposed, its various elements have been adopted at the present time and have been assimilated into models of cell adhesion proposed by several investigators.

The cell-ligand hypothesis raised certain propositions which could be tested experimentally. The first proposition concerned the suggestion that tissue-specific cell-recognition might reflect qualitative differences between cell-surface components. Could one demonstrate the existence of tissue-specific surface antigens on cells from different tissues of the chick embryo? This question was studied by immunologic methods [4], and the results demonstrated conclusively the existence of tissue-specific differences in the antigenic properties of embryonic cell surfaces. These differences coincided with the tissue-specific affinities of the cells, and suggested that some of the determinants might be involved in the mechanisms of embryonic cell recognition.

The second question raised by the cell-ligand hypothesis concerned the possibility of isolating from embryonic cells materials which possessed the activity of the postulated tissue-specific cell ligands. The addition of such materials, or "factors," to cell suspensions would be expected to result in enhanced cell-reaggragation (i.e., an increased rate of cell aggregation and the formation of larger aggregates than in controls); most importantly, such enhancement should be tissue-specific.

Cell-Aggregating "Factors"

Evidence that tissue-specific cell-aggregating "factors" can be isolated from embryonic cells was first obtained in studies on embryonic neural retina [11]. Today, several such factors have been demonstrated and the one from the neural retina was purified and biochemically characterized as glycoprotein [5]. In the initial work, these factors were isolated from the supernatant medium of cell cultures. The rationale that led to this experimental approach was that, when dispersed embryonic cells were maintained for 1-3 days in primary monolayer culture (in serum-free medium), they released into the medium various macromolecular products. Some of these products represented components normally associated with the cell surface which, under these culture conditions, tended to become exteriorized into the medium. These exteriorized materials contained various derivatives which were not tissue-specific, but they also included certain tissue-specific products, among them the cell-aggregating factors. These materials could, of course, be radioactively labeled, fractionated by suitable procedures and assayed for biologic activity.

Using these procedures, we have isolated from the culture supernatant of embryonic neural retina cells a glycoprotein-containing fraction; its addition to suspensions of embryonic retina cells caused a striking enhancement of their aggregation, i.e., the cells aggregated at a faster rate and the resulting aggregates were significantly larger than the controls. Most importantly, the effect of this retina factor was specific for retina cells, and could not be obtained on cells from other neural or non-neural tissues. Another important point is that within the resulting massive aggregates the cells became histotypically organized. Therefore, the effect of the factor was not one of agglutinating or clumping the cells, as in the case of cell reactions with lectins; rather, it was a tissue-specific enhancement of morphogenetic cell associations, as might be expected of the postulated tissue-specific cell-ligands [5]. Using similar procedures, a cell-aggregating factor specific for embryonic chick and mouse cerebrum cells was also demonstrated [3]. (See [12] for review and further information.)

The retina-specific cell-aggregating factor was purified [5,16], and its activity was localized in a glycoprotein which had in solution a molecular weight (MW) of 50,000 $\pm$ 5,000. It is possible, however, that after this material has bound to the cell surface, multimerization or complexing with similar or other cell-surface components would be required to make it active biologically. Also, it should be pointed out that, even though the tests performed so far indicated an apparent homogeneity of this glycoprotein, they did not exclude the possible existence of herterogeneities within it of the kind that would not have been detectable by the methods used heretofore. This will be a matter for future studies.

Taken as a whole, our evidence from the above mentioned and other studies indicated that this glycoprotein derives from the retina cell membrane and that its cell-aggregating effect depended on its association with the cell surface. Therefore, our working assumption has been that this glycoprotein represents a component of the cell-ligand mechanism specific for embryonic neural retina cells. Its activity is destroyed by trypsin, but not by agents which modify the carbohydrate portion of the molecule. On the other hand, its binding to the cell surface appears to depend on the availability on the cell surface of carbohydrate-containing sites [13]. After the factor had bound to the cells, the expression of its activity required optimal temperature, protein synthesis and Ca^{++}. The exact reasons for these requirements are not yet satisfactorily understood: they may be related to (a) the processes by which the factor molecules become topographically organized and/or multimerized on the cell surface into functional cell-ligand mechanisms; (b) synthesis or provision of still other molecules required for the functional completeness of the cell-ligand mechanism; (c) other, as yet unknown processes, essential for bringing about morphogenetic cell interactions.

Isolation of Cell-Aggregating Factors from Purified Cell Membranes

Although considerable indirect evidence existed that this retina-specific protein was derived from the cell surface, more direct information concerning the cell-membrane derivation of such material was clearly needed. Furthermore, since the cell-aggregating factors obtained by us, as well as by others (e.g., [1, 20]) were isolated from cells in culture, there existed the possibility that these factors may represent products peculiar to cells *in vitro,* rather than constituents typical to cells *in vivo,* i.e., in embryonic tissues. Considering the role postulated for these factors in cell-recognition and cell assembly in embryonic tissue morphogenesis, it was obviously necessary to determine if a retina-specific cell-aggregating factor could be obtained directly from cell membranes prepared from noncultured retina tissue.

To examine this question [6], we isolated cell-membrane preparations from embryonic neural retina tissue, using a procedure developed by Hemminki [7] for isolating cell membranes from brain tissue. This procedure yielded membrane preparations with a low level of mitochondrial and microsomal contamination. The membrane preparation was then extracted with butanol. When the aqueous phase, separated from the butanol phase, was tested on cell aggregation, it was found to contain retina-specific cell-aggregating activity. Fractionation and biochemical characterization of this activity showed that it resided in a glycoprotein with a 50,000 ± 5,000 MW; by all tests done so far, this membrane-derived protein appears to be very similar, perhaps identical, to the retina-specific cell-aggregating glycoprotein previously purified from the supernatant medium of cultures of retina cells.

Using these procedures, a cell-aggregating protein specific for cerebrum cells was recently obtained from membrane preparations isolated from embryonic cerebrum; by similar procedures, a spinal cord-specific cell aggregating factor was also obtained.

DISCUSSION AND SUMMARY

Our results suggest the following working hypothesis: There exists on embryonic cell surfaces a category of tissue-specific proteins which, under the experimental conditions employed by us, appear to function as morphogenetic cell-cell ligands, and to determine embryonic cell recognition. For operational convenience we refer to such proteins as *cognins*, because of their postulated role in embryonic cell recognition; thus, retina-specific cognin, cerebrum-specific cognin, etc.

Although it might be tempting to consider various hypothetical schemes for the mode of cognin action, this should be postponed until additional information becomes available about this complex problem. Particularly useful could be results of comparative biochemical-physiologic analyses of the different cognins obtainable from different embryonic tissues; these might yield clues concerning the specificities of the effects of cognins and their mechanisms of function. Equally desirable would be information concerning the ontogeny of cognins during embryonic differentiation and thereafter; their biogenesis; and their topographic organization on the cell surface. Additionally, are there structural-functional-ontogenetic relationships between cognins and other "recognition-mechanisms" present on cell surfaces, such as immunoglobulins, histocompatibility antigens, T-locus products, lectin-binding sites, etc.?

If we assume that tissue-specific cognins mediate *tissue-specific* cell recognition, this leaves open the matter of *type-specific* cell-recognition, i.e., the mechanisms which enable different types of cells within a single tissue to become organized and positioned into their histologic patterns. One possibility is that the surface of each cell type is specified by a particular "variant" of the tissue-specific cognin. Another possibility is that type-specific cell-affinities might depend primarily on topographic arrangement of cognin arrays on cell surfaces. The latter possibility would not demand an extraordinarily large range of unique chemical differences between cell surfaces; instead, a relatively limited chemical vocabulary of tissue-specific cognins, deployed on cell surfaces in diverse topographic-temporal permutations could furnish cells with the information required for type-specific cell affinities and morphogenetic cell organization.

These are, at present, largely speculative considerations, and their exploration is not likely to be an easy task. However, what has already been accomplished seems sufficiently encouraging to justify further efforts toward unraveling these very important problems.

REFERENCES

1. Balsamo J, Lilien J: Functional identification of three components which mediate tissue-type specific embryonic cell adhesion. Nature 251: 522, 1974.
2. Ben-Shaul Y, Moscona AA: Scanning electron microscopy of aggregating embryonic neural retina cells. Exp Cell Res 95: 191–204, 1975.
3. Garber BB, Moscona AA: Reconstruction of brain tissue from cell suspensions. II. Specific enhancement of aggregation of embryonic cerebral cells by supernatant from homologous cell cultures. Develop Biol 27: 235–271, 1972.
4. Goldschneider I, Moscona AA: Tissue-specific cell-surface antigens in embryonic cells. J Cell Biol 53: 435–449, 1972.
5. Hausman RE, Moscona AA: Purification and characterization of the retina specific cell-aggregating factor. Proc Natl Acad Sci USA 72: 916–920, 1975.
6. Hausman RE, Moscona AA: Isolation of retina-specific cell-aggregating factor from membranes of embryonic neural retina tissue. Proc Natl Acad Sci USA 73: 3594–3598, 1976.
7. Hemminki K: Purification of plasma membranes from immature brain. FEBS Letters 38: 79–82, 1973.
8. Monod J: *In:* Chance and Necessity. London, Collins, 1972.
9. Moscona AA: Cell suspensions from organ rudiments of chick embryo Exp Cell Res 3: 535–539, 1952.
10. Moscona AA: Rotation-mediated histogenetic aggregation of dissociated cells: a quantifiable approach to cell interactions *in vitro*. Exp Cell Res 22: 455–475, 1961.
11. Moscona AA: Analysis of cell recombinations in experimental synthesis of tissues *in vitro*. J Cell Comp Physiol 60: 65–80, 1962.

12. Moscona AA: Surface specification of embryonic cells: lectin receptors, cell recognition and specific cell ligands. *In:* Moscona AA (Ed): The Cell Surface in Development. New York, Wiley & Sons, 1974.
13. Moscona AA: Embryonic cell surfaces: mechanisms of cell recognition and morphogenetic cell adhesion. *In:* McMahon D, Fox F (Eds): Development Biology-Pattern Formation, Gene Regulation. Menlo Park, Calif., W. A. Benjamin, Inc., 1975.
14. Moscona AA: The cell surface and cell recognition in embryonic morphogenesis. Studies on experimental synthesis of tissues from cells. *In:* Marois M (Ed): From Theoretical Physics to Biology. Amsterdam, North-Holland Publishing, 1976.
15. Moscona AA, Moscona H: The dissociation and aggregation of cells from organ rudiments of the early chick embryo. J Anat 86: 287–301, 1952.
16. Moscona AA, Hausman RE, Moscona M: Experiments on embryonic cell recognition: in search for molecular mechanisms. FEBS Meeting 38: 245–256, 1975.
17. Moscona M, Moscona AA: Inhibition of adhesiveness and aggregation of dissociated cells by inhibitors of protein and RNA synthesis. Science 142: 1070, 1963.
18. Pauling L: The future of enzyme research. *In:* Gaebler OH (Ed): Enzymes: Units of Biological Structure and Function. New York, Academic Press, 1956.
19. Revel J-P: Some aspects of cellular interactions in development. *In:* Moscona AA (Ed): The Cell Surface in Development. New York, John Wiley & Sons, 1974.
20. Rutishauser U, Thiery J-P, Brackenbury R et al: Mechanisms of adhesion among cells from neural tissues of the chick embryo. Proc Natl Acad Sci USA 73: 577–581, 1976.
21. Sheffield JB, Moscona AA: Electron microscopic analysis of aggregation of embryonic cells: the structure and differentiation of aggregates of neural retina cells. Develop Biol 23: 36–61, 1970.
22. Shimada Y, Moscona AA, Fishman DA: Scanning electron microscopy of cell aggregation: cardiac and mixed retina-cardiac cell suspension. Develop Biol 36: 428–446, 1974.
23. Sperry RW: Visuomotor coordination in the Newt (Triturus viridscens) after regeneration of the optic nerve. J Comp Neurol 79: 33–55, 1943.
24. Weiss P: The problem of specificity in growth and development. Yale J Biol Med 19: 235–278, 1947.

The Evolution of Multigene Families

L. Hood, M.D., Ph.D.

Recent studies of the immune system demonstrate that the families of genes coding for antibody molecules display unusual evolutionary features that are shared by other multigene families as diverse in character as the ribosomal RNA genes, the histone genes, and the tRNA genes, and even the simple sequences coding for the DNA satellites. We shall define a *multigene family* as a group of nucleotide sequences or genes that demonstrate four properties—multiplicity, close linkage, sequence homology, and related or overlapping phenotypic functions (Fig. 1). There is a broad spectrum of potential multigene families. Indeed, one category of multigene families, that in which the antibody genes fall, may code for some of the more complex aspects of phenotype in higher organisms. Strategies employed by antibody gene families for information handling may well be utilized, at least in part, by other complex gene families such as the immune response genes and perhaps even systems such as the genes coding for the wiring diagram of the nervous system. The multigene family may well be a unit of chromosomal organization encompassing a broad spectrum of gene families and coding both simple and complex phenotypic traits. Let us begin by considering the organizational and evolutionary features of antibody genes and then proceed to analysis of other multigene systems.

ANTIBODY GENE FAMILIES

Organization of Antibody Genes

Three antibody families [22]. The mammalian germ cell appears to contain at least three families of antibody genes, lambda (λ), kappa (κ), and H, each located on a different autosome (Fig. 2). The λ and κ families code for light chains, whereas the H family codes for heavy chains. Thus antibody genes constitute a multigene system comprised of three unliked multigene families.

From the Division of Biology, California Institute of Technology, Pasadena, Calif.

Acknowledgments: This paper is adapted from a longer review by L. Hood, J. H. Campbell and S. R. Elgin in Annual Review of Genetic 9: 305, 1975. This work was supported by grants from NIH and NSF.

FIG. 1. Model of a multigene family. From [20].

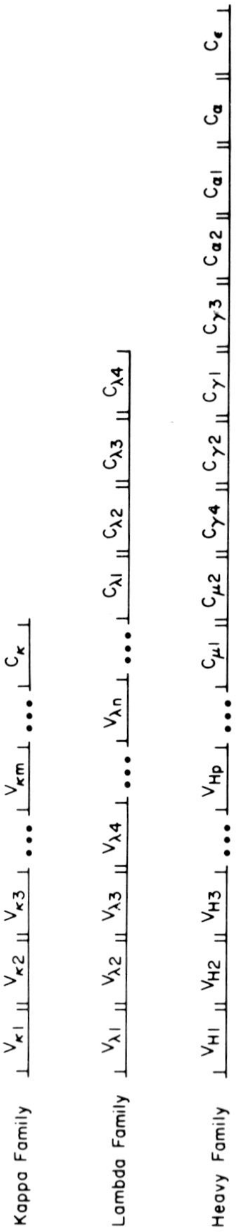

FIG. 2. A model of the genes encoding the three antibody families of man. From [20].

Two types of immunoglobulin genes [15,29]. Each immunoglobulin chain can be divided into two distinct portions, a variable and a constant region, which are encoded by separate genes in the zygote or germ line DNA (Fig. 2). Recent evidence suggests that the V and C genes are joined at the DNA level during the differentiation of the antibody-producing cell. Indeed, this joining of antibody V and C genes may be a fundamental mechanism for the differentiation or commitment of each antibody-producing cell to the synthesis of one molecular species of antibody molecule.

Number of V genes. The controversy as to the number of V genes that exist in the vertebrate organism has been debated for the past 75 years [8,16]. Recent evidence gathered from studies at the genetic, protein sequence, and nucleic acid levels suggests that the average antibody family must contain at least a hundred V genes in the germ line. Studies employing the hybridization of nucleic acid probes suggest that in certain systems somatic mutation may amplify this germ line information [29]. Irrespective of whether somatic mutation is an important mechanism for antibody diversification, it appears that the average antibody family contains multiple V genes and thus is a multigene family by the criteria discussed above.

Evolution of Antibody Genes

General scheme. Antibody polypeptides can be divided into homology units about 110 residues in length on the basis of amino acid sequence similarities (Fig. 3). The heavy chain in Figure 3 has four homology units (V_H, C_H1, C_H2, C_H3) and the light chain has two (V_L and C_L). The V_H and V_L homology units exhibit extensive sequence homology with one another, as do the C_H, C_H1, C_H2, and C_H3 homology units [13]. In spite of their lack of homology at the primary sequence level, X-ray crystallographic studies demonstrate that the V and C homology units have very similar tertiary configurations [24]. There is, of course, structural variation in the antigen-binding crevise. These homologous features suggest that all immunoglobulins were derived during evolution from a common precursor about the size of one homology unit. A hypothetical scheme for the evolution of the immunoglobulin gene families is diagrammed in Figure 4. The hypothetical precursor gene duplicated at a very early time to produce the ancestral V and C genes. Subsequently, the V gene duplicated many times over to produce a primordial multigenic family. This original family may have coded for primitive membrane receptor molecules. This multigene family in turn was duplicated either by polyploidization or by duplication and translocation of a chromosomal segment to produce the primitive antibody gene family. Subsequent duplication of this primitive antibody family may have produced the three families that have evolved to become the

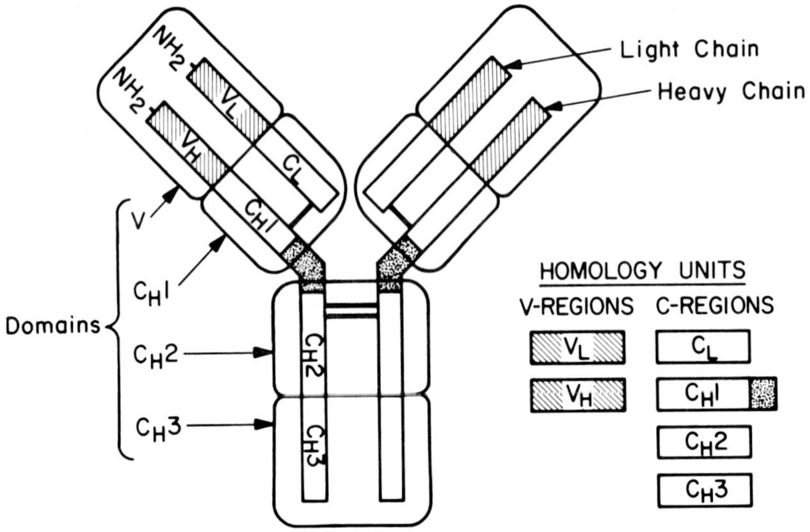

FIG. 3. A model depicting the basic structure of the immunoglobulin (IgG) molecule. From [20].

contemporary lambda, kappa, and H families. Contiguous or fused gene duplication led to C_H genes comprised of three or four homology units in the heavy chain family [13,25]. Accordingly, the evolution of antibody genes employed all of the major mechanisms of gene evolution—point mutation, discrete duplication, contiguous duplication, polyploidization and/or translocation.

Unusual evolutionary features. Antibody V genes exhibit two evolutionary features that are not generally expected in gene evolution (Fig. 5). The first feature is that the V gene families can expand or contract their gene numbers rapidly in terms of evolutionary time (Fig. 5a). For example, some mammalian species may express many V_κ and few V_λ genes, whereas others express many V_λ and few V_κ genes [18,19]. This suggests that information required to encode the light chains of diverse antibodies can be contained either in the λ or κ light chain families, and that dramatic shifts in the information content of each family can occur over the 75 million years of mammalian divergence [16].

The second unusual evolutionary feature is that the V genes of an antibody family in one species appear to have evolved structural changes together or in parallel (Fig. 5b). We designate this process as coincidental evolution. In portions of the V regions that are highly conserved, coincidental evolution is reflected by species-associated residues at certain positions that distinguish most of the immunoglobulin chains of one species from those of a second [7,17]. For example, most V_κ regions of rabbit are

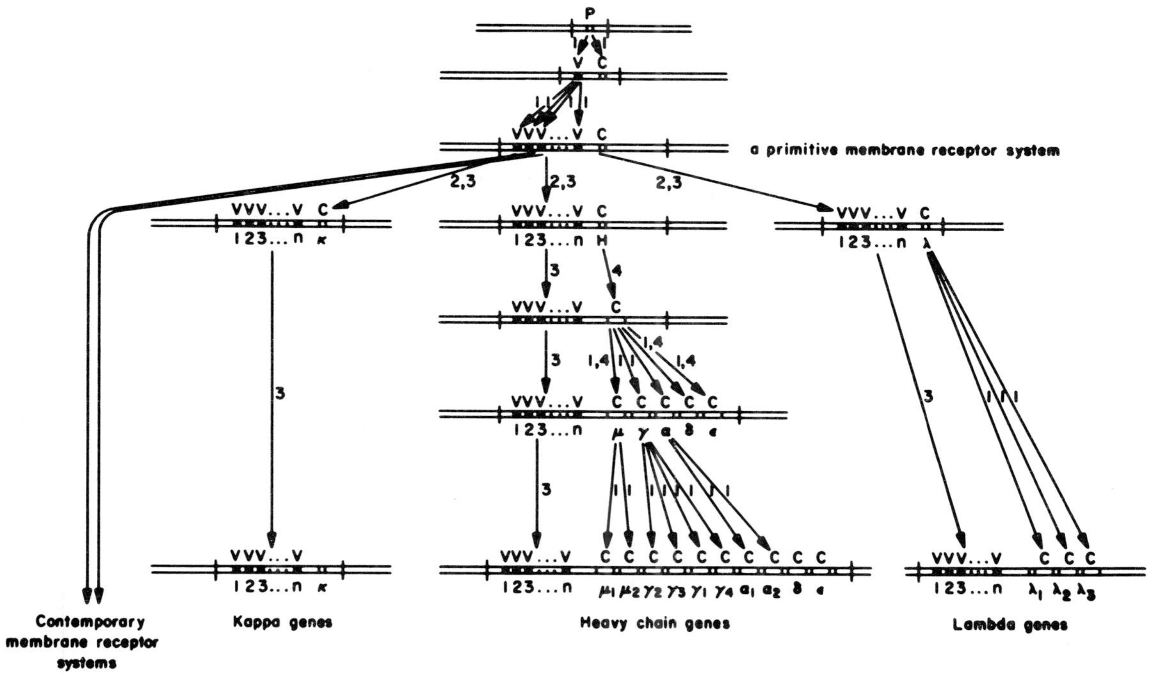

FIG. 4. A hypothetical scheme for the evolution of the antibody families. The order of gene duplication events is unknown. A number of genetic mechanisms seem to be employed in the evolution of these families as indicated by numbers adjacent to arrows. These are: 1. discrete gene duplication, 2. gene duplication by polyploidization or chromosomal translocation, 3. contiguous gene duplication, and 4. coincidental evolution of multiple genes. Mechanisms 1 and 4 may be identical (see text). Adapted from [12].

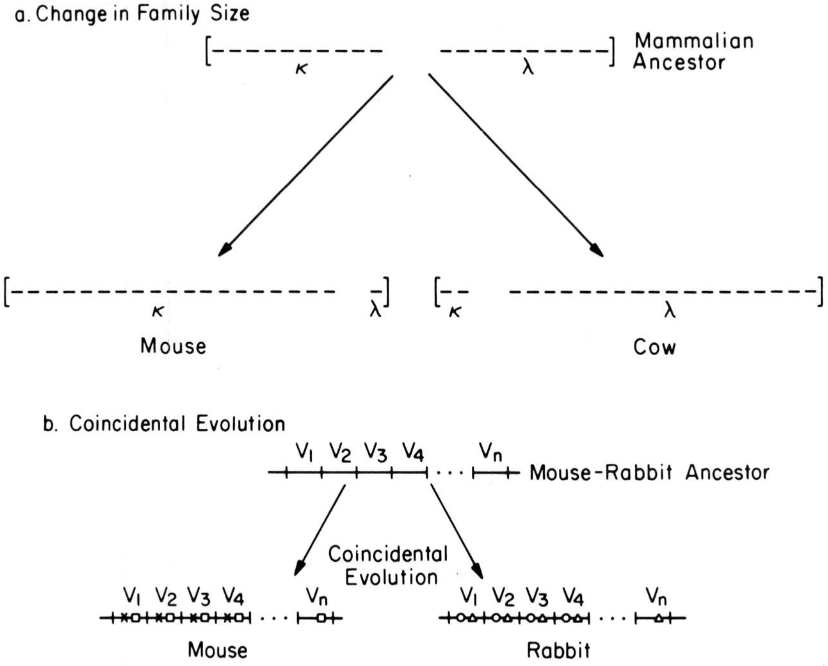

FIG. 5. Two unusual evolutionary features of multigene families. a. Diagrammatic representation of the expansion and contraction of the λ and κ gene families in two mammals. b. Diagrammatic representation of coincidental evolution during the divergence of mouse and rabbit evolutionary lines. X, □, O and Δ represent coincidental changes in the amino acid codons of the respective evolutionary lines. From [20].

distinguished from their counterparts in the mouse by the presence of valine and glycine at positions 11 and 17, respectively, rather than a leucine and glutamic acid [17]. Thus, V genes or sets of V genes from a particular family appear to have evolved together or in a coincidental fashion.

Coincidental evolution could be due to the unprecedented parallel evolution of hundreds to perhaps thousands of V genes. However, we feel that it is more likely to reflect some other, more fundamental genetic property of the multigene family. Indeed, other multigene families very different in character from that of antibodies, including ribosomal RNA genes, tRNA genes, histone genes, and DNA satellites, also exhibit these same evolutionary characteristics. We suggest that all of these very different types of multigene families share one or more common evolutionary mechanisms.

THREE CATEGORIES OF MULTIGENE FAMILIES

Multigene families can be divided into three general categories by several criteria [20]. This classification is presented in Table 1.

TABLE 1. Classification and Properties of Multigene Families*

Category	Gene products	Multiplicity	Gene or protein homology	Information content	Examples of coincidental evolution	Examples of change in family size
Simple-sequence satellites	None known	10^3–10^6	80–100%		Different satellites of *Drosophila*	Mouse satellite
Multiplicational 18S-28S RNA	RNA, protein	100–600	97–100%	One unit	Spacer regions of *X. laevis* and *X. mulleri*	
5S RNA		100–1200	97–100%	One unit	Spacer regions of *X. laevis* and *X. mulleri*	Gene number in *X. mulleri* and *X. laevis*
tRNA		6–400		One unit		
histones		10–1200	87–99%	Few units	Histone mRNAs of two species of sea urchin	Gene number in different sea urchin species
Informational antibodies	Proteins	100S	80–100%	Many units	Rabbit and mouse κ chains	V_λ and V_κ in mammals
hemoglobins		10	75–100%	Few units	Human and cow δ chains	Human and rabbit β-like genes

* In all instances, characteristics refer to a single closely linked set of genes. From [20].

The simple sequence multigene family. The simple sequence multigene family includes those gene sequences generally known as satellites which are found at the centromeric regions of most eukaryotic chromosomes (Fig. 6). These families encompass segments of DNA of a short fundamental sequence, generally 6–15 nucleotides in length, which can be repeated from 10^3 to 10^7 times [26]. The simple sequence families are not transcribed or translated, nor is their function known.

The multiplicational multigene family. The multiplicational multigene family is generally comprised of 10 to several hundred nearly identical gene repeats (Fig. 6) [4, 5, 21]. These genes may be duplicated at the DNA level because they are required in large quantities at certain stages of the cell cycle (histones) or during certain stages of development (ribosomal RNA).

Informational multigene families. The informational multigene families has individual members that can differ markedly in sequence from one another, although all are homologous and obviously share an evolutionary ancestry. The antibody genes are the only well documented example of a complex informational multigene family. This system, which can generate 10^5 to 10^7 or more molecules, has evolved to express one unit of information (an antibody molecule) on each lymphocyte and to permit environmental stimuli (antigens) to trigger expansion of this information by clonal selection and amplification. Thus a vast library of information can be selectively employed. There are also informational gene families with relatively small numbers of genes, perhaps as few as 5 or 6. For example, the beta-like hemoglobin genes of man are a family of multiple, closely linked genes with sequence homology and overlapping function (Fig. 7). Thus, informational multigene families may be simple as well as complex.

In summary, well documented examples from all three categories of multigene families exhibit four fundamental characteristics of multigene families: multiplicity, close linkage, sequence homology, and overlapping function. In addition, these examples all exhibit two unusual evolutionary features—coincidental evolutional (Fig. 6) and change in family size (Table 1).

EVOLUTIONARY MECHANISMS

Two classes of mechanisms could lead to coincidental evolution (Fig. 8). First, each gene in a family could be transmitted faithfully from parent to progeny, with coincidental evolution being imposed by natural selection (parallel evolution) or by a gene correction process (Fig. 8A). Alternatively, coincidental evolution could some about as a result of a repeated duplication of some gene in the family and the loss of other member genes during evolution (Fig. 8B). Homologous but unequal crossing over and saltatory

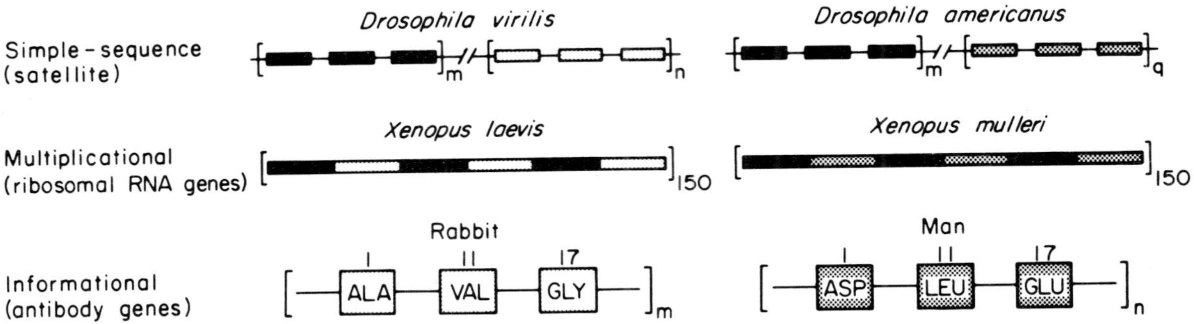

FIG. 6. Coincidental evolution of multigene families of the simple-sequence, multiplicational, and informational categories. The open and shaded areas represent regions undergoing coincidental evolution in two species, whereas the black regions are maintained virtually identical. From [20].

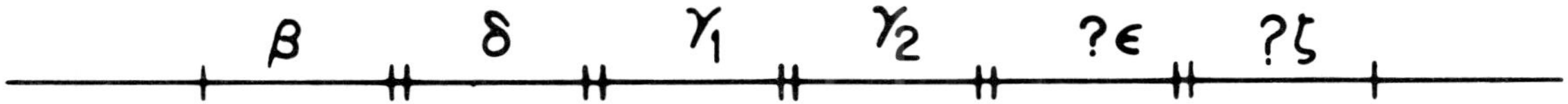

FIG. 7. A model of the gene family encoding the β-like hemoglobins of man. The linkage of the ϵ and ζ genes to the other β-like genes is uncertain. From [20].

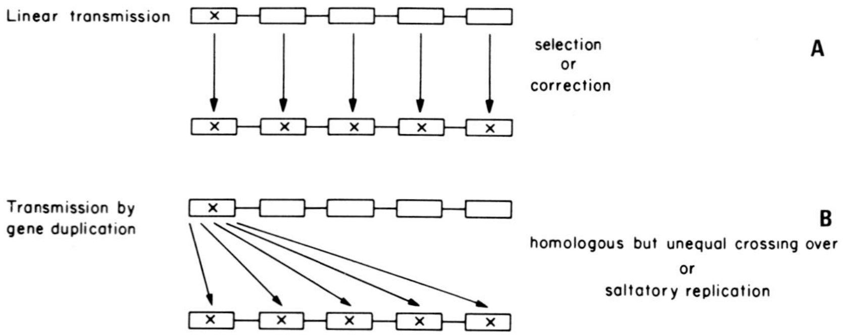

FIG. 8 (at left). Two models for coincidental evolution (see text). From [20].

FIG. 9 (below). A model of homologous but unequal crossing-over. From [20].

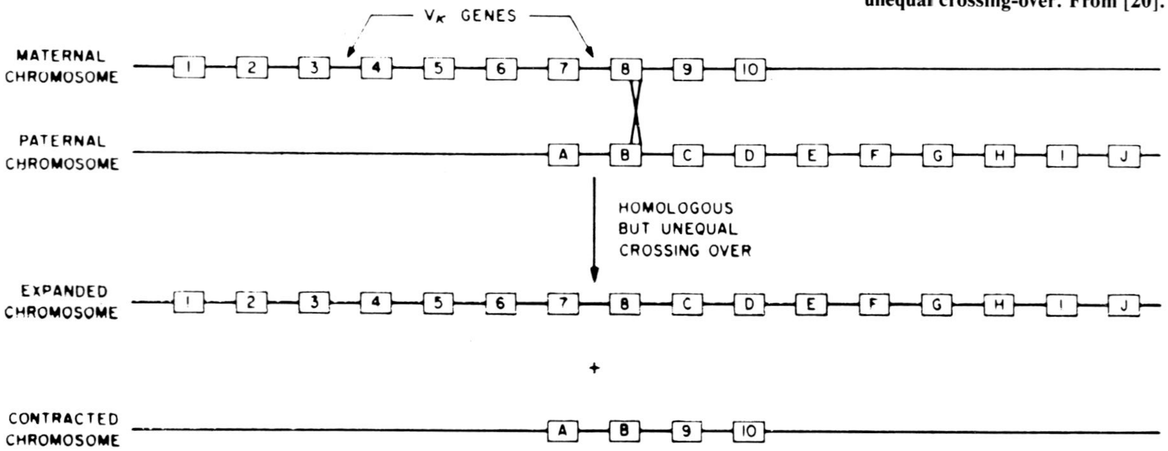

replication are two possible mechanisms of gene duplication. Mechanisms of selection, gene correction, and saltatory replication have been discussed elsewhere [4,6] and will not be considered further here.

Homologous but unequal crossing over. We feel an important mechanism for explaining the two unusual evolutionary features exhibited by various multigene families—gene expansion and contraction, and coincidental evolution—is homologous but unequal crossing over. Unequal crossing over occurs when chromosomes carrying closely linked homologous genes mispair, cross over, and yield one chromosome with an increased number of genes and a second with a decreased number of genes (Fig. 9). The repeated genes need not be identical but only similar enough to promote unequal pairing and exchange. Thus, one can see intuitively how gene expansion and contraction might occur. If there were selective pressures for an expanded (or contracted) number of genes, those chromosomes with increased (decreased) numbers of genes would be favorably selected in a population and then in time the gene family size would increase.

It is a little more difficult to visualize how coincidental evolution might be accounted for by crossing over. The population dynamics of variation within a multigene family undergoing repeated crossing over are formally analogous to population dynamics of neutral alleles in a gene pool. One form of the gene can increase in number at the expense of other genes in the family. Coincidental evolution occurs when a variant gene replaces all other genes (is fixed) in a particular evolutionary line (Fig. 10). Elegant computer simulations of the crossing over process have been carried out in computer models [2,28] and preliminary attempts have been made to analyze this process mathematically [23]. These studies readily demonstrate that crossing over can explain both gene expansion and contraction as well as coincidental evolution.

SELECTION IN MULTIGENE FAMILIES

When many genes in a multigene family have overlapping or identical functions, natural selection cannot act at the level of the single gene. Instead, it must act on blocks of genes or the functional adequacy of the gene family as a whole. As an illustration, natural selection obviously could not detect one nonfunctional 28S RNA gene among 500 functional copies. Natural selection can act on multiplicational families only when a significant number of genes become nonfunctional through mutation. Thus, for natural selection to favor a variant sequence, the number of copies of that variant must first be amplified nonselectively so that the variant can affect function of the family as a whole. Accordingly, the mechanism for coinci-

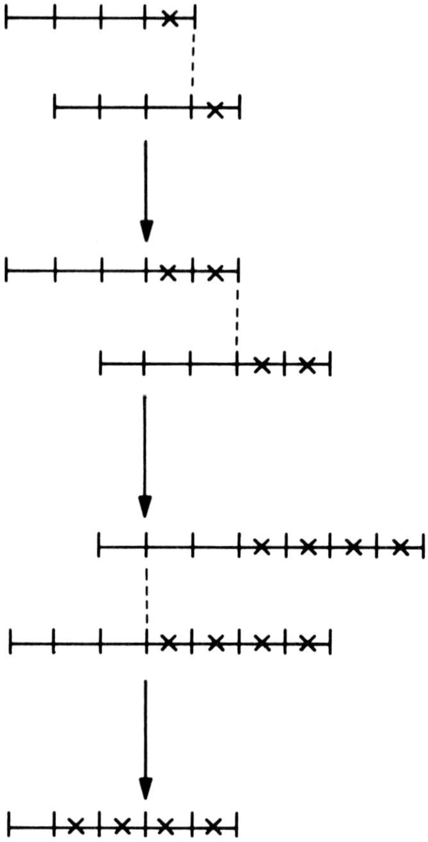

FIG. 10. A model of gene fixation by crossing-over. From [20].

dental evolution represents a fundamental requirement for the adaptive evolution of multigene families because it allows variants—useful, neutral, or deleterious—to be amplified in the family so that natural selection may operate on them.

The degree to which individual genes in an informational family are sheltered from natural selection depends on the size of the family and the extent to which the phenotypic functions overlap. In a small multigene family, such as the beta-like hemoglobin genes, some gene products clearly have different though related functions (i.e., γ and β genes) and probably are acted upon by natural selection almost as unique genes. Other genes have identical functions (i.e., $\gamma 1$ and $\gamma 2$ genes) and to some extent must be sheltered from selection. The manner in which selection operates on individual genes in a more complex informational family, such as the antibodies, is more difficult to evaluate. Certainly if an organism has 50 antibodies against an influenza strain, the loss of one would not necessarily be significant. In general, the response of most antigens is sufficiently heterogeneous to suggest that selection acts on blocks on antibody V genes with overlapping functions rather than on individual V genes.

LINKAGE REQUIREMENTS FOR A MULTIGENE FAMILY

All the multigene families known to share evolutionary features of coincidental evolution and change in family size have closely linked homologous genes. Linkage probably arises by that mechanism which generates gene families (e.g., gene duplication by saltatory replication or crossing over). However, linkage must be maintained by selection. Two types of selective pressures may preserve linkage. First, the evolutionary mechanisms for coincidental evolution may be able to operate only on linked families of genes. Second, various novel control mechanisms that operate in multigene families may function only in closely linked arrays of genes. Accordingly, tandem linkage may be a basic and necessary feature of multigene families.

It must be noted that there are families of related genes that are dispersed throughout the eukaryotic genome [10]. The evolutionary features of these dispersed multigene families as well as their functional roles are entirely unknown.

EVOLUTION OF MULTIGENE FAMILIES

Multigene families are found in eukaryotes, but not prokaryotes. As eukaryotic chromosomes became sophisticated organelles for information storage, replication, transmission, and evolution, large amounts of information could be transmitted from one generation to the next. Simple sequence families may have evolved to play a role in some of these organelle functions. Multiplicational systems probably arose to permit greater synthesis of gene products as the size of the cell increased. Informational multigene families may have evolved in response to the demands of cellular differentiation. For example, as single-celled eukaryotes evolved into metazoa, more and more informational multigene families became useful to permit discrete organ systems to express distinct cell types (e.g., the immune system, the nervous system, hormonal systems, etc.).

New multigene families can arise in two ways. The family may arise by duplication from a single gene. Alternatively, duplication and translocation of all or part of the existing multigene family may occur. For example, the hemoglobin gene family appears to have evolved during vertebrate evolution from a single gene. In contrast, the three antibody gene families almost certainly diverged from a more primitive multigene family (Fig. 4).

Newly duplicated informational multigene families can probably assume three types of function: (a) interaction with gene products of the old family to evolve more sophisticated molecules, (b) regulation of expression of the old family, and (c) acquisition of new function. Multigene families in the immune system illustrate each of these types of function. First, the three antibody gene families, lambda, kappa, and H, obviously evolved from a common primordial family (Fig. 4). The light and heavy chains interact with one another to form a molecular machine of striking sophistication (Fig. 3). Second, the immune response genes of the major histocompatibility complex are a multigene family by all criteria except that their sequence homology relationships are unknown [3]. The products of these genes regulate in some unknown fashion the immune response to a wide variety of different antigens and may share a common evolutionary origin with antibodies. Third, a number of serum proteins with diverse functions exhibit putative amino acid sequence homology with immunoglobulins—serine proteases [14], beta-2 microglobulin [9], alpha-1 acid glycoproteins [29], and haptoglobins [1]. When the evolutionary and functional relationships of these various gene families related to the antibodies become known, they will constitute a fascinating example of molecular evolution.

POTENTIAL MULTIGENE FAMILIES

The categories of multigene families described in Table 1 are useful because they present a logical basis for deducing other phenotype traits likely to be encoded by multigene families. Simple sequence families can often be isolated as satellites from the bulk of the cellular DNA. Any gene product that is required in large amounts during a certain stage of the cell cycle or during a particular developmental stage (especially in a large cell) might be a logical candidate to be encoded by a multiplicational multigene family. Informational multigene families may code for a variety of complex traits in higher organisms, as illustrated by a partial list of potential informational multigene families in Table 2. Some of these families may be small informational families, whereas others are extremely large and complex. The ever increasing sophistication of approaches for studying multigene families of proteins, RNA, and DNA levels suggests that many new multigene families will be described in the near future.

SUMMARY

The multigene family is a unit of chromosomal organization. Its gene members are closely linked, homologous in sequence, and have overlapping

TABLE 2. Possible Informational Multigene Systems

Structural proteins
Actin*
Keratins*
Collagens*
Chorionic proteins*
Wool proteins*
Serum proteins
Hemagglutinins*
Serine esterases*
Complement components*
Hemocyanins
Developmental systems
T allele
Transcriptional signals
Translational signals
Membrane systems
Immune response genes
Membrane receptors (hormone, cAMP)
Membrane transport systems
Embryonic antigens
Blood cell antigens
Nervous system
Membrane molecules mediating specific cell-cell interactions
Information storage

* These proteins have been found to exist in multiple structural forms whose significance is generally unknown. From [20].

functions. Multigene families can be divided into three categories—simple-sequence, multiplicational, and informational—by a variety of structural and functional criteria. Multigene families exhibit two novel evolutionary features—coincidental evolution and rapid change in family size—that suggest that they all share one or more evolutionary mechanisms. Natural selection cannot act directly upon individual genes in a family because of their identical or overlapping functions; hence, selection must operate upon the family as a whole or upon blocks of genes within the family. The mechanism(s) for coincidental evolution expands out variant genes with a family so that they can be acted upon by natural selection and, accordingly, permits multigene families to evolve adaptively.

The close linkage of the genes in a family appears to be a consequence of the fact that their control and evolutionary mechanisms may operate only on tandemly linked genes. New multigene families may evolve from a single gene or from other multigene families. In addition to evolving new functions, the latter mode of evolution generates a new multigene family whose members are preadapted to interact with those of the old family. These family interactions can lead to the evolution of more sophisticated molecular machines or to the regulation of one family by the second. Multigene families may be large or small. The three categories of multigene families allow potential multigene families to be identified, and they suggest specific experimental approaches for the study of new families.

Some of the most interesting genetic systems under investigation today are known or potential informational multigene families. This is not fortuitous, in that many of the most interesting aspects of phenotype are complex ones, with correspondingly complex genetic, evolutionary, and regulatory requirements. One of the frontiers in modern genetics is the identification, characterization, and understanding of informational multigene families.

REFERENCES

1. Black J, Dixon GH: Amino acid sequence of alpha chains of human haptoglobins. Nature 218: 736–741, 1968.
2. Black JA, Gibson D: Neutral evolution and immunoglobulin diversity. Nature 250: 327–328, 1974.
3. Bodmer WF: Evolutionary significance of the HL-A system. Nature 237: 139–145, 1972.
4. Brown CD, Sugimoto K: The structure and evolution of ribosomal and 5S DNAs in *Xenopus laevis* and *Xenopus mulleri*. Cold Spring Harbor Symp Quant Biol 38: 501–506, 1973.
5. Brown DD, Sugimoto K: 5S DNAs of *Xenopus laevis* and *Xenopus mulleri:* Evolution of a gene family. J Mol Biol 78: 397–415, 1973.
6. Brown DD, Wensink PC, Jordon E: A comparison of the ribosomal DNAs of *Xenopus laevis* and *Xenopus mulleri*: The evolution of tandem genes. J Mol Biol 63:57–73, 1971.
7. Capra JD, Wasserman RW, Kehoe JM: Phylogenetically associated residues within the V_HIII subgroup of several mammalian species: Evidence for a "Pauci-Gene" basis for antibody diversity. J Exp Med 138: 410–427, 1973.
8. Cohn M: A rationale for ordering the data on antibody diversification. *In:* Brent L, Holborow J, (Eds): Progress in Immunology II: 261–284. New York & Amsterdam, Elsevier & North-Holland, 1974.
9. Cunningham BA, Berggard I: Structure, evolution and significance of β_2-microglobulin. Transplant Rev 21: 3–14, 1974.
10. Davidson EH, Hough BR, Amenson CS, Britten RJ: General interspersion of repetitive with nonrepetitive sequence elements in the DNA of *Xenopus*. J Mol Biol 77: 1–23, 1973.
11. Edelman GM, Gally JA: Arrangement and evolution of eukaryotic genes. *In:* Schmitt FO (Ed): The Neurosciences: Second Study Program, 962–972. New York, Rockefeller Univ. Press, 1970.

12. Edelman GM, Gally J: The genetic control of immunoglobulin synthesis. Ann Rev Genetics 6: 1–46, 1972.
13. Edelman GM, Cunningham BA, Gall W et al: The covalent structure of an entire γG immunoglobulin molecule. Proc Natl Acad Sci USA 63: 78–85, 1969.
14. Erhan S, Greller LD: Do immunoglobulins have proteolytic activity? Nature 251: 353–355, 1974.
15. Hood L: Two genes: one polypeptide—fact or fiction? Fed Proc 31: 177–187, 1972.
16. Hood L: The genetics, evolution and expression of antibody molecules. Stadler Genet Symp 5: 73–142, 1973.
17. Hood L, Eichman K, Lackland H et al: Rabbit antibody light chains and gene evolution. Nature 228: 1040–1044, 1970.
18. Hood L, Grant JA, Sox HC: On the structure of normal light chains from mammals and birds: Evolutionary and genetic implications in developmental aspects of antibody formation and structure. *In:* Sterzl (Ed): Developmental Aspects of Antibody Structure and Formation 1: 283–309. Prague, Academia, 1970.
19. Hood L, Barstad P, Loh E, Nottenburg C: Antibody diversity: an assessment. *In:* Sercarz E, Williamson A, Fox C (Eds): The Immune System: Genes, Receptors, Signals, 119–139. New York & London, Academic Press, 1974.
20. Hood L, Campbell JH, Elgin SCR: The organization, expression, and evolution of antibody genes and other multigene families. Ann Rev Genet 9: 305–353, 1975.
21. Kedes LH, Birnstiel M: Reiteration and clustering of DNA sequences complementary to histone messenger RNA. Nature [New Biol] 230: 165–169, 1971.
22. Mage R, Lieberman R, Potter M, Terry, WD: Immunoglobulin allotypes. *In:* Sela M (Ed): The Antigens, I: 300–377. New York & London, Academic Press, 1973.
23. Ohta T: Simple mode for treating evolution of multigene families. Nature 363: 74–76, 1976.
24. Poljak R, Amzel L, Avey H, et al.: Three-dimensional structure of the Fab' fragment of a human immunoglobulin at 2.8 Å resolution. Proc Natl Acad Sci USA 70: 3305–3310, 1973.
25. Putnam F: Comparative structural study of human IgM, IgA, and IgG immunoglobulins. *In:* Brent L, Holborow J (Eds): Progress in Immunology, II: 25–38. New York and Amsterdam, Elsevier & North Holland, 1974.
26. Rae PMM: The distribution of repetitive DNA sequences in chromosomes. Adv Cell Mol Biol 2: 109–149, 1972.
27. Schmid K, Kaufmann H, Isemura S et al: Structure of α_1-acid glycoprotein. The complete amino acid sequence, multiple amino acid substitutions, and homology with the immunoglobulins. Biochemistry 12: 2711–2724, 1973.
28. Smith GP: Unequal crossover and the evolution of multigene families. Cold Spring Harbor Symp Quant Biol 38: 507–514, 1973.
29. Tonegawa S, Hozumi N, Matthyssens G, Schuller R: Somatic changes in content and context of immunoglobulin genes. Cold Spring Harbor Symp Quant Biol 41: in press.

Tumor-Specific Transplantation Antigens of Chemically Induced Tumors

E. S. Lennox, Ph.D. and K. Sikora, M.B., B.Ch.

The examination of the surface of tumor cells by means of the immune response to their cell membrane antigens has provided tools to study neoplastic transformation and differentiation.

Surface antigens have been used not only to mark out steps in the differentiation of cells of the immune system itself but also of cells of other organs as well. In all cases one examines the surface of the chosen cells by means of the immune response in some properly selected animal. Usually the formed antibodies are used, although the response of cytotoxic thymus derived lymphocytes may serve as a useful assay.

The success of the analysis by serology of the differentiation antigens expressed by tumors of the lymphoid system continues to encourage attempts at similar analysis of the large collection of apparently non-cross-reacting antigens of chemically induced sarcomas. In this case, the diversity of antigens is revealed by transplantation studies, but by and large, the molecular basis of this diversity has not been resolved by serologic and chemical analysis.

We will present here some of the puzzling aspects of these chemically induced tumors and the attempts that have been made to characterize their antigens. (See [1] for a general review.) The apparently endless variety of these antigens has encouraged speculation that they are an expression of a highly polymorphic system. Since the best characterized polymorphic membrane antigens are those of the major histocompatibility locus, attempts have been made to associate the diverse tumor antigens with them. In addition, since such a wide variety of antigens has been associated with differentiating systems, one wonders whether the antigens of the chemically induced sarcomas could be related in some way to them. Thus, each of the sarcoma antigens might be a single representative of a wide set of antigens normally expressed on different stages of differentiating cells. We review below the extent to which these associations and expectations have so far been realized.

From the M.R.C. Laboratory of Molecular Biology, Cambridge, England, (Dr. Lennox, Salk Institute for Biological Studies, San Diego, Calif.)

One of the most striking aspects of the tumors induced in inbred mice and rats by polycyclic hydrocarbons, for example by methylcholanthrene (MCA), is that each independently arising tumor expresses a unique antigen and the number of possible antigens seems very large. The definition of these antigens and a measure of their quantity is done by transplantation assays. To do this assay, one preimmunizes an animal in one of several ways, e.g., by injection of X-irradiated tumor cells or by injection of live cells and subsequent surgical excision of the resulting tumor. Animals of the same genetic background as the tumor donor are used, to avoid interference of alloantigens. Such immunized animals, when later challenged with varying numbers of live tumor cells, display a specific resistance to the immunizing tumor by shifting markedly the tumor incidence as a function of challenge dose. Other parameters of the tumor growth after challenge may also be measured, e.g., time before appearance of tumor and subsequent growth rate (tumor size) after appearance of tumor. The assay is reproducible but it is not possible to extract from it quantitative measures of amount of antigen per tumor cell. A set of tumors can be graded on the basis of comparative immunogenicity but the rank of a tumor is probably a result of the complex interplay of amount of antigen, the kind of immune response generated, and susceptibility to immune attack [18]. The comparative response of a strongly immunogenic tumor (B10MC 6A) with a weak one (B10MC 4) is shown in Figure 1. In any event, by means of this assay, independently arising tumors show distinct non-cross-reacting antigens. In fact, tumors arising in the same animal but at two different sites of MCA injection are not cross-reacting (Fig. 2). It is this apparently endless diversity of tumor antigens that fascinates and begs to be explained in terms of genes and their products.

How large is the set of non-cross-reacting antigens? The most systematic attempts to answer this question were the experiments of Basombrio [3], who tested 25 tumors in various combinations—10 against each other in all pairings and the other 15 in various multiple sets. No reproducible cross-reactions were found between any two tumors. The experience of Basombrio has been repeated on a small scale by many investigators. While it is not easy to assign a maximum number to the size of the antigen set, at least 100 would not be an unreasonable guess.

Does the large number of antigens arise from an inherent heterogeneity in the cell population on which the carcinogen acts? Since it is possible that each event is due to the amplification into a clone of a preexisting heterogeneity in the underlying carcinogen-sensitive population, two groups of investigators have examined this possibility. In one case [8] transformation of mouse prostatic cells *in vitro* was studied. Using an untransformed, nontumorigenic line of cells cloned just prior to exposure to carcinogen, several transformed tumorigenic lines were obtained. Most were immunogenic, none cross-reacted. In another case [4] cells of a nontumorigenic line were cloned by an *in vivo* procedure and tumors induced in different samples from the same clone—again with similar results. Independent isolates of transformed tumorigenic cells from the same clone are antigenically distinct. These experiments imply that generation of new antigens accompanies the transformation event.

Chemically induced sarcomas share antigens with embryos. Sera from multiparous mice or rats or those raised by injection of tissue from midterm embryos into adult syngeneic aninals react with mouse or rat sarcomas respectively. This leads to the question whether tumor specific transplantation antigens (TSTA's) are embryonic antigens. Extensive experiments with rat sarcomas [2] show that this cannot be the case. Not only do independent tumors, non-cross-reacting in transplantation assays, show extensive cross-reaction with antiembryonic sera but also immunization with embryonic tissue does not protect against subsequent tumor challenge, except in rare cases.

Do genes in or near the major histocompatibility locus influence susceptibility to tumorigenesis by MCA? Since susceptibility to tumorigenesis by RNA tumor viruses is known to be under the influence of these genes, one might suspect that similar effects would be seen with MCA induction. We failed to find this, using groups of congeneic mice differing at H-2, after

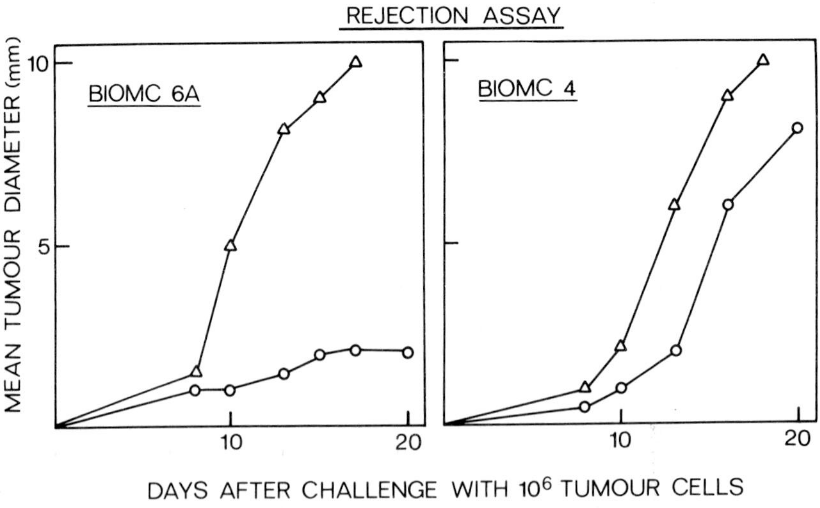

FIG. 1. Comparison of a strongly immunogenic (6A) and a weakly immunogenic (4) tumor, each derived from a B10 mouse. Mice were immunized by injection of 10^6 tumor cells subcutaneously. Ten days later, the resulting tumors were excised. After a further 10 days, immunized mice (-O-) and normal mice (-Δ-) were challenged with 10^6 tumor cells and mean tumor diameter was measured.

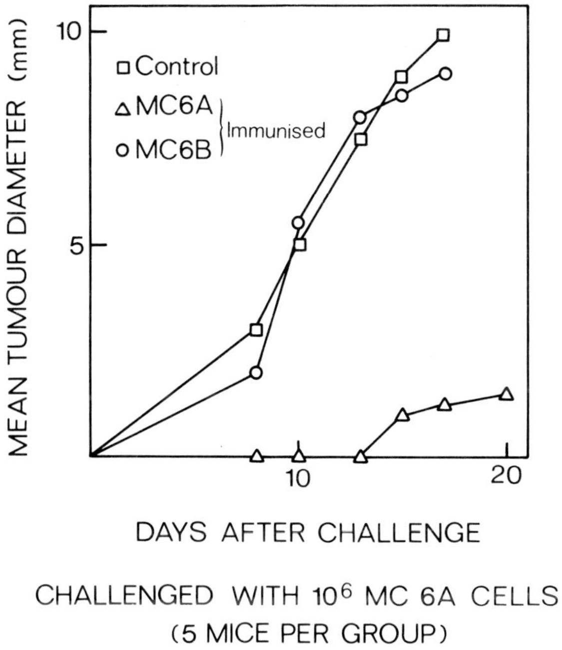

FIG. 2. Non-cross-reacting specificities of two tumors (MC 6A and MC6B) arising in the same B10 mouse. Mice were immunized with 6A or 6B and challenged as in Figure 1 with 10^6 6A or 6B cells. Only the results of 6A challenge are shown. Results with 6B are comparable.

injecting a single dose of 0.5 mg MCA/mouse, and the time to appearance of tumor was measured. The data are shown in Figure 3. In addition, growth rates of the tumors were similar after first appearance. It is possible that had we used smaller doses of MCA, we might have observed effects of the H-2 locus.

Is there a relation between the tumor-specific transplantation antigens and the major histocompatibility antigens? Four kinds of evidence are sought to answer this question: (1) quantitative relations between the amounts of TSTA and histocompatibility antigens (H-2); (2) genetic linkage of TSTA and H-2 in somatic segregation of tumor cell variants; (3) immunologic cross-reactions of H-2 and TSTA: (4) chemical comparison of purified antigens.

Using a group of tumors induced in congeneic mice, an inverse relationship was shown [12] between the amount of histocompatibility antigen assayed with unispecific antisera and the amount of TSTA as defined in a transplantation test. In a rough way we have confirmed this,* using

* Further experiments make our original observations suspect, for they reveal contamination of host cells in the animal grown tumors.

multispecific anti H-2ᵇ sera and a set of recently induced MCA tumors in B10 mice (Fig. 4). While this may imply that TSTA in some way replaces H-2 perhaps because it is an altered form of H-2 no longer serologically recognizable, there are other explanations and we shall offer one below.

In two experiments of rather different design, investigations were carried out to study the possible genetic association in murine tumors of TSTA with the chromosome known to carry the locus for H-2 expression—chromosome 17. Rather different conclusions were reached in the two cases. In each set of experiments, starting with a tumor cell line expressing two H-2 types, selection was made of sub-lines carrying one H-2 type or the other, and the possible cosegregation of TSTA with H-2 was then assayed in the selected variants.

In one set of experiments [14], the starting tumor was a fusion product of a strongly immunogenic MCA-induced murine sarcoma of type H-2ˢ with a

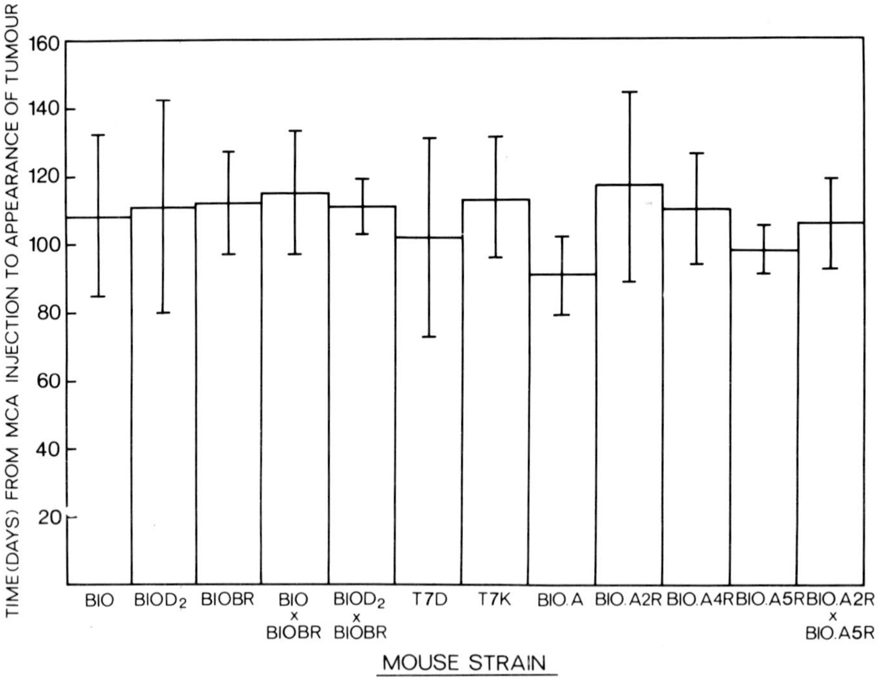

FIG. 3. Time of appearance of tumors following a single injection of MCA into congeneic mice of different H-2 types. Mice were given a single injection of 0.5 mg MCA in trioctanoin intramuscularly into each thigh. Average times to appearance of palpable tumor with standard deviations is shown.

The strains are all B10 congeneic mice or F1 hybrids of them. T7D and T7K are F1 hybrids of B10D2 × B10BR with the T7 chromosome marker carrying respectively H-2ᵈ or H-2ᵏ.

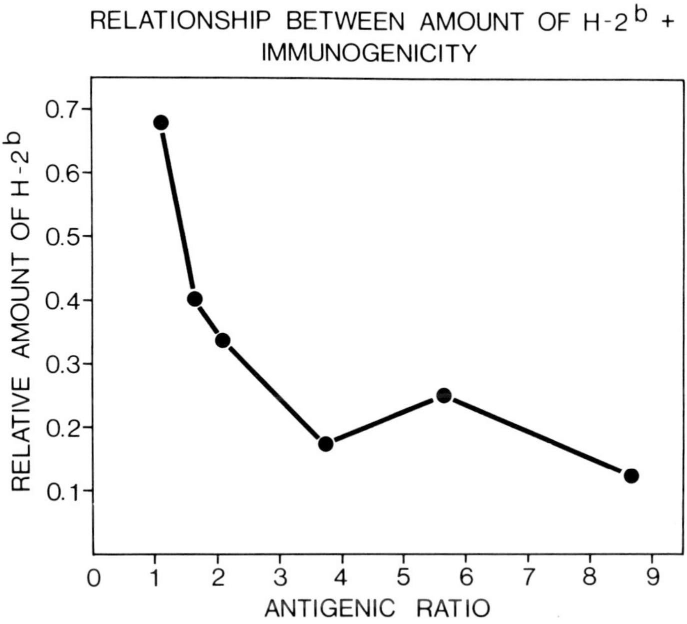

FIG. 4. Relative amount of H-2^b on B10 sarcomas vs amount of tumor antigen measured as antigenic ratio. Amount of H-2^b is measured in terms of absorbing capacity of an anti-H-2^b serum by the same numbers of tumor cells. Antigenic ratio is defined as the ratio of the tumor size 20 days after challenge in control mice to that in immunized mice.

mammary carcinoma of type H-2^a. The hybrid cell line retained the TSTA of the sarcoma line. In addition, the two chromosomes carrying H-2^a could be distinguished cytologically from the two carrying H-2^s. The hybrid cell line grew well and was immunogenic in H-2^s $\times$ H-2^a Fl heterozygous mice. By passage in H-2^s or H-2^a parental mice, lines homozygous for H-2^s or H-2^a respectively were obtained and these were assayed for presence of the original TSTA. Cytologic examination also revealed that the loss of H-2^s or H-2^a was accompanied by the loss of the corresponding pair of chromosomes 17. If the production of TSTA were controlled in the same chromosome as H-2^s, then selection against this haplotype should select against the TSTA, while selection against H-2^a should leave the expression of the TSTA unaffected. This was not the result. Both classes of variants showed reduced immunogenicity when measured in H-2^s mice challenged with the parental H-2^s sarcoma. While the series of tumors of both classes of variants differed in their immunogenicity, these differences were unrelated to whether H-2^s or H-2^a had been selected. Thus, the authors concluded that since TSTA was not completely lost when the H-2^s chromosome was eliminated, it was not synthesized or controlled by genes of the H-2 complex. Nevertheless, they suggested that the expression of TSTA was influenced by chromosomes 17.

In the other study [16] a tumor was induced in a mouse heterozygous at H-2, i.e., in an H-2^f × H-2^b F1 hybrid. Variants with H-2 loss were selected by passage in H-2^f of H-2^b parental mice and these were then tested for the presence of TSTA in F1 mice. The interesting finding is that the H-2^f and the H-2^b variants are each immunogenic and do not cross-react. Unfortunately, the starting F1 tumor is not itself immunogenic nor is there any cross-reaction with either of the two variants with H-2 loss; thus, the two new TSTA appearing in them cannot be related to antigens in the F1 tumor. Moreover, the fact that the original F1 tumor was not cloned prior to H-2 selection weakens the interpretation in terms of the segregation of TSTA with H-2. This, however, may not be a serious objection, for the comparison of the karyotypes of the F1 tumor with the two variants with the H-2 loss seems to indicate that both of them arose from the same heterozygous clone.

Several kinds of evidence have been presented that suggest cross-reaction between TSTA and major histocompatibility antigens. In one case [6] TSTA-containing material was isolated and purified from a chemically induced rat hepatoma (D23) of a Wistar rat. By several criteria, including separation on isoelectric focusing gels with or without detergent, only a single protein was detected. By using this material as antigen in rabbits, an antiserum was raised which, after absorption with normal cell membranes from rats of the same inbred strain, was then specific for hepatoma D23. Similar sera could be prepared with the same material in an allogenic strain of rats. These results suggest the presence of a D23-specific antigen, probably the TSTA in the preparation.

Finally, a serum prepared in an allogeneic strain of rats against normal cell membranes of the Wistar strain and bound to sepharose would specifically absorb ^{125}I-labeled D23 tumor antigen preparation. Since this serum is assumed to be specific for Wistar histocompatibility antigens, this is taken as evidence for its cross-reactions with the D23 tumor-specific transplantation antigen.

Earlier experiments [9] on antigen from serum of mice carrying a spontaneous lymphoma had also indicated association of H-2 antigens and the tumor-specific antigen on the same molecules, under nondissociating conditions. Each kind of antigen was recognized by an appropriate antiserum.

In other experiments the TSTA of a chemically induced sarcoma was shown to cross-react in protection experiments with alloantigens [13]. A Balb/c sarcoma induced by MCA was immunogenic in Balb/c and in F1 hybrids of Balb/c with C3H and C57B1/6 mice. The tumor apparently had a tumor-specific antigen for it did not cross-react with other Balb/c sarcomas. Cross-reaction of the TSTA of this tumor with normal mouse alloantigens was indicated by the observation that Balb/c mice grafted with C3Hf normal tissue (but not that of C57B1 or W tissue) showed increased resistance to the sarcoma. Extensive comparison with other immunogenic sarcomas was not done. While the results were not entirely straightforward, these experiments hinted at possible cross-reaction of a TSTA with the major histocompatibility antigens. Later experiments [17] did not support this connection for they indicated that the antigens of normal tissue responsible for immunizing against the tumor were not those of the major histocompatibility complex but were minor alloantigens.

The chemical analyses of TSTA and detailed comparison with H-2, have been inhibited mainly by the rather primitive state of TSTA chemistry. This is partly because by and large, even the crude material released from tumor cell membranes and used as starting material for purification has been only weakly immunogenic in the transplantation assays that define TSTA. In addition, tumor-specific antisera are usually very weak and not suitable as reagents to identify TSTA during purification. Attempts to make stronger specific sera have not been successful. Our failure, illustrated in Figure 5, is an example.

Observations coming from an unexpected direction [7] may be relevant to experiments seeking relationships between TSTA and major histocompatibility antigens.

These recent observations emphasize a predominant role played by the major cellular histocompatibility antigens as specific targets for immune attack by cytotoxic lymphocytes (CTL). Even when the immunogenic cell differs from those of the responding host in many other surface proteins, e.g., mouse cells immunogenic to human cells, the responding human CTL recognize mainly the mouse major histocompatibility antigens [15]. Moreover, if the surface antigens of a mouse cell are altered by appearance of new antigens, e.g., by virus infection, and CTL are stimulated in a host having the same H-2 antigens as the uninfected cell, then recognition of this new antigen by CTL seems to involve simultaneous recognition of H-2. While the mechanism underlying this phenomenon has not been brought to light, several features are clear. One is the basic finding: If cytotoxic lymphocytes are stimulated by a cell having a certain H-2 and a non H-2 antigen, then their specificity is determind both by that H-2 and the non H-2 antigen. In the case of virus-infected cells, they are thus specific for the virus and for the H-2 background of the infected cell, not separately but for their presence together on the same cell. The other is that this phenomenon occurs not only for a wide variety of dissimilar viruses, but also for a wide variety of other stable components of the cell surface [5].

As for basic mechanisms, several have been proposed, one of which is that in some way H-2 and the new antigen interact [5,19] and that the result of this interaction is recognized. While this may not be correct, it at least

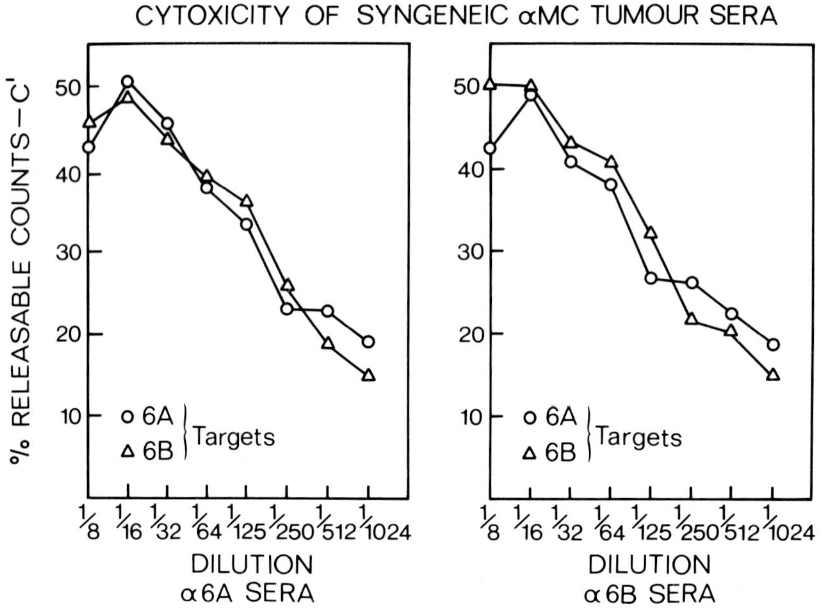

FIG. 5. Lack of specificity of antisarcoma sera prepared in syngeneic mice. Antisera were prepared by repeated injection of X-irradiated cells of the independent B10 tumors 6A and 6B into B10 mice. Cytotoxicity of the sera with added rabbit complement was measured by release of ^{51}Cr from labeled target cells. Counts releasable by detergent is taken as 100%. Background release with complement (C′) alone is subtracted. Cytotoxicity of each serum on cells of the immunizing tumor and of the independent tumor is shown.

helps remember the basic feature, i.e., that both H-2 and the new antigen are required for recognition.

Where might TSTA fit into this picture? The first answer might be—nowhere, since we always measure a set of non-cross-reacting TSTA's in the same H-2 context; that is, H-2^b tumors are assayed for TSTA in H-2^b mice, syngeneic to the one in which the tumor arose. But since there is some evidence from virus-infected cells that the amount of normal H-2, serologically recognized, decreases upon infection, one might seek here the origin of the apparent inverse relation between amounts of H-2 and TSTA. There are, of course, other explanations.

Further, if the observed TSTA is a result of the interaction with H-2 of new proteins generated in the carcinogen-induced transformation and if these proteins could interact with each other as well as with H-2, a small number of them might generate a large number of non-cross-reacting specificities.

In this connection, there is evidence that some new H-2 specificities appear and some old ones disappear on virus infected cells [10] and tumor cells [11].

In genetic segregation experiments with tumors heteroyzgous for two H-2 types and examining TSTA cosegregation with H-2, the possible association of TSTA with H-2 proteins could make complications in interpretation. In this case one would expect to see modifications of the TSTA on removal of one H-2 type or the other and these modifications might mask attempts to seek genetic linkage of TSTA to H-2. One could also imagine complex situations where removal of one H-2 type, by forcing associations of surface proteins with the other type, could lead to the appearance of new TSTA.

In summary, the source of the diversity of tumor-specific transplantation antigens still remains to be found. The answer to the simple question of whether they are members of a polymorphic set is not yet known. Their possible relation to major histocompatibility antigens is suggested by some experiments and denied by others. The evidence is against their relationship to embryonic antigens recognized so far. Their diversity continues to provide an intriguing puzzle to be solved.

REFERENCES

1. Baldwin RW: Immunological aspects of chemical carcinogenesis. Adv Cancer Res 18: 1–75, 1973.
2. Baldwin RW, Embleton MJ, Price MR, Vose BM: Embryonic antigen expression in experimental rat tumours. Transplant Rev 20: 77–99, 1974.
3. Basombrio MA: Search for common antigenicities among twenty-five sarcomas induced by methylcholanthrene. Cancer Res 30: 2458–2462, 1970.
4. Basombrio M, Prehn RT: Antigenic diversity of tumors chemically induced within the progeny of a single cell. Int J Cancer 10: 1–8, 1972.
5. Bevan MJ: The major histocompatibility complex determines susceptibility to cytotoxic T cells directed against minor histocompatibility antigens. J Exp Med 142: 1349–1364, 1975.
6. Bowen JG, Baldwin RW: Tumour-specific antigen related to rat histocompatibility antigens. Nature 258:75–76, 1975.
7. Doherty PC, Blanden RV, Zinkernagel RM: Specificity of virus-immune effector T cells for H-2K or H-2D compatible interactions: Implications for H-antigens diversity. Transplant Rev 29: 89–124, 1976.
8. Embleton MJ, Heidelberger C: Antigenicity of clones of mouse prostate cells transformed *in vitro*. Int J Cancer 9: 8–18, 1972.
9. Fujimoto S, Chen CH, Sabadini E, Sehon AH: Association of tumor and histocompatibility antigens in sera of lymphoma-bearing mice. J Immunol 111: 1093–1100, 1973.
10. Garrido F, Schirrmacher V, Festenstein H: H-2-like specificities of foreign haplotypes appearing on a mouse sarcoma after vaccinia virus infection. Nature 259: 228–230, 1976.
11. Garrido F, Festenstein H, Schirrmacher V: Further evidence for derepression of H-2 and Ia-like specificities of foreign haplotypes in mouse tumor cell lines. Nature 261: 705–707, 1976.
12. Haywood GR, McKhann CF: Antigenic specificities on murine sarcoma cells. J Exp Med 133: 1171–1187, 1971.

13. Invernizzi G, Parmiani G: Tumour-associated transplantation antigens of chemically induced sarcomata cross reacting with allogeneic histocompatibility antigens. Nature 254: 713–714, 1975.
14. Klein G, Klein E: Are methylcholanthrene-induced sarcoma-associated, rejection-inducing (TSTA) antigens, modified forms of H-2 or linked determinants? Int J Cancer 15: 879–887, 1975.
15. Lindahl KF, Bach FH: Human lymphocytes recognize mouse alloantigens. Nature 254: 607–609, 1975.
16. Oth D, Berebbi M, Meyer G: Tumor-associated antigens in isoantigenic variants of a 3-methylcholanthrene-induced sarcoma. J Natl Cancer Inst 55: 903–908, 1975.
17. Parmiani G, Invernizzi G: Alien histocompatibility determinants on the cell surface of sarcomas induced by methylcholanthrene. Int J Cancer 16: 756–767, 1975.
18. Prehn, RT: Analysis of antigenic heterogeneity within individual 3-methylcholanthrene-induced mouse sarcomas. J Natl Cancer Inst 45: 1039–1045, 1970.
19. Schrader JW, Edelman GM: Participation of the H-2 antigens of tumor cells in their lysis by syngeneic T cells. J Exp Med 143: 601–614, 1976.

Effects of Steroid Hormone Receptors on Gene Transcription

Bert W. O'Malley, M.D., William A. Coty, Ph.D.,
Robert J. Schwartz, Ph.D. and
William T. Schrader, Ph.D.

The immature chick oviduct responds to estrogen administration with dramatic cellular proliferation and differentiation [14]. One major step in this process is the expression of new genes coding for the synthesis of eggwhite proteins, including ovalbumin [15,24]. The presence of these specific proteins as biochemical markers for cell differentiation makes the chick oviduct an excellent system to study the process of steroid hormone action. Studies in this laboratory have been directed toward understanding steroid hormone control of gene expression at the molecular level.

Estrogen-induced oviduct growth and differentiation are accompanied by profound changes in the transcription of DNA. Competition hybridization experiments have demonstrated qualitative changes in nuclear RNA transcribed from repetitive DNA sequences [23], and increased transcription of unique sequence DNA [17]. These changes are reflected in an increase in sequence complexity of cellular mRNA [22]. These new mRNA species include biologically active ovalbumin mRNA, which accumulates prior to the appearance of ovalbumin protein in the oviduct [6,9,29]. In addition to these changes in messenger RNA synthesis, estrogen induces new ribosomal components in order to increase the over-all capacity of the cells to synthesize protein [21].

Upon withdrawal of estrogen treatment, there is a reduction in the over-all level of RNA and protein synthesis, and the synthesis of ovalbumin mRNA ceases [6,9,28]. Readministration of either estrogen or progesterone (secondary stimulation) results in a rapid increase in the production of ovalbumin mRNA and the induction of other eggwhite proteins [28]. Thus, we can assess the effect of steroid hormone on gene expression in the chick oviduct by the measurement of a specific product of a single gene.

From the Department of Cell Biology, Baylor College of Medicine, Houston, Texas.
Acknowledgments: This work was supported by National Institutes of Health Grants HD-8188, HD-7857, HD-7495, and a Ford Foundation Grant for the Cell Biology Department, Baylor College of Medicine, Houston, Texas.

Our investigations into the mechanism of steroid-hormone action in the chick oviduct have proceeded along two lines. First, we have studied the various steps in RNA synthesis *in vitro,* using purified components, in order to determine the site of hormone regulation. Secondly, we have purified and characterized steroid-hormone receptors from the chick oviduct in an effort to elucidate the role that these important proteins play in gene regulation. The results of these two lines of investigation now allow us to assess directly the effect of pure steroid hormone-receptor complex on gene expression *in vitro*. In this review, we will discuss our recent studies on progesterone-receptor regulation of transcription of chick oviduct chromatin *in vitro*.

Measurement of Over-all Gene Expression *in vitro*

The capacity of oviduct chromatin to serve as a template for *E. coli* RNA polymerase has previously been shown to increase following estrogen administration to the immature chick [8,40]. While chromatin template activity measurements may generally reflect the amount of DNA sequences made available to RNA polymerase, the components of such a reaction are so complex that these experiments shed little light on the biochemical mechanisms of hormone-induced alterations in gene transcription. In order to determine the effect of steroids on gene transcription in the chick oviduct, it appeared necessary to monitor and control all of the parameters involved in RNA synthesis. The process of initiation of RNA synthesis received special attention, since this step appeared to be regulated by steroids [37].

Procedures for measuring the initiation of RNA chains *in vitro* were adopted from studies initially carried out in bacterial and bacteriophage systems [1,18]. The initiation of RNA synthesis can be divided into two basic processes. First, RNA polymerase binds randomly and reversibly to DNA to form a series of nonspecific complexes. However, after sufficient time and at a proper temperature, RNA polymerase binding proximal to an initiation region forms a stable binary complex with DNA [39, 48]. This complex has undergone a transition involving the local destabilization of the DNA duplex structure [43], and is now capable of rapidly initiating an RNA chain [43]. The second process is the actual initiation step in which RNA polymerase catalyzes the formation of the first phosphodiester bond between two nucleoside triphosphates.

The existence of stable RNA polymerase-DNA initiation complexes was elucidated through the use of the drug rifampicin [11, 46]. Rifampicin is a competitive inhibitor of RNA synthesis, which acts prior to the formation of the first phosphodiester bond but which has no effect on RNA chain elongation. RNA chain initiation by RNA polymerase bound to DNA in a stable initiation complex is so rapid that it can occur in the presence of the drug [12]. The fraction of RNA polymerase in stable complexes can be determined following the simultaneous addition of a mixture of rifampicin and the four ribonucleoside triphosphates [5]. RNA polymerase molecules which are free in solution or randomly bound to DNA will be inhibited by rifampicin. Under these conditions, reinitiation of RNA transcription will be completely inhibited and thus each RNA polymerase in a stable preinitiation complex can synthesize one and only one RNA chain. By measuring the number of RNA chains made, the rifampicin challenge assay provides a method for quantification of the number of RNA polymerase initiation sites on a given template [7,37,44,45].

Acute Progesterone Stimulation of Initiation of RNA Synthesis

Although incapable of inducing eggwhite protein synthesis in the oviduct during primary stimulation, progesterone not only reestablishes ovalbumin protein synthesis but appears generally to mimic estrogen as a secondary stimulant to hormonally withdrawn chicks [20,28]. Furthermore, the similarity in the extent of eggwhite protein synthesis induced by progesterone suggests that the induction of RNA synthesis during secondary stimulation is similar to estrogen and not influenced by the cytodifferentiation of new tubular gland cells [28]. Therefore, it was of interest to understand the manner by which progesterone can substitute for estrogen in hormone-withdrawn chicks.

Estrogen-treated chicks were withdrawn from hormone treatment for 12 days and then restimulated with a single injection of progesterone (2.0 mg). Oviduct chromatin was isolated following secondary stimulation and assayed for initiation sites by the rifampicin-challenge technic. The number of RNA chains initiated was calculated as described above. As shown in Table 1, withdrawn chromatin had the capacity to support the initiation of 8,600 RNA chains/pg of DNA. Following a single injection of progesterone, a rapid increase in the number of initiation sites was found. Within a half-hour of hormone treatment, the number of initiation sites had nearly doubled to a level of 15,900 sites. After one hour of progesterone stimulation, a maximum of 23,000 initiation sites was detected. Thereafter, the number of initiation sites declined, so that by 24 hours after progesterone administration, 13,500 sites were observed. During this time of intense transcriptional activity, the parameters of RNA chain propagation rate ($\sim$7.5 nucleotides/second) and the number average chain length of the RNA product ($\sim$770 nucleotides) did not vary significantly from withdrawn oviduct chromatin (Table 1). Therefore, the progesterone-induced increase in chromatin transcription in the oviduct was due mainly to an increase in the number of available RNA polymerase initiation sites.

To investigate further the mechanism by which progesterone induces increased chromatin transcription in withdrawn chicks, we tested the possibility that progesterone acted at sites on chromatin which were identical to those effected by estrogen. We compared the dose dependence of initiation-site stimulation by progesterone (Table 2) and estrogen [44]. Either hormone gave approximately the same maximal extent of stimulation, from 9,100 to approximately 23,000 sites/pg DNA at the same dose (1.25 mg) of either steroid.

We then tested the possibility that these steroids induced the initiation of transcription at the same sites. Both DES (1.25 mg) and progesterone (1.25 mg) were administered together to withdrawn chicks. If estrogen and progesterone acted at different chromatin RNA initiation sites, then the stimulation of RNA synthesis by the two hormones would be additive. The results in Table 2 show that the number of RNA chain initiations was not additive when both hormones were simultaneously administered. These data suggest that in the withdrawn chick oviduct, estrogen and progesterone may regulate transcription at similar sites in chromatin.

TABLE 1. The Effect of Progesterone Administration *in Vivo* on RNA Polymerase Transcription of Withdrawn Oviduct Chromatin *in Vitro*

Hours after Progesterone Administration[a]	Size of RNA Product (Nucleotides)	Initial Elongation Rate (Nucleotides/sec)[b]	pmol of RNA Chains Initiated per 5μg DNA	Initiation Sites/pg DNA[c]
0	820	7.2	.072	8,600
0.5	730	7.0	.132	15,900
1.0	750	7.8	.191	23,000
2.0	750	6.7	.134	16,100
6.0	800	8.0	.123	14,800
24.0	750	8.0	.112	13,500

[a] Chicks were stimulated for 12 days with DES, withdrawn from hormone treatment for 10 days, and then injected with 2 mg of progesterone.

[b] Initial elongation rate was determined for 1 min of chain propagation as previously described [37].

[c] The number of initiated chains was calculated from the total nucleotides incorporated at the transition point of RNA polymerase titration curves, assuming that RNA contained 25% UMP, and divided by the number average chain size of 770 nucleotides as described [37].

TABLE 2. Effect of the Combined Administration of Progesterone and Diethylstilbestrol on Initiation Sites for RNA Synthesis in Oviduct Chromatin

Dose of Hormone[a]	RNA Chains Initiated (pmols/5 μg DNA)	Initiation Sites/pg DNA[b]
Control	.076	9,100
0.25 mg Progesterone	.135	16,200
1.25 mg Progesterone	.187	22,500
2.0 mg Progesterone	.196	23,600
1.25 mg DES	.191	23,000
1.25 mg Progesterone + 1.25 mg DES	.180	21,600

[a] Chicks were stimulated for 12 days with DES, withdrawn from hormone treatment for 10 days and then injected with various doses of hormone.

[b] The number of initiated chains was calculated as described in the legend to Table 1.

We also investigated the effect of progesterone on transcription of the ovalbumin gene. We have previously shown that estrogen stimulation of ovalbumin mRNA synthesis is achieved through regulation of transcription of the ovalbumin gene [10]. Since secondary progesterone stimulation induces a net accumulation of ovalbumin mRNA [28], it was of interest to determine whether this stimulation was also at the level of gene transcription.

To test this hypothesis, withdrawn chicks were given a single injection of progesterone (2 mg), and oviduct chromatin was isolated and transcribed *in vitro,* using *E. coli* RNA polymerase. A qualitative measure of the levels of ovalbumin mRNA sequences ($mRNA_{ov}$) in *in vitro* chromatin transcripts was obtained by hybridization with [^{3}H]DNA complementary to ovalbumin mRNA ($cDNA_{ov}$) [10,47]. From the kinetics of reassociation of *in vitro* synthesized RNA with $cDNA_{ov}$ at various RNA concentrations, the concentration of $mRNA_{ov}$ sequences was determined.

The data in Table 3 show that prior to progesterone treatment, no detectable ovalbumin mRNA sequences were present in the transcript. Progesterone administration induced a rapid appearance of $mRNA_{ov}$ sequences, indicating that progesterone, like estrogen [10], acts to regulate the availability of the ovalbumin gene for transcription by RNA polymerase. A measurable increase occurred after 4 hours, which preceded the progesterone-induced accumulation of ovalbumin mRNA [36].

These experiments demonstrate that during secondary hormonal stimulation of withdrawn chicks, progesterone is an adequate substitute for estrogen with respect to its effect on gene expression.

TABLE 3. *In Vitro* Synthesis of $mRNA_{ov}$ from Oviduct Chromatin Templates

Source of Chromatin[a]	pg $mRNA_{ov}/\mu g$ DNA[b]
Withdrawn	0
2 hr × progesterone	10.5
6 hr × progesterone	16.2
24 hr × progesterone	30.3

[a] Chicks were stimulated for 12 days with DES, withdrawn from hormone treatment for 10 days, and then injected with 2 mg of progesterone.

[b] Bulk RNA synthesis and hybridization with $cDNA_{ov}$ was performed as previously described [10].

Purification and Properties of Chick-Oviduct Progesterone Receptor

Receptor proteins are likely to mediate steroid-hormone control of gene expression for two reasons. First, these proteins bind hormone in target tissue with high affinity and with a specificity for hormone structure consistent with *in vivo* steroid potency. Secondly, receptors are translocated into the nucleus upon hormone administration at a time prior to other hormone-dependent responses. Since the initial identification of the crude oviduct progesterone receptor [38], we have sought to characterize and purify this important regulatory protein.

A major form of progesterone receptor in crude cytoplasmic extract is a 6S molecule [32] of approximately 200,000 molecular weight, containing two nonidentical hormone-binding subunits [32,34]. Occupancy of these sites by progesterone renders the 6S dimer metastable and promotes dissociation into subunits [31].

The two receptor subunits designated A and B are both taken up in the nucleus [27,35], but have distinct binding specificities toward nuclear components: the A subunits binds only to DNA, while the B subunit binds only to chromatin [35]. Furthermore, the B subunit binding to chromatin is target tissue-specific [13,26,41,42]. Although both DNA and chromatin-binding subunits are present in the intact 6S dimer, only the activity of the B subunit is expressed. The intact dimer is able to bind to chromatin, but dissociation into subunits is required for DNA binding activity (W. A. Coty, W. T. Schrader and B. W. O'Malley, submitted for publication). These properties of the progesterone receptor, which are summarized in Figure 1, led us to propose that the two receptor subunits might act together

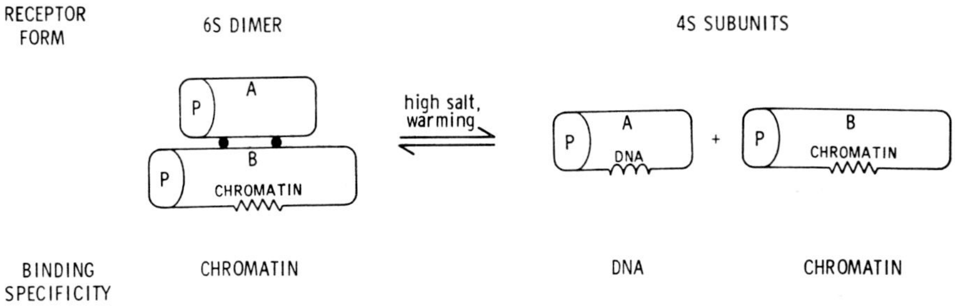

FIG. 1. Subunit structure of chick oviduct progesterone receptor.

in regulating gene transcription [25,32,35]. The B subunit could act as the specifier by virtue of its affinity for a tissue-specific class of chromatin nonhistone (acidic) proteins. This would then direct the effector A subunit toward a regulatory region on the DNA, resulting in altered gene transcription.

We have recently succeeded in purifying to homogeneity A and B subunits and the intact progesterone receptor by a variety of technics, including steroid-affinity chromatography [3] and conventional ion-exchange and gel filtration technics [16]; W. A. Coty et al., submitted for publication). These purified receptor forms retain all the physical and biologic activities of crude receptor preparations.

Effect of Purified Progesterone Receptor on *in vitro* Initiation of Transcription

The availability of purified receptor components enabled us to test our proposed model by examining whether receptors have a direct effect on transcription *in vitro*. The use of purified receptor in these studies is essential, in order to avoid artifacts caused by interfering enzyme activities [3] and to be able to relate directly any observed effect to the action of the receptor.

We first examined whether receptor could stimulate *in vitro* the number of initiation sites on oviduct chromatin. Chromatin from withdrawn chicks was incubated for 30 minutes at 22° with varying concentrations of intact receptor dimer purified by a modification of the affinity chromatography procedure of Kuhn et al. [16]. Saturating levels of RNA polymerase (15 μg) were then added, and further incubated for 30 minutes to allow the formation of stable preinitiation complexes. Finally, the RNA synthesis reaction was started by the simultaneous addition of rifampicin, nucleoside triphosphates and heparin, and RNA synthesis was measured. Progesterone receptor stimulated rifampicin-resistant RNA synthesis approximately 75% in a dose-dependent manner (Fig. 2). Half-maximal stimulation occurred at 5×10^{-9}M receptor, in excellent agreement with chromatin binding affinities [13]. The kinetics for the stimulation of chromatin transcription *in vitro* revealed a $T_{½}$ of 15 min [36]. This value is close to the optimal time required for receptor binding to chromatin at 22°C and similar to the kinetics of receptor appearance in nuclei *in vivo* and *in vitro* following progesterone administration [2,27].

An additional control was performed in which the receptor was com-

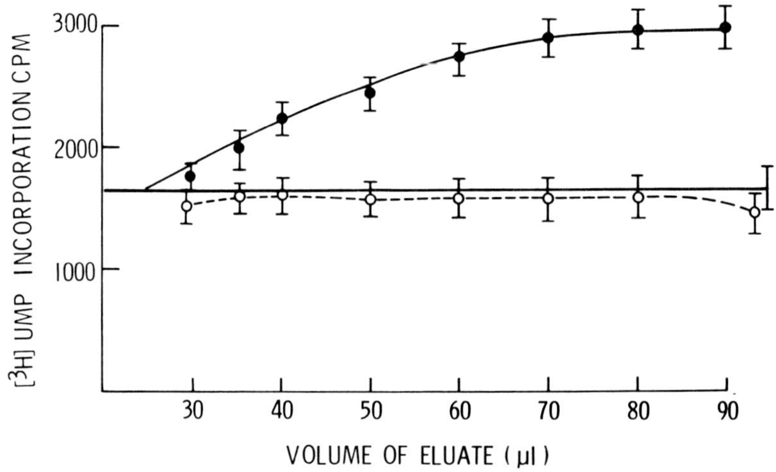

FIG. 2. Effect of purified progesterone receptor complexes on oviduct chromatin RNA synthesis *in vitro*.

Chick oviduct cytosol was precipitated with ammonium sulfate [34] and divided into two parts. Progesterone was added to one half (O---O), while the other half remained untreated (●---●). Both fractions were separately incubated with a deoxycorticosterone hemisuccinate-bovine serum albumin-Sepharose affinity resin and eluted as described [16]. The eluted fraction pretreated with progesterone did not contain any progesterone receptor, while the untreated fraction contained progesterone receptor [1.5×10^{-8} M). Various amounts of the eluted fractions in 10 mM Tris-HCl (pH 7.4), 1 mM Na_2 EDTA, 12 mM 1-thioglycerol and 50 mM KCl were added to a maximum volume of 100 µl to 25 µl of a withdrawn oviduct chromatin suspension (5 µg of DNA) and incubated for 30 minutes at 22°C. A saturating amount of *E. coli* RNA polymerase (15 µg) was then added together with the following components with 20 µmol Tris-HCl (pH 7.9), 1.0 µmol $MnCl_2$, 0.5 µmol β-mercaptoethanol and 100 µg of bovine serum albumin and incubated for an additional 30 minutes at 22°C. At the end of the preincubation period, RNA synthesis was started by the addition of 37.5 nmol of ATP, GTP, CTP and UTP and 10 µCi [^{3}H]UTP, 10 µg of rifampicin, 0.1 µmol potassium phosphate (pH 7.0) and 200 µg of heparin in a final volume of 250 µl for 15 minutes at 37°C. Synthesized RNA was precipitated with 5% trichloroacetic acid, collected on glass fiber filters and counted in scintillation fluid at an efficiency of 19.5%. The straight line represents the level of constitutive RNA synthesis in the absence of added receptors.

TABLE 4. Effect of Progesterone-Receptor Complexes on RNA Synthesis from Oviduct Chromatin *in Vitro*

Components Added to Assay	[^{3}H]UMP Incorporated into RNA/5 μg Chromatin DNA[a] CPM	Percentage Activity
Background:		
RNA polymerase only	230	8
Progesterone-receptor complex plus RNA polymerase	400	14
Chromatin alone	200	7
Control:		
Chromatin plus RNA polymerase	2730	100
Chromatin plus boiled progesterone-receptor complex	2860	105
Chromatin plus 10^{-8} M progesterone alone	2430	89
Experimental:		
Chromatin plus RNA polymerase plus 10^{-8} M		
Progesterone-receptor complex	4100	150
plus α-Amanitin (10 μg)	3740	137
plus Actinomycin D (10 μg)	320	12

[a] RNA synthesis was assayed in the presence of nucleotides and rifampicin as described in Figure 1.

plexed with progesterone prior to affinity chromatography. This prevented binding of the receptor to the steroid affinity resin, and subsequent elution to the resin failed to yield any detectable receptor. This control preparation was subsequently inactive in stimulating initiation of transcription (Fig. 2). This experiment strongly indicates that stimulation of chromatin transcription is due to progesterone receptor and not to a minor contaminant adsorbed nonspecifically to the affinity column.

We have previously reported that crude receptor preparations can often spuriously stimulate RNA synthesis by a template-independent process which is unrelated to hormone receptor [3]. To rule out such an effect in the present study, we investigated the sensitivity of rifampicin-resistant incorporation of [^{3}H]UMP into RNA to various inhibitors of RNA synthesis. Table 4 shows that RNA synthesized in the presence of receptor and chromatin was completely dependent upon template and inhibited by actinomycin D. Furthermore, none of the components alone contained significant synthetic activity, and hence neither the chromatin, polymerase nor receptor were contaminated with other enzymes which were capable of

TABLE 5. Effect of Progesterone Receptor on the Size and Rate of RNA Chain Synthesis on Oviduct Chromatin

Concentration of Progesterone Receptor	Total Incorporation of Nucleotide[a] (pmoles)	Initial Elongation Rate[b] (Nucleotides/sec)	Number Average Chain Size[c] (Nucleotides)	RNA Chains Initiated (pmol/5 μg) Chromatin DNA	Initiation Sites/pg DNA
No Receptor	62	7.9	739	.083	10,000
1×10^{-8} M	91	7.8	720	.126	15,200

[a] RNA was synthesized in the presence of rifampicin and nucleotides as in Figure 1 and corrected for base composition of RNA, assuming 25% UMP.

[b] Measured in first minute of synthesis.

[c] Measured by sucrose gradient analysis of RNA synthesized in 15 min.

incorporating [^{3}H]UTP into acid-insoluble material. In these experiments, the purified progesterone receptor preincubated with chromatin in the presence of *E. coli* RNA polymerase increased rifampicin-resistant RNA synthesis 50% over control values. With the addition of α-amanitin, a potent inhibitor of oviduct RNA polymerase II, there was little effect on the progesterone receptor-directed stimulation of RNA transcription. Thus, the stimulation of RNA synthesis was not due to the activation of endogenous RNA polymerase in the chromatin preparations but rather was due to transcription by the added *E. coli* enzyme. Nevertheless, it should be noted that receptor-mediated stimulation of RNA synthesis also occurred in the presence of exogenously added oviduct RNA polymerase II (R. J. Schwartz, M. J. Tsai, W. T. Schrader and B. W. O'Malley, in preparation).

The data in Table 4 also show that the increased RNA synthesis was dependent upon the native structure of the hormone-receptor complex, since neither free progesterone nor boiled receptor was effective in stimulating RNA synthesis. Importantly, the purified progesterone receptor does not contain any detectable proteolytic or nuclease activity which would spuriously affect the chromatin template [36]. Thus, the stimulation of [^{3}H]UMP incorporation is derived through a chromatin template-dependent process which is directly mediated by an intact progesterone-receptor complex. In addition, calculations of the number of RNA chains synthesized (Table 5) show that receptor stimulates initiation of RNA chain synthesis, rather than increasing the rate of chain elongation or the chain length of the RNA product.

The stimulatory effect of the progesterone receptor on RNA chain initiation on chromatin was next verified by a totally independent assay. Quantitation of the number of 5′-termini present in the population of *in vitro* synthesized RNA chains was measured by the incorporation of [γ-^{32}P]GTP into RNA [19]. We have previously shown that 57% of the initial 5′-tetraphosphate nucleosides (A + G) from the RNA synthesized from oviduct chromatin is guanosine [37]. As shown in Table 6, the number of 5′-tetraphosphate nucleosides incorporated also increased 50% in the presence of progesterone receptor. The increased incorporation of [γ-^{32}P]GTP into the 5′-termini of *in vitro* synthesized RNA substantiates our observations that the progesterone receptor stimulates initiation of transcription in the cell-free assay by increasing the number of chromatin sites for initiation of RNA synthesis available to RNA polymerase. Previous studies have demonstrated a quantitative tissue specificity for binding of oviduct progesterone-receptor complexes to nuclei and to chromatin [2,13]. We therefore investigated the tissue specificity of receptor stimulation of transcription. At receptor concentrations which produced maximal stimulation, there was an increase of 4,700 initiation sites per pg of DNA in oviduct chromatin but only 750 and 300 additional receptor-induced sites in liver and erythrocyte chromatin respectively. It thus appears that the

TABLE 6. Effect of Progesterone Receptor on the Incorporation of [γ-^{32}P]GTP at the 5′-end of Rifampicin-Resistant RNA Chains

Treatment	[^{32}P]GTP (pmol)[a]	5′ Termini (pmol)[b]	RNA Chains/ pg of DNA
Control	.048 ± .002	.086	10,300
1 × 10^{-8} M Progesterone Receptor	.073 ± .003	.131	15,700

[a] Withdrawn chick oviduct chromatin (5 μg) and 1 × 10^{-8} M progesterone receptor complex were preincubated for 30 min at 22°C. RNA polymerase (15 μg) and the assay components described in Figure 1 were added for an additional 30 min. RNA synthesis was then started by the addition of 85.2 μCi of [γ-^{32}P]GTP(5000 CPM/pmol), together with nucleotides, rifampicin and heparin. After 10 sec. at 37°C the transcription reaction was stopped by pipetting a 125 μl sample on DEAE filter paper and the filters were washed according to the procedure of Roeder [30]. The incorporated [γ-^{32}P]GTP was identified as guanosine tetraphosphate after degradation of the RNA product with alkali [19,36].

[b] Total number of initiated chains was estimated by the value of guanosine 5′-tetraphosphate nucleosides multiplied by 1.8, a factor for total chain starts.

progesterone receptor-mediated effect is at least partly dependent on the presence of tissue-specific proteins in the oviduct chromatin. These proteins may be related to the nonhistone chromosomal proteins which convey quantitative tissue specificity for receptor binding and comprise a vital part of the chromatin "acceptor sites" for steroid hormone-receptor binding [41,42].

The model for receptor action described above would predict that the intact 6S dimer is necessary for maximal receptor activity. If the A subunit is the "effector" protein, it could have reduced activity, whereas the B or "specifier" subunit would have no effect alone on transcription. The effects of the various receptor forms on transcription [4] are in agreement with these predictions. The 6S dimer is most active, while the A subunit can elicit a similar response but at 10-fold higher concentration, and the B subunit has no stimulatory effect. These results are consistent with the concept that the two functionally distinct receptor subunits are both necessary for receptor action.

The Effect of Progesterone Receptor on *in vitro* Transcription of the Ovalbumin Gene

The data presented above strongly support our proposal that direct transcriptional control is the primary locus of steroid-hormone action.

TABLE 7. ***In Vitro*** **Synthesis of Ovalbumin mRNA from Chick Oviduct Chromatin in the Absence or Presence of Purified Progesterone-Receptor Complexes**

Source of Chromatin	Progesterone Receptor (1×10^{-8} M)	Chromatin in Reaction (μg DNA)	RNA Synthesized (μg)	Percent $mRNA_{ov}$ in RNA	pg of mRNA Synthesized ($\times 10^{-3}$)	pg $mRNA_{ov}/\mu g$ DNA
Withdrawn Oviduct	–	400	125	0.0015	1.9	0–4.8
Withdrawn Oviduct	+	400	135	0.015	20.9	50.0

Chromatin was preincubated with progesterone receptor for 30 min at 22°C. Bulk RNA synthesis was performed at 22°C as described by Harris et al. [10]. The purified RNA was hybridized with $cDNA_{ov}$ (1.5 μg) as described previously [10].

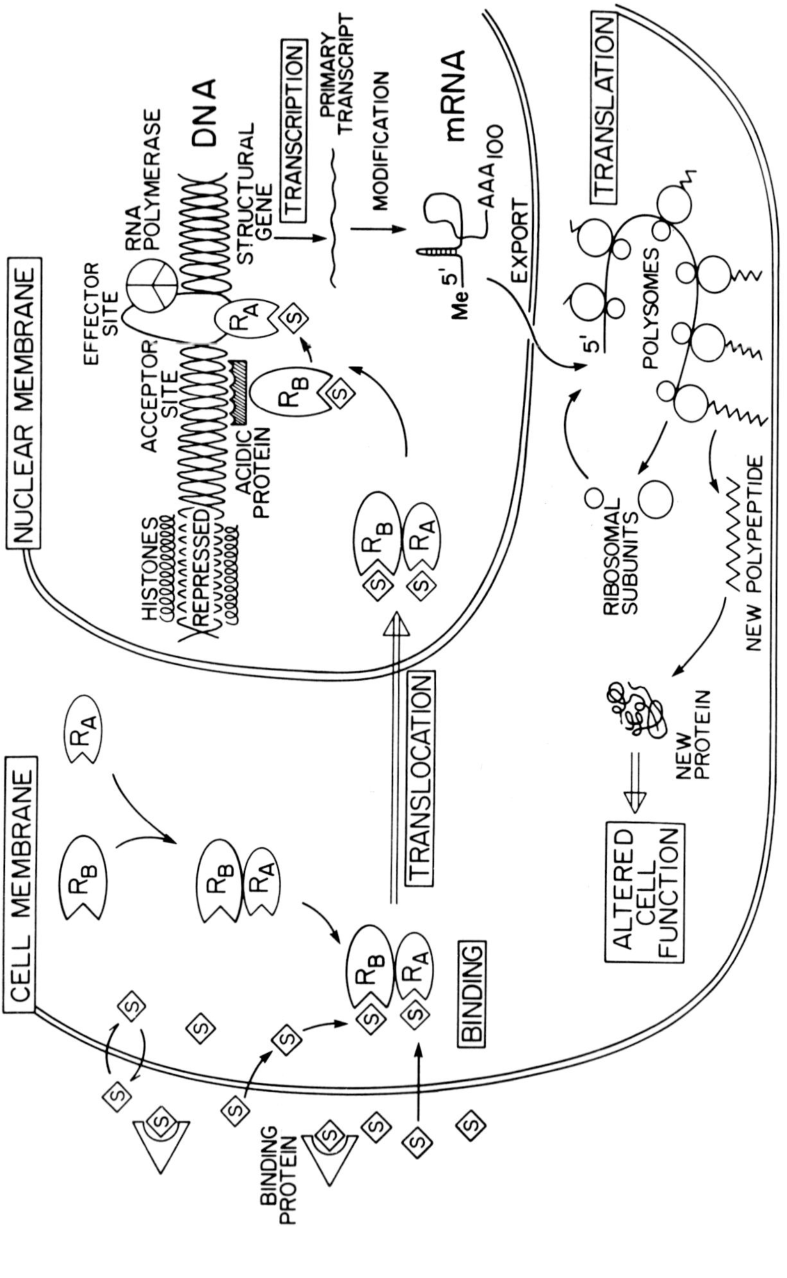

FIG. 3. A hypothetical mechanism for steroid hormone action in the chick oviduct.

Since progesterone can induce ovalbumin in withdrawn chicks, it was conceivable that the purified progesterone receptors could induce transcription of the ovalbumin gene in isolated oviduct chromatin. Such a demonstration would strongly reaffirm the notion that the *in vitro* receptor-chromatin interaction mimics the events *in vivo*. To test this assumption, we asked whether the purified progesterone receptor complex could directly stimulate the transcription of the ovalbumin gene in a cell-free system. The results of this experiment are shown in Table 7. Bulk amounts of RNA were synthesized from both withdrawn chromatin and from withdrawn chromatin incubated in the presence of progesterone receptor (1×10^{-8} M). Both RNA preparations were assayed for ovalbumin mRNA sequences by hybridization with [^{3}H]cDNA$_{ov}$. The RNA synthesized in the presence of receptor-hormone complex contained at least a 10-fold enrichment of ovalbumin mRNA sequences as compared with that found in untreated, withdrawn chromatin controls [10]. It thus appears that a steroid-receptor complex may act directly on chromatin to both enhance the number of initiation sites for RNA synthesis and also to induce the synthesis of specific mRNAs for induced proteins.

DISCUSSION

The data presented in this paper are in direct support of the concept that steroid hormones act by exerting a direct positive regulation of gene transcription. This can be most clearly seen by measurements of *in vitro* transcription of a specific hormonally regulated gene, such as the ovalbumin gene. These results are inconsistent with models of steroid-hormone action which postulate a requirement for the induction of RNA or protein intermediates which then exert secondary or feedback effects on transcription.

We have also demonstrated directly the regulation of transcription of a hormone-dependent gene by purified steroid-hormone receptor *in vitro*. This regulatory process can be related to the subunit structure of the receptor; the active receptor form contains equal amounts of DNA and chromatin binding components.

Our results support the over-all model for steroid-hormone action illustrated in Figure 3. Steroid hormone entering the cell binds to the intact receptor dimer in the cytoplasm. This dimer then moves into the nucleus where it binds to tissue-specific nonhistone proteins at or near a hormonally regulated gene. The DNA binding A subunit is then released near its site of action, so that it has a high probability of encountering the correct region of DNA. The interaction of the A subunit with the DNA then facilitates the transcription of a previously dormant gene. The primary transcript is then processed, transported to the cytoplasm and translated to yield a new protein species, resulting in altered cell function.

While this proposed model remains speculative, we can now test its predictions using pure receptor components in a completely *in vitro* system for the regulation of transcription in a eukaryotic cell.

REFERENCES

1. Bautz EKF, Bautz FA: Initiation of RNA synthesis: the function of σ in the binding of RNA polymerase to promoter sites. Nature 226: 1219–1222, 1970.
2. Buller RE, Schrader WT, O'Malley BW: Progesterone-binding components of chick oviduct. IX. The kinetics of nuclear binding. J Biol Chem 250: 809–818, 1975.
3. Buller RE, Schwartz RJ, O'Malley BW: Steroid hormone receptor fraction stimulation of RNA synthesis: a caution. Biochem Biophys Res Commun 69: 106–113, 1976.
4. Buller RE, Schwartz RJ, Schrader WT, O'Malley BW: Progesterone-binding components of chick oviduct: *in vitro* effect of receptor subunits on gene transcription *in vitro*. J Biol Chem 251: 5178–5186, 1976.
5. Chamberlin MJ, Ring J: Studies of the binding of *Escherichia coli* RNA polymerase to DNA. V. T7 RNA chain initiation by enzyme-DNA complexes. J Mol Biol 70: 221–237, 1972.
6. Chan L, Means AR, O'Malley BW: Rates of induction of specific translatable messenger RNAs for ovalbumin and avidin by steroid hormones. Proc Natl Acad Sci USA 70: 1870–1874, 1973.
7. Cox RF: Transcription of high-molecular-weight RNA from hen oviduct chromatin by bacterial and endogenous form-B RNA polymerases. Eur J Biochem 39: 49–61, 1973.
8. Cox RF, Haines ME, Carey NH: Modification of the template capacity of chick-oviduct chromatin for form B RNA polymerase by estradiol. Eur J Biochem 32: 513–524, 1973.
9. Harris SE, Rosen JM, Means AR, O'Malley BW: Use of a specific probe for ovalbumin messenger RNA to quantitate estrogen induced gene transcripts. Biochemistry 14: 2072–2081.
10. Harris SE, Schwartz RJ, Tsai M-J et al: Effect of estrogen on gene expression in the chick oviduct. *In vitro* transcription of the ovalbumin gene in chromatin. J Biol Chem 251: 524–529, 1976.
11. Hartmann H, Honikel KO, Knüsel F, Nüesch J: The specific inhibition of the DNA-directed RNA synthesis by rifampicin. Biochem Biophys Acta 145: 843–844, 1967.
12. Hinkle DC, Chamberlin MJ: Studies on the binding of *Escherichia coli* RNA polymerase to DNA. I. The role of sigma subunit in site selection. J Mol Biol 70: 157–185, 1972.
13. Jaffe RC, Socher SH, O'Malley BW: An analysis of the binding of the chick oviduct progesterone receptor to chromatin. Biochim Biophys Acta 399: 403–419, 1975.
14. Kohler PO, Grimley PM, O'Malley BW: Estrogen induced cytodifferentiation of the ovalbumin-secreting glands of the chick oviduct. J Cell Biol 40: 8–27, 1969.
15. Kohler PO, Grimley PM, O'Malley BW: Protein synthesis: differential stimulation of cell-specific proteins in epithelial cells of chick oviduct. Science 160: 86–87, 1968.
16. Kuhn RW, Schrader WT, Smith RG, O'Malley BW: Progesterone-binding components of chick oviduct. X. Purification by affinity chromatography. J Biol Chem 250: 4220–4228, 1975.
17. Liarakos CD, Rosen JM, O'Malley BW: Effect of estrogen on gene expression in the chick oviduct. II. Transcription of chick tritiated unique deoxyribonucleic acid as measured by hybridization in ribonucleic acid excess. Biochemistry 12: 2809–2816, 1973.

18. Lill H, Lill U, Sippel A, Hartmann G: The inhibition of the RNA polymerase reaction by rifampicin. *In:* Silvestri L (Ed): RNA-Polymerase and Transcription. Amsterdam, North-Holland Publishing Co., 1970, pp. 55–64.
19 Maitra V, Hurwitz J: The role of DNA in RNA synthesis. IX. Nucleoside triphosphate termini in RNA polymerase products. Proc Natl Acad Sci USA 54: 815–822, 1965.
20. McKnight SG, Pennequin P, Schimke RT: Induction of ovalbumin mRNA sequences by estrogen and progesterone in chick oviduct as measured by hybridization to complementary DNA. J Biol Chem 250: 8105–8110, 1975.
21. Means AR, Abrass I, O'Malley BW: Protein biosynthesis on chick oviduct. I. Changes during estrogen-mediated tissue differentiation. Biochemistry 10: 1561–1570, 1971.
22. Monahan JJ, Harris SE, O'Malley BW: Effect of estrogen on gene expression in the chick oviduct: effect of estrogen on the sequence and population complexity of chick oviduct poly (A)-containing RNA. J Biol Chem 251: 3738–3748, 1976.
23. O'Malley BW, McGuire WL: Studies on the mechanism of estrogen-mediated tissue differentiation: regulation of nuclear transcription and induction of new RNA species. Proc Natl Acad Sci USA 60: 1527–1534, 1968.
24. O'Malley BW, McGuire WL, Korenman SG: Estrogen stimulation of synthesis of specific proteins and RNA polymerase activity in the immature chick oviduct. Biochem Biophys Acta 145: 204–207, 1967.
25. O'Malley BW, Schrader WT: Progesterone-receptor components: identification of subunits binding to the target-cell genome. J Steroid Biochem 3: 617–629, 1972.
26. O'Malley BW, Spelsberg TC, Schrader WT et al: Mechanisms of interaction of a hormone-receptor complex with the genome of an eucaryotic target cell. Nature 235: 141–144, 1972.
27. O'Malley BW, Toft DO, Sherman MR: Progesterone-binding components of chick oviduct. II. Nuclear components. J Biol Chem 246: 1117–1122, 1971.
28. Palmiter RD: Regulation of protein synthesis in chick oviduct. I. Independent regulation of ovalbumin, conalbumin, ovomucoid and lysozyme induction. J Biol Chem 247: 6450–6461, 1972.
29. Palmiter RD, Smith LT: Purification and translation of ovalbumin, conalbumin, ovomucoid and lysozyme messenger RNA. Mol Biol Rep 1: 129–134, 1973.
30. Roeder RG: Multiple forms of deoxyribonucleic acid-dependent ribonucleic acid polymerase in *Xenopus laevis*. Isolation and partial characterization. J Biol Chem 249: 241–248, 1974.
31. Schrader WT, Coty WA, Smith RG, O'Malley BW: Purification and properties of progesterone receptors from chick oviduct. Ann NY Acad Sci, in press.
32. Schrader WT, Heuer SS, O'Malley BW: Progesterone receptors of chick oviduct: identification of 6S receptor dimers. Biol Reprod 12: 134–142, 1975.
33. Schrader WT, Kuhn RW, O'Malley BW: Progesterone-binding components of chick oviduct. Receptor B subunit protein purified to apparent homogeneity from laying hen oviducts. J Biol Chem 252: 299–307, 1977.
34. Schrader WT, O'Malley BW: Progesterone-binding components of chick oviduct. IV. Characterization of purified subunits. J Biol Chem 247: 51–59, 1972.
35. Schrader WT, Toft DO, Sherman MR: Progesterone-binding components of chick oviduct. VI. Interaction of purified progesterone-receptor components with nuclear constituents. J Biol Chem 247: 2401–2407, 1972.
36. Schwartz RJ, Kuhn RW, Buller RE et al: Progesterone binding components of the chick oviduct: *In vitro* effects of purified hormone-receptor complexes on the initiation of RNA synthesis in chromatin. J Biol Chem 251: 5166–5177, 1976.
37. Schwartz RJ, Tsai M-J, Tsai SY, O'Malley BW: Effect of estrogen on gene expression in the chick oviduct. V. Changes in the number of RNA polymerase binding and initiation sites in chromatin. J Biol Chem 250: 5175–5182, 1975.

38. Sherman MR, Corvol PO, O'Malley BW: Progesterone-binding components of chick oviduct. I. Preliminary characterization of cytoplasmic components. J Biol Chem 245: 6085–6096, 1970.
39. Sippel AE, Hartmann GR: Rifampicin resistance of RNA polymerase in the binary complex with DNA. Eur J Biochem 16: 152–157, 1970.
40. Spelsberg TC, Mitchell WM, Chytil F et al: Chromatin of the developing chick oviduct: changes in the acidic proteins. Biochim Biophys Acta 312: 765–778, 1973.
41. Spelsberg TC, Steggles AW, Chytil F, O'Malley BW: Progesterone-binding components of chick oviduct. V. Exchange of progesterone-binding capacity from target to nontarget tissue chromatins. J Biol Chem 247: 1368–1374, 1972.
42. Spelsberg TC, Steggles AW, O'Malley BW: Progesterone-binding components of chick oviduct. III. Chromatin acceptor sites. J Biol Chem 246: 4188–4197, 1971.
43. Travers A, Baillie DL, Pedersen S: Effect of DNA conformation on ribosomal RNA synthesis *in vitro*. Nature [New Biol] 243: 161–163, 1973.
44. Tsai M-J, Schwartz RJ, Tsai SY, O'Malley BW: Effects of estrogen on gene expression in the chick oviduct. IV. Initiation of RNA synthesis on DNA and chromatin. J Biol Chem 250: 5166–5174, 1975.
45. Tsai SY, Tsai M-J, Schwartz RJ et al: Effects of estrogen on gene expression in chick oviduct: nuclear receptor levels and initiation of transcription. Proc Natl Acad Sci USA 72: 4228–4232, 1975.
46. Umezawa H, Mizuno S, Umasaki H, Hitta K: Inhibition of DNA-dependent RNA synthesis by rifampicins. J Antibiot 21: 234–235, 1968.
47. Young BD, Harrison PR, Gilmour RS et al: Kinetic studies of gene frequency. II. Complexity of globin complementary DNA and its hybridization characteristics. J Mol Biol 84: 555–568, 1974.
48. Zillig W, Zechel K, Rabussay D et al: On the role of different subunits of DNA-dependent RNA polymerase from *E. coli* in the transcription process. Sympos Quan Biol 35: 47–58, 1970.

Maternal Contributions to Embryogenesis in *Drosophila*

Ching-Hung Kuo, Ph.D. and Alan Garen, Ph.D.

The development of even as comparatively simple an organism as *Drosophila* involves an incredibly complex array of events that is genetically programmed to occur in a precise temporal and spatial order. A few general principles have emerged from this complexity. Nuclear transplantation experiments have shown that some and probably all cells in an organism contain virtually the same genetic information [1]. This conclusion has profound significance because it indicates that cell differentiation and functional specialization is based on selective gene expression. Lineage studies with genetic mosaics have shown that development proceeds clonally. Early in embryogenesis, separate clusters of cells become determined as the primordia for specific tissues, and these clusters generate clones that eventually differentiate into specialized cell types [2–5]. Although the importance of early events in determining the subsequent course of development has long been recognized, the molecular basis of the process remains a mystery.

During the first two hours after fertilization, the *Drosophila* embryo develops as a syncytium in which the nuclei undergo rapid mitoses and eventually move from the interior to the surface. Cell formation begins with the appearance, in the polar region at the posterior end of the egg, of a cluster of about 40 pole cells which are the primordia for the gametes, and shortly afterwards a confluent surface layer of about 7,500 cells is formed [6]. The resulting cellular blastula is a highly determined structure, in the sense that the developmental fates of the cells are predictable from the positions occupied in the blastula [2–4]. Once these positions are established, the cells appear to function as autonomous units of development, since the developmental fates are not changed when the cells are rearranged [7,8].

The egg has a major role in the control of the early stages of *Drosophila* embryogenesis, as shown by studies of maternal effect mutants [9–12]. The

From the Department of Molecular Biophysics and Biochemistry, Yale University, New Haven, Conn.

Acknowledgments: This research was supported by grants from the National Science Foundation (Developmental Biology) and the National Institutes of Health (Child Health and Human Development). C.-H.K. was supported by postdoctoral fellowships from The Damon Runyon Cancer Fund and U.S. Public Health Service.

homozygous females of these mutants produce defective eggs which, after fertilization by normal sperm, cannot support the subsequent development of the genetically competent heterozygous embryos. Abnormalities occur early in the development of the embryos, usually during the preblastula, blastula or beginning gastrula stages, indicating that these early stages depend on maternal gene functions required for the production of normal eggs. For two maternal effect mutants, *deep orange* [13] and *rudimentary* [14], the egg defects causing abnormal embryogenesis could be repaired by injecting cytoplasm from normal eggs into the young embryos. The injection results provide direct evidence that egg cytoplasm is the source of at least some of the essential components for early embryogenesis. Possibly all of these components are provided by the egg, since the only mutants in which abnormalities of early embryogenesis have been detected are the maternal effect mutants.

It was recently shown that one of the early embryonic functions controlled by *Drosophila* egg cytoplasm is the determination of the pole cells in the blastula [15]. In these experiments cytoplasm from the posterior region of unfertilized eggs was injected into the anterior region of preblastula embryos; the host embryos continued to develop to the blastula stage but, unlike a normal blastula, contained functional pole cells in the anterior as well as posterior regions. Thus, polar cytoplasm from an unfertilized egg is capable of inducing pole cell formation in a region that does not normally produce those cells. Although this remarkable experiment has as yet been tried only with polar cytoplasm, it should be feasible to do similar tests for other regions. A reasonable expectation is that cytoplasm from any region of the egg will be capable, after injection into another region, of inducing the formation of the same types of determined blastula cells that normally are formed in the region from which the cytoplasm is obtained.

The developmental program of *Drosophila* is not entirely determined by the egg, as illustrated by the development of the gametes. Although the polar cytoplasm of the egg determines the formation of the pole cells from which the gametes develop, the sexual phenotype of the mature gametes is determined by the genotype of the zygote. This dual control of developmental specificity, initially by localized egg components specified by maternal genes and later by cellular components specified by zygotic genes, probably applies not only to the gametes but also to the somatic tissues.

One of the egg components involved in embryogenesis is the maternal messenger-RNA that is transcribed during oogenesis and translated after fertilization [16]. We have examined the maternal RNA in stage-14 oocytes of *Drosophila melanogaster*, which is the last stage before the oocyte enters the uterus where fertilization occurs, using the method of molecular hybridization of the RNA with labeled complementary cDNA [17]. This method involves an affinity chromatography step on oligo(dT)cellulose, in order to isolate poly(A)RNA that provides the templates for the enzymatic synthesis of the cDNA. The poly(A)RNA fraction contains many of the messenger-RNA species present in eukaryotes, but not all [18]. The hybridization curve of the oocyte poly(A)RNA with cDNA, shown in Figure 1, has three distinct phases, corresponding to three major classes of poly(A)RNA that hybridize at relatively fast, intermediate and slow rates, respectively. The fast hybridizing class is estimated to contain two molecular species of poly(A)RNA, the intermediate class 240 species, and the slow hybridizing class 7500 species (Table 1). The fast hybridizing species are not translated *in vitro,* and these have been identified as the two ribosomal RNA species in *Drosophila* mitochondria [17]. The other poly(A)RNA of the oocyte contains maternal messenger-RNA that can be efficiently translated *in vitro,* producing a complex mixture of discrete proteins [17]. The maternal messenger-RNA is probably the only RNA available for translation during the preblastula stage of embryogenesis, since transcription of the zygotic nuclei does not begin until the blastula stage [19]. Therefore the extensive protein synthesis that occurs during the preblastula stage can be attributed to the translation of maternal rather than zygotic messenger-RNA. Translation of maternal messenger-RNA is considered to be a possible control mechanism of early embryogenesis, which is activated by fertilization [16]. However, unfertilized *Drosophila* eggs, shortly after being oviposited, synthesize protein at the same rate as preblastula embryos [19,20], and also appear to form the same products as indicated by the autoradiographic patterns obtained after electrophoresis of [^{35}S]methionine-labeled eggs and embryos on SDS-acrylamide gels [20]. Thus, translation of the maternal messenger-RNA in *Drosophila* proceeds as effectively in unfertilized eggs as in young embryos and is therefore not dependent on fertilization or embryogenesis. In virgin females, stage-14 oocytes can be stored for as long as nine days at 20°C without losing the capacity to support normal embryogenesis after the females are mated [21]. It remains to be shown whether or not the maternal messenger-RNA is preserved in the stored oocytes.

A *Drosophila* egg contains nonhistone DNA-binding proteins that can be isolated by affinity chromatography on DNA-cellulose. Although initially located in the egg cytoplasm, some of these proteins probably become associated with the zygotic nuclei early in embryogenesis and might have a role in regulating zygotic gene activity. Of particular interest are DNA-binding egg proteins that are not general components of all *Drosophila* cells and are therefore either unique to the egg or also present only in certain other cells. We are attempting to isolate such proteins by comparing eggs with a cultured line of somatic cells, in order to identify DNA-binding egg proteins that are not common to the cells. An immunochemical procedure has been developed for this purpose that uses affinity chromatography on

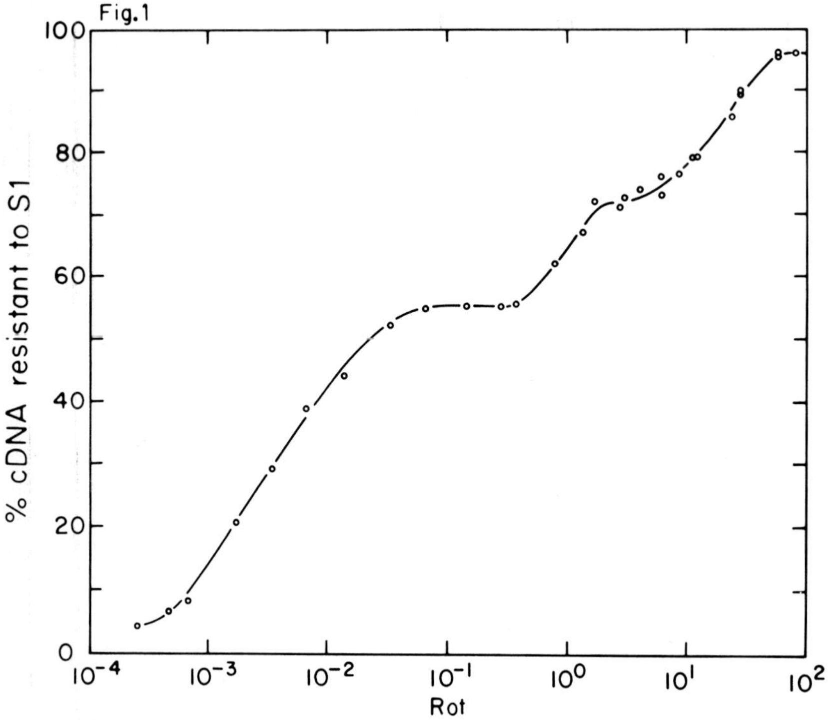

FIG. 1. Hybridization of oocyte [^{3}H]cDNA to oocyte poly(A)RNA. The hybridization reaction was done in a solution containing 240 mM sodium phosphate pH 6.8, 1.0 mM EDTA and 0.1% $NaDodSO_4$, to which was added about 10^6 cpm/ml [^{3}H]cDNA and poly(A)RNA at a concentration of either 0.55 μg/ml, 11 μg/ml or 274 μg/ml, depending on the Rot value to be determined. The preparation of the cDNA and poly(A)RNA samples is described in another report [17]. The reactants were mixed at 25°, and 5 μl aliquots were sealed in glass capillary tubes, which were placed in boiling water for 10 minutes and afterwards incubated at 70°C for the desired period of hybridization. The hybridization reaction was stopped by rapidly cooling the samples in crushed ice, opening the tubes and diluting the samples to a final volume of 200 μl in a solution of 50 mM CH_3COONa pH 4.5, 100 mM of NaCl, 1 mM NaCl, 1 mM $ZnSO_4$, and 5 μg denatured calf thymus DNA. Each sample was divided into two 100 μl portions, and 1 μg Sl-nuclease was added to one of the portions. Both portions were incubated for 30 min at 45°C, after which was added Cl_3CCOOH to a final concentration of 5%, 20 mM $Na_2P_2O_7$ and 200 μg bovine serum albumin carrier. The precipitates were collected on glass-fiber filter papers and the [^{3}H] activity was counted. The ordinate for each figure is calculated as the percentage of precipitable [^{3}H] counts in the portion treated with Sl-nuclease, relative to the counts in the untreated portion of the same sample. The abscissa is the product of RNA concentration and reaction time; and Rot value of 1.0 is defined as equivalent to an RNA concentration of 2.0 absorbance units at 260 nm reacting for 3600 sec.

TABLE 1. Three major Classes of Maternal poly(A)RNA in Mature Oocytes

Hybridization class	% of the hybridizeable cDNA	% of the poly(A)RNA	$Rot_{1/2}$		Number of species of poly(A)RNA	ng poly(A)RNA in an oocyte	Average number of molecules/species in an oocyte
			measured	adjusted			
Fast	58	48	0.003	0.0014	2	4.9	3.7×10^9
Intermediate	17	21	0.85	0.18	240	2.2	1.1×10^7
Slow	25	31	18.	5.6	7500	3.2	5.1×10^5

The three classes correspond to the three phases of the hybridization curve in Figure 1, each class presumably containing distinct species of poly(A)RNA that hybridize at relatively fast, intermediate and slow rates. The % hybridizeable cDNA is the plateau value for each class normalized to 96% total hybridizeable cDNA. The % poly(A)RNA is the % cDNA adjusted for the average molecular weight of each class, which was estimated to be 4×10^5 for the fast hybridizing class and 5×10^5 for the other two classes [17]. The measured $Rot_{1/2}$ values are the points at which 50% of the cDNA in each class is hybridized; the adjusted $Rot_{1/2}$ values are corrected for the % poly(A)RNA in each class. In order to estimate the number of species in each class, an $Rot_{1/2}$ value for rabbit globin poly(A)RNA was first determined in the same way as described for Figure 1. Since the globin preparation contained equal amounts of two species of molecular weight 2×10^5, the $Rot_{1/2}$ value obtained, 6×10^{-4}, was used as the reference value for a single species of molecular weight 4×10^5.

antibody and antigen columns, and involves the following steps:

1. An anti-egg serum is prepared against the DNA-binding egg proteins.
2. An egg antigen column is prepared by coupling to Sepharose the DNA-binding egg proteins, and a cell antigen column is prepared by coupling a corresponding fraction of DNA-binding cell proteins.
3. The anti-egg serum is absorbed to the cell antigen column and the bound antibodies are eluted. These are anti-egg antibodies that cross-react with the cell proteins.
4. The flowthrough from step 3 is absorbed to the egg antigen column and the bound antibodies are eluted. These are anti-egg antibodies that do not cross-react with the cell proteins.
5. Two antibody columns are prepared by coupling to Sepharose the antibodies purified in step 3 for the first column, and the antibodies purified in step 4 for the second column.
6. The egg DNA-binding protein sample is absorbed to the first antibody column to bind antigens that are common to the cells, and the flowthrough is absorbed to the second antibody column to bind antigens that are not common to the cells. The antigens bound to the second column are eluted.
7. The egg antigens recovered from the second antibody column can be fractionated to provide individual antigens for preparing monospecific sera. The flowthrough from this column, which probably contains additional antigens that failed to elicit antibodies in the first serum preparation, can be used to prepare new sera.

From a serum prepared against the DNA-binding egg proteins, an antibody that shows no detectable cross-reaction with the cultured cell line has been isolated by the above procedure. The protein that binds to this antibody has been purified and characterized as a single species with a molecular weight of 80,000 daltons, which is present in the egg at a relative concentration of about 0.1% of the total soluble protein. Immunohistochemical studies are now in progress, using peroxidase-labeled antibody to trace the synthesis and distribution of this protein during oogenesis and its subsequent fate during embryogenesis.

REFERENCES

1. Gurdon J: The Control of Gene Expression in Animal Development. Cambridge, Mass., Harvard Univ. Press, 1974.
2. Garcia-Bellido A, Merriam JR: Cell lineage of the imaginal discs in *Drosophila melanogaster*. J Exp Zool 170: 61–76, 1969.
3. Hotta Y, Benzer S: Mapping of behavior in *Drosophila* mosaics. Nature 240: 527–535, 1972.
4. Wieschaus E, Gehring W: Clonal analysis of priomordial disc cells in the early embryo of *Drosophila melanogaster*. Develop Biol 50: 249–263, 1976.

5. Mintz B: Gene control of mammalian differentiation. Ann Rev Genet 8: 411–470, 1974.
6. Turner FR, Mahowald AP: Scanning electron microscopy of *Drosophila* embryogenesis. I. The structure of the egg envelopes and the formation of the cellular blastoderm. Develop Biol 50: 95–108, 1976.
7. Chan LN, Gehring W: Determination of blastoderm cells in *Drosophila melanogaster.* Proc Natl Acad Sci USA 68: 2217–2221, 1971.
8. Illmensee K: Private communication.
9. Rice TB: Isolation and characterization of maternal-effect mutants: An approach to the study of early determination in *Drosophila melanogaster.* Ph.D. Thesis, Yale University, 1973. Advisor, Dr. A. Garen.
10. Bakken AH: A cytological and genetic study of oogenesis in *Drosophila melanogaster.* Develop Biol 33: 100–122, 1973.
11. Rice TB, Garen A: Localized defects of blastoderm formation in maternal effect mutants of *Drosophila.* Devel Biol 43: 277–286, 1975.
12. Zalokar M, Audit C, Erk I: Developmental defects of female-sterile mutants of *Drosophila melanogaster.* Develop Biol 47: 419–432, 1976.
13. Garen A, Gehring W: Repair of the lethal developmental defect in deep orange embryos of *Drosophila* by injection of normal egg cytoplasm. Proc Natl Acad Sci USA 69: 2982–2985, 1972.
14. Okada M, Kleinman IA, Schneiderman HA: Repair of a genetically-caused defect in oogenesis in *Drosophila melanogaster* by transplantation of cytoplasm from wild-type eggs and by injection of pyrimidine nucleosides. Develop Biol 37: 55–62, 1974.
15. Ilmensee K, Mahowald AP: Transplantation of posterior polar plasm in *Drosophila.* Induction of pole cells at the anterior pole of the egg. Proc Natl Acad Sci USA 71: 1016–1020, 1974.
16. Gross PR: *In:* Moscona AA, Monroy A (Eds): Current Topics in Developmental Biology. 2: 1–46, 1967.
17. Kuo C-H, Garen A: In preparation.
18. Brawerman G: Eukaryotic messenger RNA: Ann Rev Biochem 43: 621–642, 1974.
19. Zalokar M: Autoradiographic study of protein and RNA formation during early development of *Drosophila* eggs. Develop Biol 49: 425–437, 1976.
20. Beckendorf S, Garen A: Unpublished results.
21. Wyman R: Unpublished results.

The Differentiation of Cartilage

Albert Dorfman, M.D., Ph.D., Pei-Lee Ho, B.S.,
Charles M. Strom, B.S., Barbara M. Vertel, Ph.D.
and William B. Upholt, Ph.D.

Cartilage, primarily a component of the skeletal system, has evolved physical properties appropriate for its specific physiologic functions. It is highly probable that these physical properties depend on the presence of the macromolecules chondroitin sulfate proteoglycan and Type II collagen $[\alpha 1(II)]_3$, which are synthesized by the specific cell type designated as the chondrocyte.

Cartilage represents a useful model for the study of differentiation for the following reasons: (1) Avian chondrocytes are readily propagated in cell culture, and under appropriate conditions synthesize matrix materials characteristic of the phenotype of the intact tissue; (2) chondrocytes differentiate from primitive mesenchyme in cell culture; (3) chondroitin sulfate proteoglycan may be readily isolated and reasonably characterized; (4) the biosynthetic pathway of chondroitin sulfate is well established and methods are available for the study of the enzymic steps required for formation of this complex polymer; (5) Type II collagen is well characterized and may be readily identified.

For these reasons, investigations have been carried out to define the events that result in the conversion of mesenchyme cells to chondrocytes [28]. Such studies have also involved the analysis of mechanisms of control of synthesis of these macromolecules in differentiated chondrocytes. Differentiating chick limb bud mesenchyme cells and embryonic chick epiphyseal and sternal chondrocytes have been utilized for these studies. The development of the chondrocyte has been largely monitored by the capacity to synthesize a specific chondroitin sulfate proteoglycan and, to a limited extent, Type II collagen. More extensive studies of the synthesis of collagen by cultured chondrocytes have been carried out in other laboratories [33,34,36].

From the Departments of Pediatrics and Biochemistry and the Joseph P. Kennedy, Jr., Mental Retardation Research Center, Pritzker School of Medicine, University of Chicago, Chicago, Ill.

Acknowledgments: C.M.S. is a Medical Scientist Trainee supported by NIGMS Grant 5-T05 GMO 1939. This investigation was supported by USPHS Grants AM-05996, HD-04583, HD-09402 and a grant from the Home for Destitute and Crippled Children.

Chondroitin sulfate proteoglycan is a macromolecule with a molecular weight of approximately 2.5×10^6 daltons. It consists of a protein core which makes up 8% of the molecule and is covalently attached to chondroitin sulfate chains of molecular weight varying between 10–25,000 daltons [32]. Each chain is linked by way of a galactosylgalactosylxylosyl linkage to the hydroxyl groups of serine residues in the core protein [43]. Additionally, keratan sulfate, a sulfated polymer of N-acetyl-glucosamine and galactose, is attached to the protein O-glycosidically to serine and threonine residues. Keratan sulfate chains are smaller than chondroitin sulfate chains. Most recent studies suggest that the keratan sulfate chains are clustered in a region of the molecule close to that involved in the binding to hyaluronic acid (see below) [21]. The proteoglycan subunit forms large aggregates with hyaluronic acid, which are stabilized by one or more glycoproteins which have been designated link proteins. Current concepts suggest that multiple proteoglycan subunits are bound by a specific binding region of the core protein to the hyaluronic acid and link proteins [16–18].

Biosynthesis of the proteoglycan subunit is well characterized [43,54]. It proceeds by synthesis of the core protein on the rough endoplasmic reticulum, followed by addition of xylosyl residues transferred from UDP-xylose to serine by the enzyme, UDP-xylosyl:core protein xylose transferase. Following xylosylation, two galactosyl residues are added from UDP-Gal by two distinct galactosyltransferases (I and II). Initiation of alternating glucuronic acid and N-acetylgalactosamine residues occurs by transfer of a glucuronyl residue from UDP-GlcUA by a unique glucuronic acid transferase. Extension of the chain proceeds by alternate group transfer from UDP-GalNAc and UDP-GlcUA by two additional specific transferases. Completion of the polysaccharide chains results from transfer of sulfate from 3-phosphoadenosine 5′-phosphosulfate to either the 4 or 6 positions of the N-acetylgalactosamine units. It has been proposed that coordination of these reactions occurs on a multienzyme complex [23]. Initiation probably depends on the binding of the xylosyltransferase to core protein and the second enzyme of the complex, galactosyltransferase I. The specific interaction of xylosyltransferase and galactosyltransferase I has been demonstrated [50]. At present it is unknown whether xylosylation occurs before or after the termination of synthesis of the core protein. There is evidence that sulfation occurs as polysaccharide chain elongation proceeds [12,40]. The mechanism of chain termination has not yet been elucidated. No data are available concerning the keratan sulfate biosynthesis.

IN VITRO DIFFERENTIATION OF LIMB BUD

There exists an extensive literature on the differentiation of cartilage from embryonic mesenchyme, particularly in the chick. Whereas a large number of studies have been concerned with the development of vertebral cartilage from somites, this system is complicated by the fact that the notochord and neural tube play a role as inducers of cartilage [57]. More recently, substantial information has been accumulated on the development of limb bud cartilage, a phenomenon readily demonstrable *in vitro* and uncomplicated by the problem of specific inducers [28].

Studies presented in this paper have been carried out utilizing stage 24 chick embryonic limb buds. Details of the methods of culture have been reported elsewhere [27]. Briefly, stage 24 limb buds are trypsinized and cultured either over agar or at 25×10^6 cells (high density) per 60 mm falcon tissue culture dish. Differentiation to chondrocytes is measured by the appearance of a metachromatic matrix, high levels of $^{35}SO_4$ incorporation into chondroitin sulfate proteoglycan, and cell morphology. If mesenchyme cells are cultured at low density, 5×10^5 cells per 60 mm dish, they continue to divide and synthesize protein, DNA and RNA at a normal rate, but even after subculture, differentiation does not occur. In contrast, if cells are grown over agar for 48 hours and then subcultured at low density, cartilage nodules are formed. When the high density mesenchyme cultures which normally will be readily converted to a sheet of cartilage were grown in the presence of 3.2×10^{-5}M 5-bromo-2′-deoxyuridine (BrdUrd) for the first 48 hours, differentiation was blocked. This irreversible blockage of cartilage formation by BrdUrd also occurs if the analogue is present in low density cultures over agar. The above observations have been the basis of further study for the mechanism of cartilage differentiation.

During the conversion of mesenchyme to chondrocytes there is a shift in the type of collagen formed. It has been shown [30,31] that as cartilage develops within the limb bud, Type II collagen appears. In collaboration with Barbara Smith and George Martin we have shown that a similar transformation occurs in limb bud cell culture (Smith, B., Martin, G., Ho, P.-L., Levitt, D. and Dorfman, A., unpublished results). When cartilage differentiation is inhibited by the presence of BrdUrd, Type I collagen $[\alpha 1(I)]_2\alpha 2$ is produced. We have not yet ascertained whether collagen produced by the initial mesenchyme is Type I or Type III, $[\alpha 1(III)]_3$.

The failure of cells treated with BrdUrd to produce large amounts of proteoglycan may be due to a variety of disturbances of synthesis which can be enumerated as follows: (1) interference with production of core protein, (2) interference with synthesis of uridine nucleotide precursors and (3) diminished levels of transferases required for synthesis of chondroitin sulfate chains. Measurement of uridine nucleotide levels showed no significant decreases in BrdUrd-treated cells, while levels of two of the essential transferases, xylosyltransferase and N-acetylgalactosamine transferase, indicated only a slight dimininution, probably insufficient to account for the marked diminution in synthesis of chondroitin sulfate proteoglycan [27].

Chromatography on Bio-Gel A50M of the $^{35}SO_4$-labeled material produced by cultured cells showed two distinct peaks of radioactivity. The highest molecular weight material was present in large amounts in cultures that differentiated to cartilage and only to a very limited extent in BrdUrd-treated cultures. In contrast, there was little difference between the amounts of a lower molecular weight material in the two cultures. These findings indicated that two different types of proteoglycan may exist. It was suggested that the larger type may be cartilage-specific, the synthesis of which is inhibited by prior treatment with BrdUrd, whereas the smaller is present even after BrdUrd treatment [29]. About this time, Palmoski and Goetinck [38] reported similar observations for BrdUrd-treated chondrocytes and nanomelic chicks. Recently, Pennypacker and Goetinck [39] found that the smaller size macromolecule represented 90% of the proteoglycan synthesized by the nanomelic chicks.

Subsequently, another tool became available for the study of the capacity of cells to synthesize chondroitin sulfate. Brett and Robinson [5] showed that the addition of *D*-xylose to chondrocytes reverses the inhibition by puromycin of chondroitin sulfate synthesis. Subsequently Okayama et al. [37] and Robinson et al. [42] showed that a variety of xylosides, including p-NO_2-phenyl-β-*D*-xyloside and 4-methyl-α-*D*-xyloside, were even more effective in the stimulation of synthesis of chondroitin sulfate. In several studies from this laboratory, it has been shown that the xylosides permit the initiation of synthesis of free chondroitin sulfate chains in chondrocytes and in a variety of other cell types, including nonconnective tissue cells [15,48]. Apparently, the first galactosyl residue may be transferred to the β-*D*-xylosides or free *D*-xylose to initiate chondroitin sulfate chain synthesis, obviating the requirement for xylosylated core protein. Recently, Schwartz, Ho and Dorfman [49] have shown that the xyloside treatment decreases production of chondroitin sulfate proteoglycan but does not inhibit the synthesis of core protein.

The availability of the β-*D*-xylosides as probes permits the determination of the capacity of cells to synthesize chondroitin sulfate chains independently of synthesis of xylosylated core protein. When these probes were applied to *in vitro* differentiation of limb bud cartilage, it was readily demonstrated that both predifferentiated mesenchyme and post-BrdUrd-treated cells have a high capacity for synthesis of chondroitin sulfate chains [29]. Since only moderate reduction of xylosyltransferase activity was observed, it seemed likely that the lesion in post-BrdUrd-treated cells involved synthesis of core protein.

DETERMINATION OF CORE PROTEIN

The result described above indicated that synthesis of core protein is a critical step in differentiation. For many reasons it was desirable to develop a method for measurement of core protein independent of chondroitin sulfate chains. Available core protein isolation procedures do not lend themselves to quantitative detection of the small amounts of material present in cell culture. The immunology of chondroitin sulfate proteoglycans has been previously studied by a number of investigators [14,24,31,46,47]. They have demonstrated that antibodies may be obtained to the protein portion of the subunit but not to the chondroitin sulfate chains. These antibodies were thought to exhibit cross-reactivity between proteoglycans of certain species. The antigenic site of the peptide chains has not been elucidated. Additionally, antibodies have been obtained for link proteins which appear to be more species-specific. Difficulties in defining the purity of the proteoglycans have prevented the development of specific quantitative methods, although a radioimmune assay utilizing iodinated core protein has been reported [24].

We decided to reinvestigate this problem, utilizing a new approach [22]. An assumption was made that any protein in cartilage which contains covalently linked ester sulfate may be defined as a core protein. Previous experience had indicated that the intact proteoglycan subunit was not convenient for immunologic studies because of the high anionic charge of the chondroitin sulfate chains which leads to nonspecific ionic interactions. Accordingly, proteoglycan was prepared from chick embryonic epiphyses and treated with highly purified testicular hyaluronidase to remove the bulk of the chondroitin sulfate chains. The material obtained by that method was purified and used to immunize rabbits by conventional technics.

As a test antigen, proteoglycan was obtained from differentiated high density limb bud cultures which were labeled for a short period of time with $^{35}SO_4$ to obtain proteoglycan of high specific activity. The proteoglycan was purified and then subjected to testicular hyaluronidase digestion. Although approximately 95% of the radioactivity was removed as oligosaccharides by this treatment, sufficient radioactivity remained associated with keratan sulfate and/or the stubs of chondroitin sulfate attached to core protein to permit the use of such preparations for the development of a radioimmune assay. Utilizing the $(NH_4)_2$ SO_4 method of Minden and Farr [35], it was possible to show a linear relationship between percentage of labeled antibody and log of antibody concentration. Subsequent studies have been carried out with antisera prepared against highly purified proteoglycan subunit obtained from embryonic epiphyses and adult sternae. [^{3}H]Acetate-labeled proteoglycan has also been utilized to circumvent the rapid decay of $^{35}SO_4$. Reasonably reproducible results are obtained with different lots of antisera as well as antigen. Thus far, no more than approximately 75% precipitation of radioactivity has been achieved. It is not yet clear whether this is due to the low avidity of antibodies or to the presence of multiple antigens. The radioimmune assay is readily converted to an inhibition assay which permits

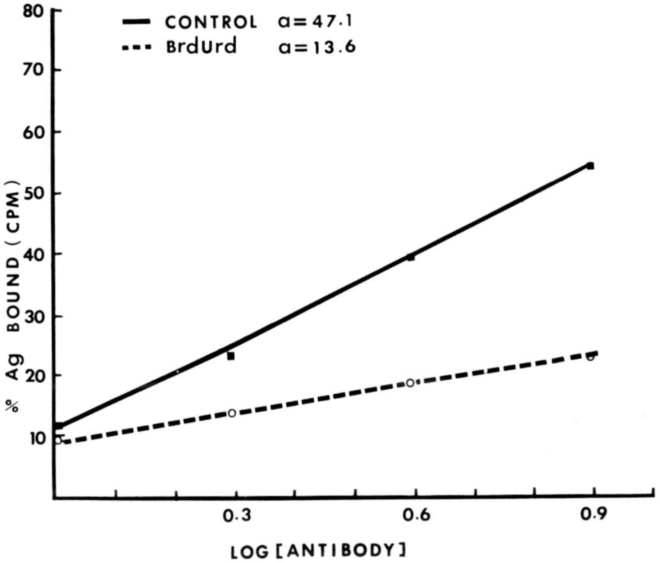

FIG. 1. Direct precipitation curves of control and BrdUrd preparations. Antibody concentration is expressed as log of serum concentration. Final serum dilutions used were 1:12 to 1:96.

the detection of nonradioactive core protein at μg levels and the examination of the identity of various antigens.

Utilizing the radioimmune assay it has been possible to reinvestigate the nature of the BrdUrd-lesion in limb bud cultures. Figure 1 illustrates the measurement of core protein in cultures which have differentiated as compared to the material produced by a comparable number of BrdUrd-treated cells. There is a striking difference between the two slopes which could be caused either by presence of a partially cross-reactive material or by presence of an excess of an antigen with low specific activity. (The amount of antigen used in this method is defined only in terms of radioactivity.) These possibilities can be distinguished by the inhibition method which determines more directly the amount of antigen present. When inhibition studies were carried out, the results illustrated in Figure 2 and summarized in Table 1 were obtained. The material isolated from the BrdUrd cultures contained an order of magnitude less proteoglycan. Whether the results shown in Figure 1 indicate the presence of an additional cross-reactive antigen requires further investigation. Utilizing the inhibition method, it was also possible to demonstrate that the antigen produced by differentiated cultures is identical with highly purified proteoglycan subunit isolated from epiphyses.

Studies using the Ouchterlony double diffusion method confirmed the identity of the principal antigen produced by differentiated cultures with that of the core protein of highly purified subunit obtained from chick embryonic epiphyses. There was also an indication of the presence of an additional antigen in both BrdUrd-treated and differentiated cultures [22].

Taken together, these data provide direct evidence for the hypothesis that BrdUrd treatment of limb bud mesenchyme inhibits the acquisition of the capacity to synthesize the core protein of a cartilage-specific chondroitin sulfate proteoglycan. There appears to exist another proteoglycan which is synthesized by differentiated chondrocytes as well as by mesenchyme cells whose differentiation is inhibited by treatment with BrdUrd.

The availability of an antiserum to the protein core has made possible a number of new approaches to the study of cartilage differentiation. Purified IgG has been prepared and used to develop an immunohistochemical method for the localization of proteoglycan. For this purpose, indirect immunoprecipitation has been employed, utilizing goat anti-rabbit IgG which had been coupled to either fluorescein or hemocyanin. Figure 3 illustrates a series of micrographs obtained with the scanning electron microscope. Easily visualized in Figure 3a is the presence of 500Å hemocyanin molecules which identify proteoglycan. In contrast, cultures treated with normal rabbit IgG in Figure 3b and post-BrdUrd cultures treated with anti-core protein IgG in Figure 3c do not show hemocyanin molecules.

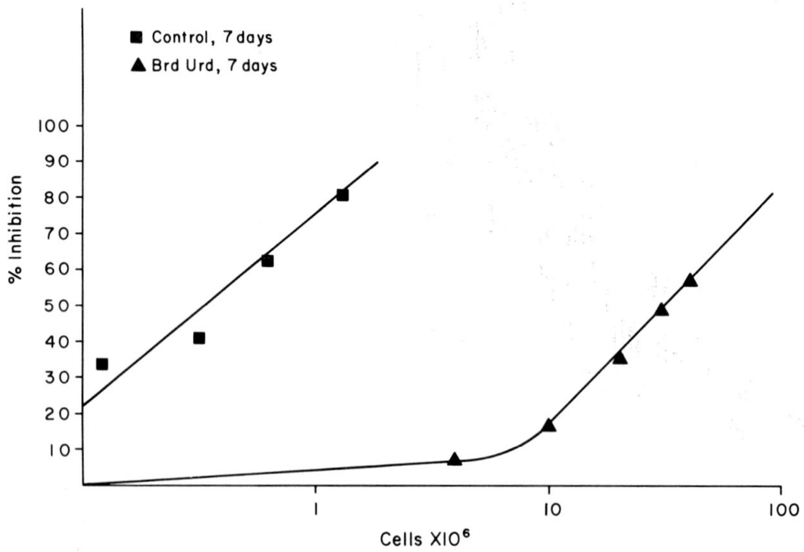

FIG. 2. Inhibition assay using $^{35}SO_4$-labeled limb bud control culture core protein as radioactive antigen. Cold antigens compared were from 7-day control and BrdUrd-treated limb bud cultures.

TABLE 1. Immunoreactive Protein in Control and BrdUrd-Progeny Limb Bud Cells

Sample	Inhibition	Immuno-reactive Protein	Cells Needed For Inhibition	Immuno-reactive Protein
	%	μg	No. $\times 10^{-6}$	$\mu g/10^6$ cells
Control, 7 days	33.0	29	0.12	241.7
	41.5	40	0.32	125.0
	63.0	96	0.64	150.0
	80.5	200	1.30	153.9
BrdUrd, 7 days	7.4	<10	4	<2.5
	16.9	14.5	10	1.4
	36.0	32.5	20	1.6
	49.0	55.0	30	1.8
	57.0	76.0	40	1.9

We have now undertaken a series of studies aimed at delineating more specifically the biosynthesis of the core protein with the hope that the molecular events of differentiation and the mechanism of its inhibition by BrdUrd can be elucidated. It has been possible to isolate polysomes from differentiating cartilage and to utilize specific antibodies to examine the *in vitro* biosynthetic product. Preliminary results indicate that nascent core protein can be identified on polysomes. Furthermore, a wheat germ system appears to synthesize this protein *in vitro*. It is hoped that with the isolation of core protein mRNA, more detailed information concerning its structure and the control of its synthesis may be obtained.

THE MECHANISM OF ACTION OF BrdUrd

An extensive literature has appeared concerning the effects of BrdUrd on both procaryotic and eucaryotic cells. Of special interest for the purpose of these studies has been the demonstration in a number of cell types that treatment with this analogue results in the inhibition of expression of the differentiated phenotype [45,59]. There is ample evidence to indicate that the effect of BrdUrd requires incorporation of the analogue into the genome [7,58]. In differentiated cells there is some evidence that this effect is reversible, but when cells in the process of differentiating are treated, the inhibition of phenotypic expression is irreversible [7,8,27]. Originally we proposed that the specific effect of BrdUrd may result from its incorporation into regions of the genome which are relatively rich in thymidine and may regu-

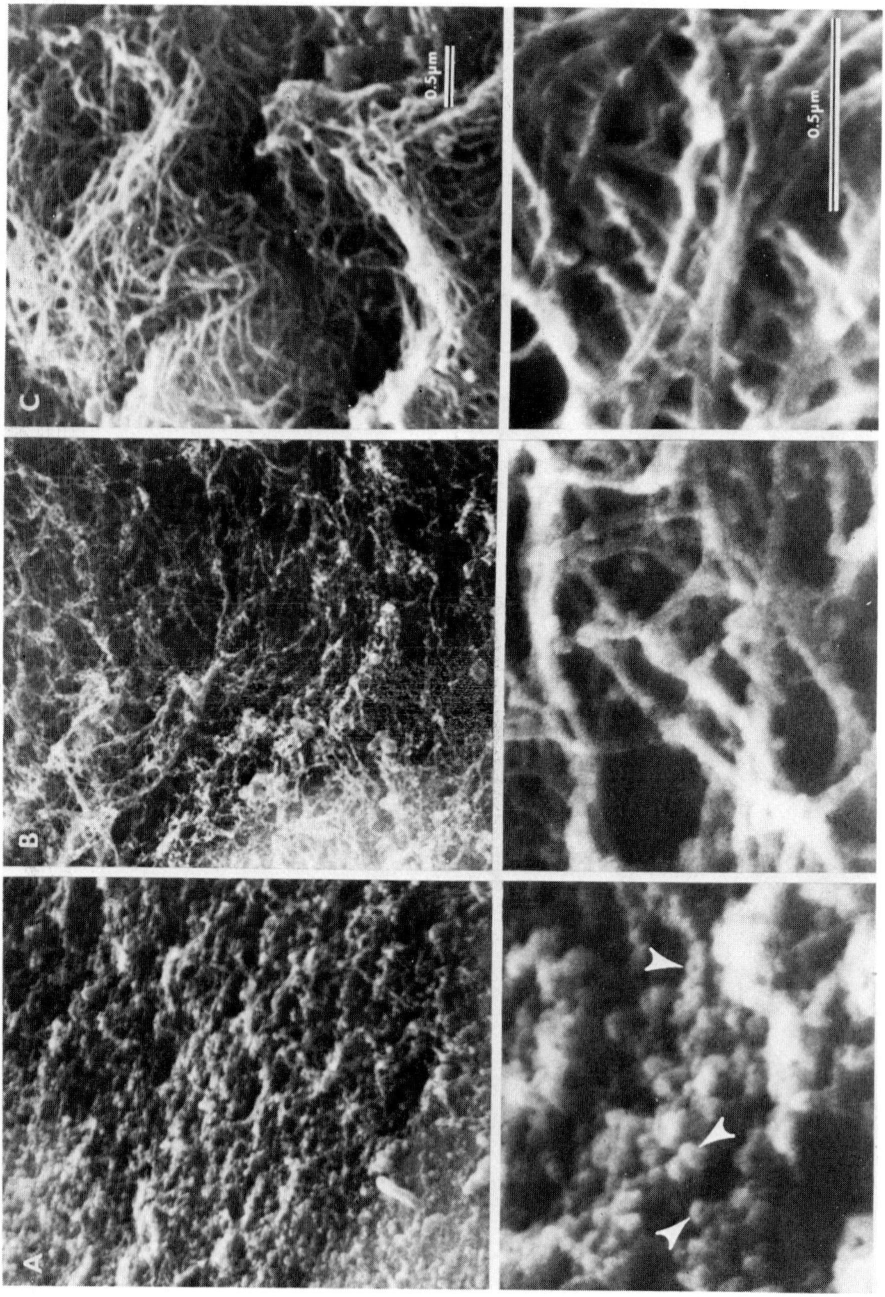

FIG. 3. See legend, facing page

late the activity of genes required for the expression of the differentiated phenotype [28]. However, Schwartz and Kirsten [52] demonstrated that BrdUrd is preferentially incorporated (with respect to thymidine) into moderately-repetitive DNA sequences in rat fibroblasts exposed to low levels of the analogue. Schwartz [51] has also demonstrated the presence of BrdUrd-rich regions vulnerable to attack by the single stranded specific nuclease, S_1, at low pH after probable depurination.

We have undertaken a study of the kinetic distribution of BrdUrd in differentiating chick limb bud [55]. When mesenchyme cells were incubated with [^{3}H]thymidine ([^{3}H]dT) during days 1 and 2 in culture, label was incorporated into repetitive, moderately-repetitive, and unique classes of DNA. In contrast, when [^{3}H]BrdUrd was added during the first 48 hours in culture (in the presence of 32 μM BrdUrd), the label was preferentially incorporated into a late moderately-repetitive region. Simultaneous incubation of unlabeled BrdUrd and [^{3}H]dT revealed a selective inhibition of [^{3}H]dT incorporation in the moderately-repetitive region. Cultures incubated during days 3 and 4 with [^{3}H]dT incorporated label in all three classes of DNA, however, when [^{3}H]dT was present during days 3 and 4 in cultures previously incubated with BrdUrd during the first 48 hours, the [^{3}H]dT was preferentially incorporated into the late moderately-repetitive region. These experiments suggest that the analogue not only replaces dT but that there is a preferential incorporation into a specific portion of the genome. Further studies are being carried out to determine more exactly the nature of the DNA which contains this relatively increased concentration of BrdUrd. For this purpose [^{3}H]BrdUrd in the presence of 32 μM BrdUrd was added to developing mesenchyme during the first two days of culture and the total DNA was isolated and sheared to produce fragments which sediment as a single 5-6S peak in a neutral linear sucrose gradient. Palindromic DNA was removed by boiling the sheared DNA for 10 minutes and passing it over a hydroxyapatite column, equilibrated at 62° in 0.18M phosphate buffer. This

---

FIG. 3. Immunohistochemical reactions visualized in the scanning electron microscope.

A. High density limb bud culture (8 days) incubated with anti-core protein IgG and hemocyanin-coupled goat anti-rabbit IgG. The positive immunochemical reaction resulted in an apparent increase in filament thickness seen at low magnification (top; compare with control shown in B). At higher magnification (bottom), characteristic 500Å hemocyanin molecules (arrows) are observed completely covering the matrix material.

B. High density limb bud culture (8-days) control incubated with normal rabbit IgG and hemocyanin-coupled goat anti-rabbit IgG. The matrix appears equivalent to normal nonreacted controls and shows no evidence of immunochemical reaction. At higher magnification (bottom), filaments, probably Type II collagen and granules, presumably proteoglycan, are seen typically distributed in the matrix. Hemocyanin molecules are not seen.

C. Post-BrdUrd-treated high density limb bud culture (8 days) incubated with anti-core protein IgG and hemocyanin-coupled goat anti-rabbit IgG. Matrix synthesized by these cells does not react immunohistochemically and no hemocyanin molecules are observed.

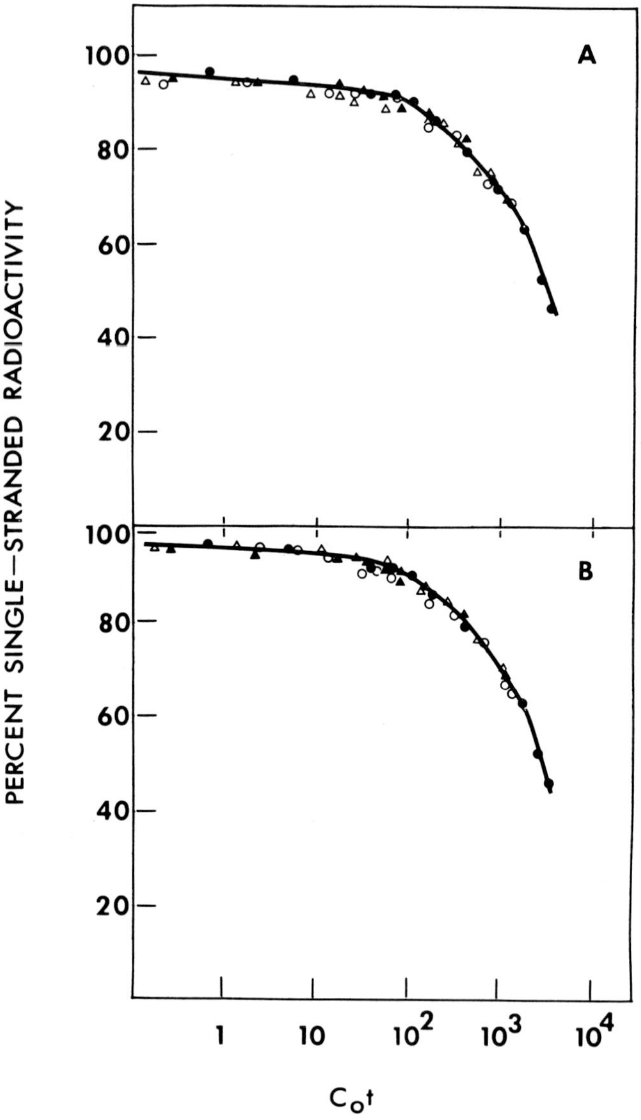

FIG. 4. Reassociation of [^{3}H]BrdUrd probe with non-cartilage DNA's.

A. ●, ▲, two preparation of liver DNA; ○, △, two preparations of limb bud DNA.

B. ●, ▲, same as in 4A; ○, △, two preparations of 8-day BrdUrd-blocked limb bud culture DNA.

material was used as a probe. Nonradioactive sheared DNA's were prepared from 13-day whole embryos, 13-day embryonic livers, 17-day embryonic sternae, stage 24 limb buds, 8-day limb bud cultures which had differentiated to cartilage and 8-day limb bud cultures which had been treated with BrdUrd during the first 2 days in culture.

A series of studies were carried out with [^{3}H]BrdUrd probe to determine its reassociation with the various DNA's [56]. Figure 4 compares the reassociation of probe with DNA from limb bud and BrdUrd-blocked cultures and DNA from liver. In contrast, Figure 5 compares the reassociation of probe with differentiated cartilage DNA from 8-day limb bud cultures and embryonic sternae and DNA from liver. The results were consistent with three different preparations of probe and at least two different preparations of each type of DNA. The curves shown are composites of various preparations of each DNA with the same preparation of probe. Reassociation of probe in the presence of a 50-fold excess of sheared salmon sperm DNA was performed to control for self-reannealing of probe. Figures 4 and 5 clearly show that both sternal cartilage DNA and 8-day cultured limb bud DNA drive the reassociation reaction approximately twice as fast so do liver DNA, undifferentiated limb bud DNA or DNA from 8-day cultures blocked from differentiation by BrdUrd. In addition to driving the reassociation of probe at a slower rate, the total percentage of radioactivity which reannealed in the moderately-repetitive fraction (C_0t = 10–1000) was lower in liver (20%) than in sternal cartilage (30%) or 8-day limb bud culture DNA (30%). Unfortunately, availability of material limited the C_0t values attainable for the cartilage, limb bud and BrdUrd-blocked DNA's. Work is in progress to complete these curves.

Statistical analyses were performed by defining the parameter $C_0t_{1/3}$ as the C_0t value at which 1/3 of the probe had reannealed. Although the curves are not theoretically linear, since the moderately-repetitive regions are

TABLE 2. Statistical Summary of Probe Reassociation

DNA Type	n	$C_0t_{1/3} \pm$ Std. Error	% Moderately-Repetitive
Liver	3	1545 + 63	20
Limb Bud	3	1545 + 87	20
Limb Bud Cultures, 8-day	3	694 + 67	30
Sternae	3	803 + 50	30
BrdUrd-Blocked Limb Bud Cultures, 8-day	3	1571 + 30	20
Total Embryo	2	900 + 88	33

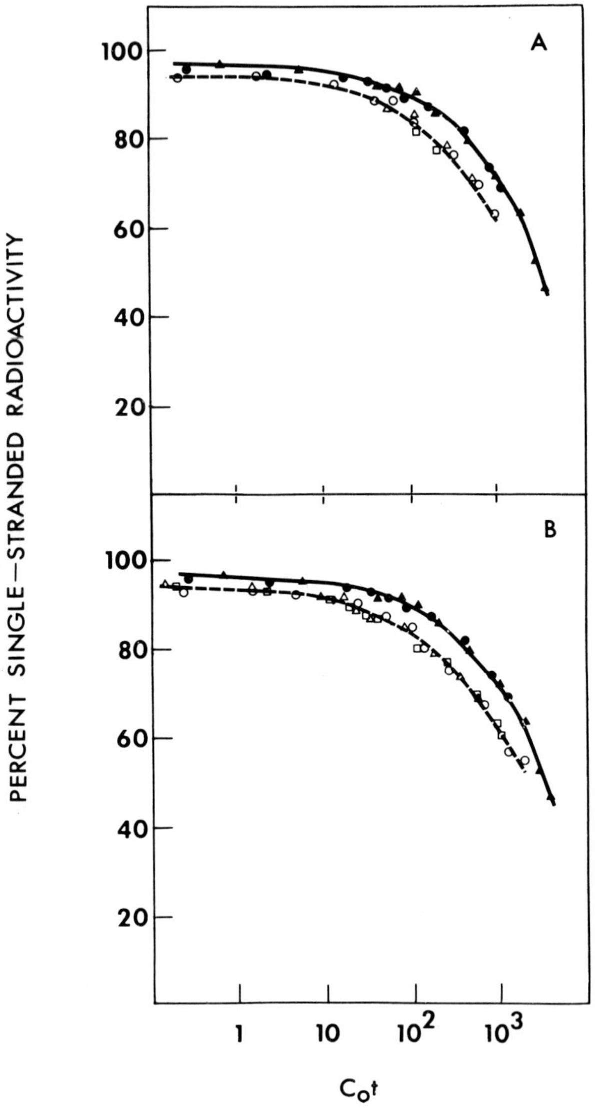

FIG. 5. Reassociation of [^{3}H]BrdUrd-probe with cartilage DNA's.

A. ●————▲, two preparations of liver DNA; ○------△, two preparations of sternal DNA.

B. ●————▲, same as in 5A; △------□------○ three preparations of 8-day limb bud culture.

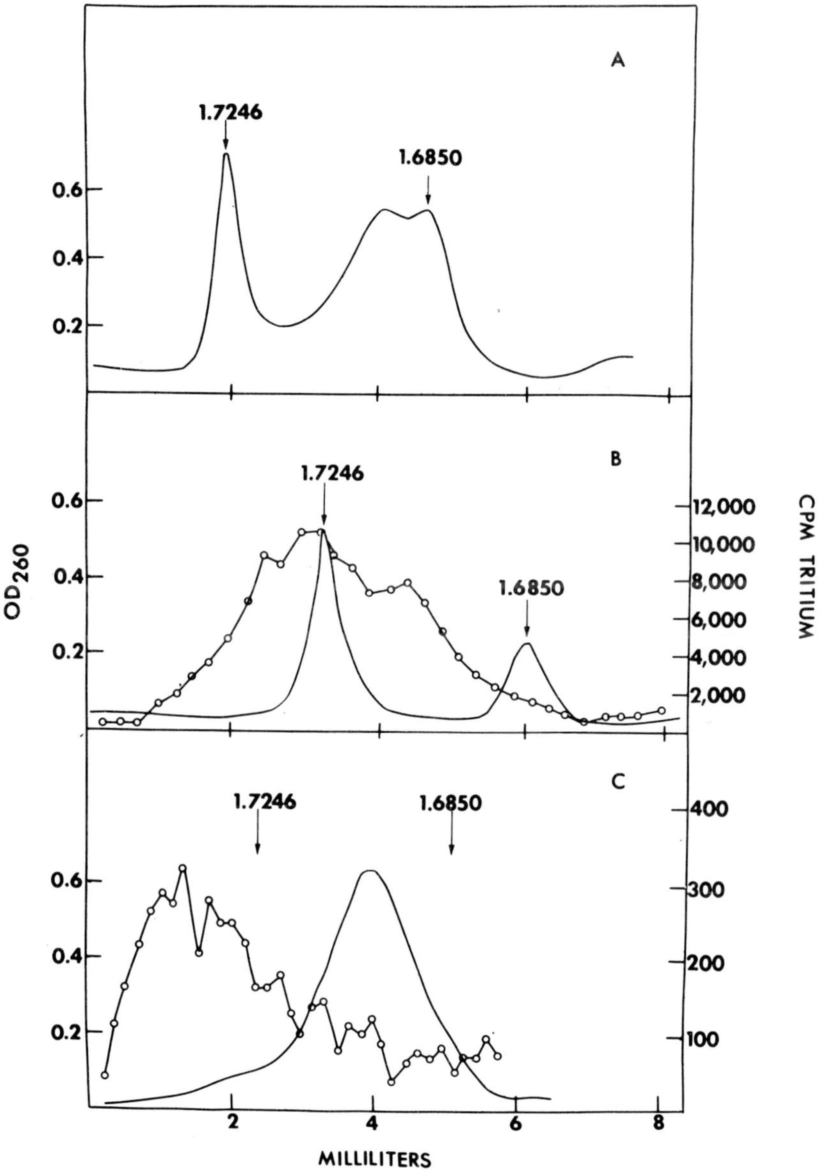

FIG. 6. CsCl density gradients: ———— OD_{260}, O————O, cpm [^{3}H].

A. 40 μg *C. perfringens* DNA (density 1.6850), 30 μg *M. lysodeikticus* DNA (density 1.7246), 30 μg unsheared total embryonic DNA.

B. 40 μg *C. perfringens* DNA, 30 μg *M. lysodeikticus* DNA, 100,000 cpm [^{3}H]BrdUrd DNA (unsheared).

C. 50 μg sheared total embryonic DNA, 10,000 cpm [^{3}H]BrdUrd DNA (sheared).

nearly linear, computer analyses were carried out to determine optimal linear regression. In all cases, the best fit line had a correlation coefficient greater than 0.97. The $C_ot_{1/3}$ of each separate experiment was obtained by linear estimate from the best fit regression line. Each determination was treated as an independent observation. Means and standard errors were calculated and are shown in Table 2. Both the sternal cartilage and cultured cartilage DNA's differ significantly ($p < 0.005$) from the three non-cartilage DNA's. The C_ot of total embryonic chick DNA was slightly higher than that of the DNA derived from sternae.

Double stranded [^{3}H]BrdUrd-DNA, both sheared and unsheared was prepared as described for probe and centrifuged in CsCl with nonradioactive, total embryonic chick DNA (sheared or unsheared), or standards. The results of these experiments are shown in Figure 6. The unlabeled, unsheared total embryonic chick DNA banded as a single peak at a density 1.699. The unsheared [^{3}H]BrdUrd-DNA reproducibly banded in a broad region with three peaks of densities 1.710, 1.727 and 1.737. These correspond to approximately 10%, 27% and 35% substitution of dT residues by BrdUrd respectively. Shearing of the probe produced a complicated pattern, with the heaviest peak occurring at a density of 1.741 (40% substitution). Calculations performed utilizing the specific activity of the [^{3}H]BrdUrd pool during incubation and the final optical density and specific radioactivity of the purified probe indicated an average total substitution of only 10% of dT residues by BrdUrd. We must conclude from these data that the BrdUrd is clustered in the chick genome. These results are consistent with the observation of Schwartz [51] of clustering of BrdUrd in rat cells.

DISCUSSION

A central problem of biology is the mechanism by which specific portions of the genome are expressed in a coordinated way in eukaryotic cells. The study of prokaryotic cells has resulted in the discovery of a number of mechanisms of regulation, but so far these have failed to offer adequate explanations for the coordinated regulation of structural genes that occur in temporal succession in the process of differentiation. Space does not permit a review of various models that have been proposed.

The differentiating limb bud may offer a model for the examination of coordination, since at least two structurally unrelated gene products, core protein and Type II collagen, are characteristic of the cartilage phenotype. Further study is required to determine whether link protein may also be specific for this phenotype. Likewise, it must be determined more definitively whether the core protein of cartilage-specific proteoglycan represents a single gene product. The findings by Heinegard [20], Baxter and Muir [1], Hascall and Heinegard [19] and Rosenberg et al. [44] of proteoglycans of different sizes require further explanation.

The studies reviewed in this paper strongly indicate that the transition of mesenchyme cells to chondrocytes is characterized by the expression of the genes specifying Type II collagen and core protein of chondroitin sulfate proteoglycan. The level of control of this expression can now be explored in view of the available methods utilizing antibodies to examine the detailed intracellular mechanisms. Difficult problems must be solved to explain the influence of environmental factors on gene expression. We do not know the mechanism by which cell density affects the specific gene expression. Switches in types of collagen synthesized apparently occur as a result of changes in culture conditions or length of time in culture [2,13,33,34,36]. Distinction between changes in phenotypic expression with environmental and those which are due to differentiation are difficult to make.

At another level the possible changes in the genome with differentiation clearly needs examination with more sensitive technics. The earlier more simplistic approaches that resulted in the conclusion that all cells of a given organism contain identical DNA which is replicated as one continuous process are clearly open to question.

In the limb bud system, the presence of BrdUrd during differentiation irreversibly blocks subsequent chondrogenesis. Hydroxyurea was added to limb bud cultures in concentrations ten-fold higher than sufficient to block 95% of [^{3}H]dT incorporation in chondrocytes. This concentration of hydroxyurea inhibited [^{3}H]dT incorporation less than 50% and did not block or delay subsequent differentiation (Strom, C. M. and Dorfman, A., unpublished results). Leder et al. [26] observed that Friend leukemia cells differentiated when hydroxyurea was present during dimethylsulfoxide induction. Taken together, these observations suggest that cell division is not a prerequisite for differentiation. In the limb bud system, the effect of BrdUrd may be observed without a background of complete replication of the cellular genome.

Since liver or undifferentiated pre-cartilage DNA exhibit fewer sequences complimentary to [^{3}H]BrdUrd probe than cartilage DNA, we suggest that during differentiation, cartilage-specific sequences are being amplified. Total embryonic chick DNA also contains a greater number of probe-complementary sequences than does liver or pre-cartilage DNA [56]. The probe probably contains sequences which are amplified in other differentiated tissues. Thus, if total embryonic chick DNA contains many different amplified species of DNA, it may be expected to reanneal with a greater percentage of moderately-repetitive probe sequences than DNA derived from undifferentiated limb buds or from a single differentiated cell type such a liver. If each differentiated tissue contains a specific class of amplified moderately-repetitive sequences, then each tissue should have similar self-annealing profiles. deJimenez et al. [11] have demonstrated this phenomenon in the adult chick.

Britten and Davidson [6] suggested that moderately-repetitive DNA sequences might be important in the regulation of gene expression. Davidson and co-workers [9] have shown that 80% of the sequences complementary to mRNA are contiguous to moderately-repetitive DNA in the sea urchin. We have demonstrated an increased amount of [^{3}H]BrdUrd probe-complementary sequences in mature cartilage as compared with its precursor cells or BrdUrd-blocked limb bud cultures. Eight-day BrdUrd-blocked cultures contain less than 10% of [^{3}H]BrdUrd initially present in DNA after a 2-day incubation (Strom, C. M. and Dorfman, A., unpublished observations). The presence in the probe of moderately-repetitive sequences not specific for cartilage would lead to an underestimate of any differences between cartilage and non-cartilage DNA's.

We propose that the determinative event of chick cartilage differentiation may be an amplification of cartilage-specific sequences present initially in multiple copies. During the first 48 hours of stage 24 limb bud cultures, this amplification is occurring without replication of the total genome.

Perhaps the selective incorporation of BrdUrd is due to a separate polymerase which has a higher affinity for BrdUrd than dT. Rizki and Rizki [41] observed a differential distribution of BrdUrd and dT in isotichs isolated from *Drosophila* larvae fed with these radioactive nucleosides. Bradshaw and Papaconstantinou [3] detected the differential incorporation of [^{3}H]BrdUrd into certain puff regions of salivary chromosomes in *Rhynchosciara*. Perhaps the BrdUrd-dependent cell line reported by Davidson and Bick [10] lacks the normal polymerase, with only the BrdUrd-preferring polymerase remaining.

In the presence of BrdUrd, most of the amplified sequences would become heavily substituted with BrdUrd and subsequently broken down. Skalko and Packard [53] have demonstrated an eleven-fold decrease in the half-life of [^{3}H]BrdUrd in substituted DNA with respect to [^{3}H]dT. Krider and Blake [25] have detected a reduction in the number of moderately-repetitive ribosomal DNA sequences in *Drosophila* larvae after feeding with the analogue.

We suggest that during the first two days of limb bud culture, BrdUrd-substituted amplified sequences are being produced and subsequently degraded. Such cells are therefore temporally beyond the critical stage of determination but lack the amplified sequences necessary to express the cartilage phenotype. Irreversible inhibition of differentiation occurs. Since various chick cell types (muscle, red cells and cartilage) and tissues of diverse eukaryotic organisms exhibit BrdUrd sensitivity of differentiated functions, this proposed amplification may be a generalized concomitant of cellular differentiation.

REFERENCES

1. Baxter E, Muir H: The nature of the protein moieties of cartilage proteoglycans of pig and ox. Biochem J 149: 657–668, 1975.
2. Binya PD, Nimni ME: The progeny of rabbit articular chondrocytes synthesize $[\alpha 1(I)]_3$ but not Type II cartilage collagen $[1(II)]_3$. Fed Proc 35: 1520 (Abstract), 1976.
3. Bradshaw WS, Papaconstinou JV: Differential incorporation of 5-bromodeoxyuridine into DNA puffs of larval salivary gland chromosome in *Rhynchosciara*. Biochem Biophys Res Commun 41: 306–312, 1970.
4. Brante K, Tsignos CP, Muir H: Immunology of pig cartilage proteoglycans. *In:* Balazs EA (Ed): Chemistry and Molecular Biology of the Intercellular Matrix. New York, Academic Press, 1970.
5. Brett MJ, Robinson HC: The effect of xylose on chondroitin sulphate biosynthesis in embryonic cartilage. Proc Aust Biochem Soc 4: 92 (Abstract), 1971.
6. Britten RJ, Davidson EH: Gene regulation for higher cells: A theory. Science 165: 349–357, 1969.
7. Coleman AW, Coleman JR, Kankel D, Werner I: The reversible control of animal cell differentiation by the thymidine analogue 5-bromodeoxyuridine. Exp Cell Res 59: 319–328, 1970.
8. Coleman JR, Coleman AW, Hartline JH: A clonal study of the reversible inhibition of muscle differentiation by the halogenated thymidine analogue 5-bromodeoxyuridine. Dev Biol 19: 527–548, 1969.
9. Davidson EH, Hough BR, Klein WH, Britten RJ: Structural genes adjacent to interspersed repetitive DNA sequences. Cell 4: 217–238, 1975.
10. Davidson RL, Bick MD: 5-Bromodeoxyuridine dependence. A new mutation. Proc Natl Acad Sci USA 70: 138–142, 1973.
11. deJimenez ES, Gonzolez JL, Dominque JL, Saloma ES: Characterization of DNA from differentiated cells. Eur J Biochem 45: 25–29, 1974.
12. DeLuca S, Richmond ME, Silbert JE: Biosynthesis of chondroitin sulfate. Sulfation of the polysaccharide chain. Biochemistry 12: 3911–3915, 1973.
13. Deshmukh K, Kline WG: Characterization of collagen synthesized by rabbit articular chondrocytes in various culture systems. Fed Proc 35: 1520 (Abstract) 1976.
14. DiFerrante N: The immunological properties of hyaluronidase-treated bovine proteoglycans. *In:* Balazs EA (Ed): Chemistry and Molecular Biology of the Intercellular Matrix. New York, Academic Press, 1970.
15. Galligani L, Hopwood J, Schwartz NB, Dorfman A: Stimulation of synthesis of free chondroitin sulfate chains by *β-D*-xylosides in cultured cells. J Biol Chem 250: 5400–5406, 1976.
16. Gregory JD: Multiple aggregation factors in cartilage proteoglycan. Biochem J 133: 383–386, 1973.
17. Hardingham TE, Muir H: The specific interaction of hyaluronic acid with cartilage proteoglycans. Biochim Biophys Acta 297: 401–405, 1972.
18. Hascall VC, Heinegård D: Aggregation of cartilage proteoglycans. I. The role of hyaluronic acid. J Biol Chem 249: 4232–4241, 1974.
19. Hascall VC, Heinegård D: The structure of cartilage proteoglycans. *In:* Slavkin HC,

Greulich RC (Eds): Extracellular Matrix Influences on Gene Expression. New York-San Francisco-London, Academic Press, 1975.

20. Heinegård D: Hyaluronidase digestion and alkaline treatment of bovine tracheal cartilage proteoglycans. Isolation and characterization of different keratan sulfate proteins. Biochim Biophys Acta 285: 193–207, 1972.
21. Heinegård D, Hascall VC: Aggregation of cartilage proteoglycans. III. Characteristics of the proteins isolated from trypsin digests of aggregates. J Biol Chem 249: 4250–4256, 1974.
22. Ho P-L, Levitt D, Dorfman A: A radioimmune study of the effect of bromodeoxyuridine on the synthesis of proteoglycan by differentiating limb bud cultures. Dev Biol 55: 233–243, 1977.
23. Horwitz A, Dorfman A: Subcellular sites for synthesis of chondromucoprotein of cartilage. J Cell Biol 38: 358–369, 1968.
24. Keiser H, DeVito J: Immunochemical studies of cartilage proteoglycan subunit fragments. Conn Tiss Res 2: 273–282, 1974.
25. Krider HM, Blake JA: Effect of 5-bromodeoxyuridine on redundancy of ribosomal genes in *Drosophila*. Nature 256: 436–438, 1975.
26. Leder A, Orkin S, Leder P: Differentiation of erythroleukemic cells in the presence of inhibitors of DNA synthesis. Science 190: 893–894, 1975.
27. Levitt D, Dorfman A: The irreversible inhibition of differentiation of limb bud mesenchyme by bromodeoxyuridine. Proc Natl Acad Sci USA 69: 1253–1257, 1972.
28. Levitt D, Dorfman A: Concepts and mechanisms of cartilage differentiation. *In:* Moscona A, Monroy A (Eds): Current Topics in Developmental Biology. New York, Academic Press, 1974.
29. Levitt D, Dorfman A: Control of chondrogenesis in limb-bud cell cultures by bromodeoxyuridine. Proc Natl Acad Sci USA 70: 2201–2205, 1974.
30. Linsenmayer TF: Temporal and spatial transitions in collagen types during embryonic chick development. II. Dev Biol 40: 372–377, 1974.
31. Linsenmayer TF, Toole BP, Trelstad RL: Temporal and spatial transitions in collagen types during embryonic chick limb development. Dev Biol 35: 232–239, 1973.
32. Mathews MB: Polyanionic proteoglycans. *In:* Connective Tissue. Macromolecular Structure and Evolution. Berlin-Heidelberg-New York, Springer-Verlag, 1975.
33. Mayne R, Vail MS, Mayne PM, Miller EJ: Changes in type of collagen synthesized as clones of chick chondrocytes grow and eventually lose division capacity. Proc Natl Acad Sci USA 73: 1674–1678, 1976.
34. Mayne R, Vail MS, Miller EJ: Analysis of changes in collagen biosynthesis that occur when chick chondrocytes are grown in 5-bromo-2′-deoxyuridine. Proc Natl Acad Sci USA 72: 4511–4515, 1975.
35. Minden P, Farr RS: The ammonium sulphate method to measure antigen-binding capacity. *In:* Weir DM (Ed): Handbook of Experimental Immunology. Philadelphia, Davis, 1967.
36. Müller P, Lemmen C, Gay S et al: Biosynthesis of collagen by chondrocytes *in vitro*. *In:* Slavkin HC, Greulich RC (Eds): Extracellular Matrix Influences on Gene Expression. New York-San Francisco-London, Academic Press, 1975.
37. Okayama M, Kimata K, Suzuki S: Influence of p-nitrophenyl *β-D*-oxyloside on the synthesis of proteochondroitin sulfate by slices of embryonic chick cartilage. J Biochem 74: 1069–1073, 1973.
38. Palmoski MJ, Goetinck, PF: Synthesis of proteochondroitin sulfate by normal, nanomelic and 5-bromodeoxyuridine-treated chondrocytes in cell culture. Proc Natl Acad Sci USA 69: 3385–3388, 1972.

39. Pennypacker JP, Geotinck PF: Biochemical and ultrastructural studies of collagen and proteochondroitin sulfate in normal and nanomelic cartilage. Dev Biol 50: 35–47, 1976.
40. Richmond ME, DeLuca S, Silbert JE: Biosynthesis of chondroitin sulfate. Microsomal acceptors of sulfate, glucuronic acid and N-acetylgalactosamine. Biochemistry 12: 3898–3903, 1973.
41. Rizki RM, Rizki TM: Distribution of 5-bromodeoxyuridine in the pyridine oligonucleotides of *Drosophila* DNA. Mutat Res 30: 343–354, 1975.
42. Robinson HC, Brett MJ, Tralaggan PJ et al: The effect of *D*-xylose, β-*D*-xylosides and β-*D*-galactoside on chondroitin sulphate biosynthesis in embryonic chicken cartilage. Biochem J 148: 25–34, 1975.
43. Rodén L: Biosynthesis of acidic glycosaminoglycans (mucopolysaccharides). *In:* Fishman W (Ed): Metabolic Conjugation and Metabolic Hydrolysis. New York, Academic Press, 1970.
44. Rosenberg L, Margolis R, Wolfenstein-Todel C et al: Organization of extracellular matrix in bovine articular cartilages. *In:* Slavkin HC, Greulich RC (Eds): Extracellular Matrix Influences on Gene Expression. New York-San Francisco-London, Academic Press, 1975.
45. Rutter WJ, Pictet RL, Morris PW: Toward molecular mechanisms of development processes. Ann Rev Biochem 42: 601–646, 1973.
46. Sandson J, Rosenberg L, White D: The antigenic determinants of protein-polysaccharides of cartilage. J Exp Med 123: 817–828, 1966.
47. Saunders AM, Mathews MB, Dorfman A: Antigenicity of chondroitin sulfate. Fed Proc 21: 26 (Abstract) 1962.
48. Schwartz NB, Galligani L, Ho P-L, Dorfman A: Stimulation of synthesis of free chondroitin sulfate chains by β-*D*-xylosides in cultured cells. Proc Natl Acad Sci USA 71: 4047–4051, 1974.
49. Schwartz NB, Ho P-L, Dorfman A: Effect of β-xylosides on synthesis of cartilage-specific proteoglycan in chondrocyte cultures. Biochem Biophys Res Commun 71: 851–856, 1976.
50. Schwartz NB, Rodén L, Dorfman A: Biosynthesis of chondroitin sulfate: Interaction between xylosyltransferase and galactosyltransferase. Biochem Biophys Res Commun 56: 717–724, 1974.
51. Schwartz SA: Localization of 5-bromodeoxyuridine distribution in rat DNA as determined by single-strand specific nucleases. Biochem Biophys Res Commun 65: 1081–1087, 1975.
52. Schwartz SA, Kirsten WH: Distribution of 5-bromodeoxyuridine in the DNA of rat embryo cells. Proc Natl Acad Sci USA 71: 3570–3574, 1974.
53. Skalko RG, Packard DS: Mechanisms of halogenated nucleoside embryotoxicity. Ann NY Acad Sci 255: 552–558, 1975.
54. Stoolmiller AC, Dorfman A: The metabolism of glycosamino-glycans. *In:* Florkin M, Stotz EH (Eds): Comprehensive Biochemistry. Amsterdam-London-New York, Elsevier, 1969.
55. Strom CM, Dorfman A: The distribution of thymidine and 5-bromodeoxyuridine in the DNA of developing chick cartilage. Proc Natl Acad Sci USA 73: 1019–1023, 1976.
56. Strom CM, Dorfman A: The amplification of moderately-repetitive DNA sequences during chick cartilage differentiation. Proc Natl Acad Sci USA 73: 3428–3432, 1976.
57. Thorp FK, Dorfman A: Differentiation of connective tissue. *In:* Moscona A, Monroy A (Eds): Current Topics in Developmental Biology. New York, Academic Press, 1967.
58. Turkington RW, Majumder GC, Riddle M: Inhibition of mammary gland differentiation *in vitro* by 5-bromodeoxyuridine. J Biol Chem 246: 1814–1819, 1971.
59. Wilt FH, Anderson M: The action of 5-bromodeoxyuridine on differentiation. Dev Biol 28: 443–447, 1972.

Control of Cell Differentiation in Normal Hematopoietic and Leukemic Cells

Leo Sachs, Ph.D.

Differentiation *in vitro* to Normal Macrophages, Granulocytes, Lymphocytes and Erythrocytes

It is indeed a privilege to be present at this meeting in honor of Ted Puck, who, in addition to his many other contributions, introduced the single cell cloning procedure for the study of mammalian cells [30]. I would like to discuss some of our studies, which include single cell cloning, on the mechanism that controls the growth and differentiation of normal hematopoietic cells and how such studies can elucidate the blocks in cell differentiation that occur during leukemogenesis. In order to carry out such studies, we first developed experimental systems in which the growth and differentiation of normal mammalian white blood cells can be studied in culture. We have shown that differentiation of normal granulocytes, macrophages, mast cells and lymphocytes can be induced in mass cultures in liquid medium [10,11,32] and that normal macrophages, granulocytes (Fig. 1) [14,28,29], and lymphocytes (Fig. 2) [4] can be cloned in a semi-solid medium such as agar [28] or methylcellulose [14]. In contrast to the formation of lymphocyte colonies [4], the formation of granulocyte and macrophage colonies does not seem to require addition of a foreign antigen.

Normal undifferentiated hematopoietic cells can be induced to form colonies of mature macrophages and granulocytes by a protein inducer [14,28,29] which we now call MGI (macrophage and granulocyte inducer) [16,33,34] that is secreted by various types of cells, including fibroblasts, and can be found in serum. MGI, which has also been referred to by others as CSF or CSA [23,38], is specific for the induction of macrophage and granulocyte colonies. There are different requirements for macrophage and granulocyte colonies [14,33], and the type of colony induced may be due to different cofactors for MGI [16,33]. Another specific inducer, erythropoietin, induces the formation of erythroid colonies [36], and colonies of lymphocytes can be induced by using an appropriate lectin such as phytohemagglutinin, and pokeweed mitogen [4] or Concanavalin A. In all cases, the inducer has to

From the Department of Genetics, Weizmann Institute of Science, Rehovot, Israel.

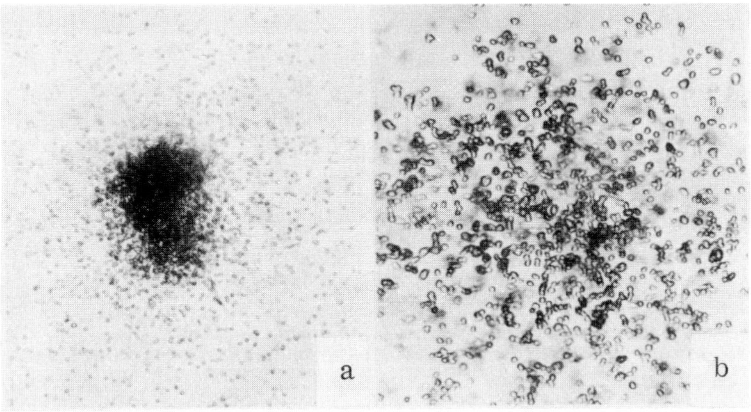

FIG. 1. Granulocyte colony (a) and macrophage colony (b) from normal mouse hematopoietic cells cloned in agar with MGI.

be present until the formation of colonies with the differentiated cells is completed. It has been shown with macrophages, granulocytes [26] and lymphocytes [9] that colonies can be derived from single cells and are therefore clones.

Normal Cell Differentiation in Myeloid Leukemic Cells

The finding of these inducers raised the question of whether an inducer required for normal cell differentiation can induce the normal differentiation of leukemic cells. In order to test this possibility, we have studied myeloid leukemic cells from different sources [5,15,19,27].

Our results have shown that there is one type of myeloid leukemic cell, which we will call $Fc^+C3^+D^+$, that can be induced by the protein inducer MGI to form surface membrane receptors for the Fc portion of immunoglobulin G and C3 component of complement (Fig. 3) [19], to synthesize and secrete lysozyme [15] and differentiate to mature granulocytes and macrophages (Fig. 4). The surface receptors induced on these myeloid leukemic cells are also found on normal mature granulocytes and macrophages and can be detected by rosette formation with sheep erythrocytes coated with antibody or with antibody and complement (Fig. 5). Induction of differentiation to mature granulocytes in these myeloid leukemic cells resulted in the loss of their leukemogenicity. The mature granulocytes induced by MGI from $Fc^+C3^+D^+$ leukemic cells behaved like normal granulocytes [6,37].

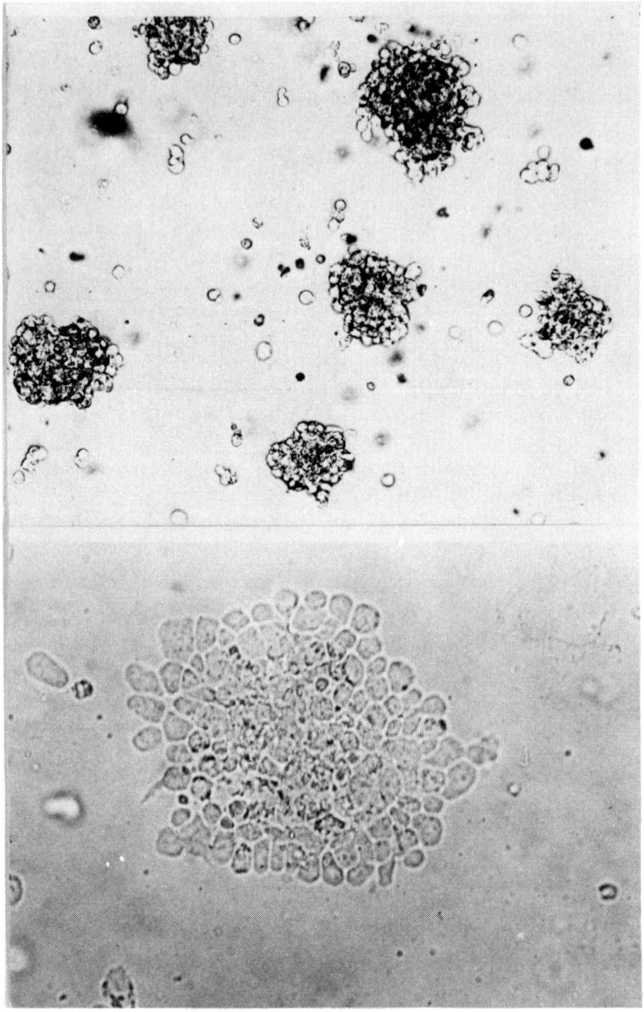

FIG. 2. Colonies of T lymphocytes from normal human peripheral blood lymphocytes cloned in agar with the lectin phytohemagglutinin. Colonies inside the agar (above) and colony on the agar (below).

Different Blocks in the Differentiation of Myeloid Leukemic Cells

In addition to $Fc^+C3^+D^+$ myeloid leukemic cells, there is another type of myeloid leukemic cell ($Fc^+C3^+D^-$) that can be induced to form Fc and C3 rosettes and lysozyme, with a lower inducibility than $Fc^+C3^+D^+$ cells, but could not be induced to differentiate to mature granulocytes or macrophages. A third type of cell ($Fc^-C3^-D^-$) could not be induced by MGI for these rosettes, lysozyme, or mature cells (Fig. 6). We have obtained tissue culture clones of these different types of myeloid leukemic cells [5,15,19]. Blast cells from all types induced myeloid leukemia *in vivo* and, as mentioned above, induction of granulocyte differentiation in $Fc^+C3^+D^+$ cells resulted in the loss of leukemogenicity. D^+ clones after the induction of differentiation can also be distinguished from D^- clones, by the migration of the differentiated cells (Fig. 7).

For comparison with the protein inducer MGI, we have studied the effects of various compounds, including steroid hormones [17,20]. These studies have shown that $Fc^+C3^+D^+$ cells were induced to form C3 but not Fc rosettes by dexamethasone, prednisolone and estradiol (Fig. 8). Induction required protein synthesis, was not inhibited by cordycepin or vinblastine, and optimum induction required the continued presence of the hormone. $Fc^+C3^+D^+$ cells were also induced by these hormones to form macrophages but not granulocytes. In contrast to the protein MGI, these

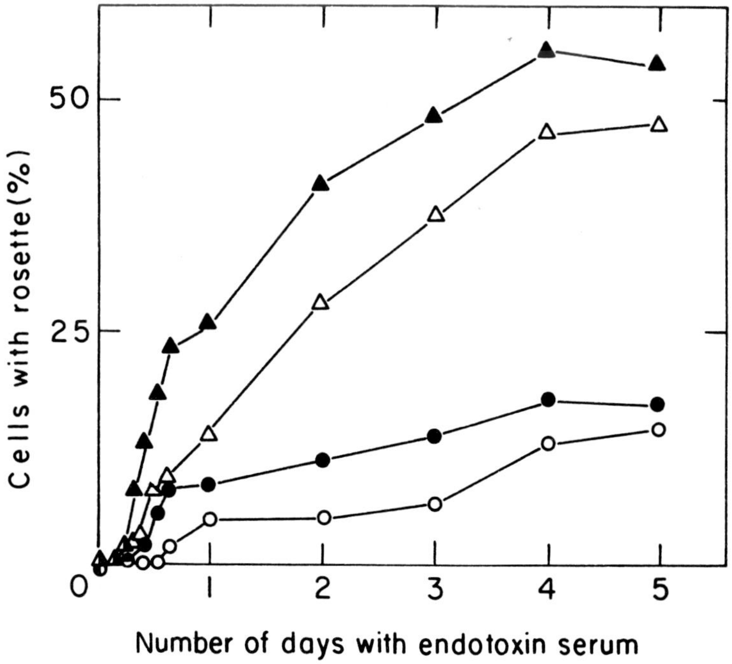

FIG. 3. Induction of Fc and C3 rosettes on $Fc^+C3^+D^+$ and $Fc^+C3^+D^-$ cells at different times after incubation with serum from mice injected with bacterial endotoxin (endotoxin serum). This serum contains MGI. Fc rosettes on $Fc^+C3^+D^+$ (●——●) and $Fc^+C3^+D^-$ (○——○) cells. C3 rosettes on $Fc^+C3^+D^+$ (▲——▲) and $Fc^+C3^+D^-$ (△——△) cells.

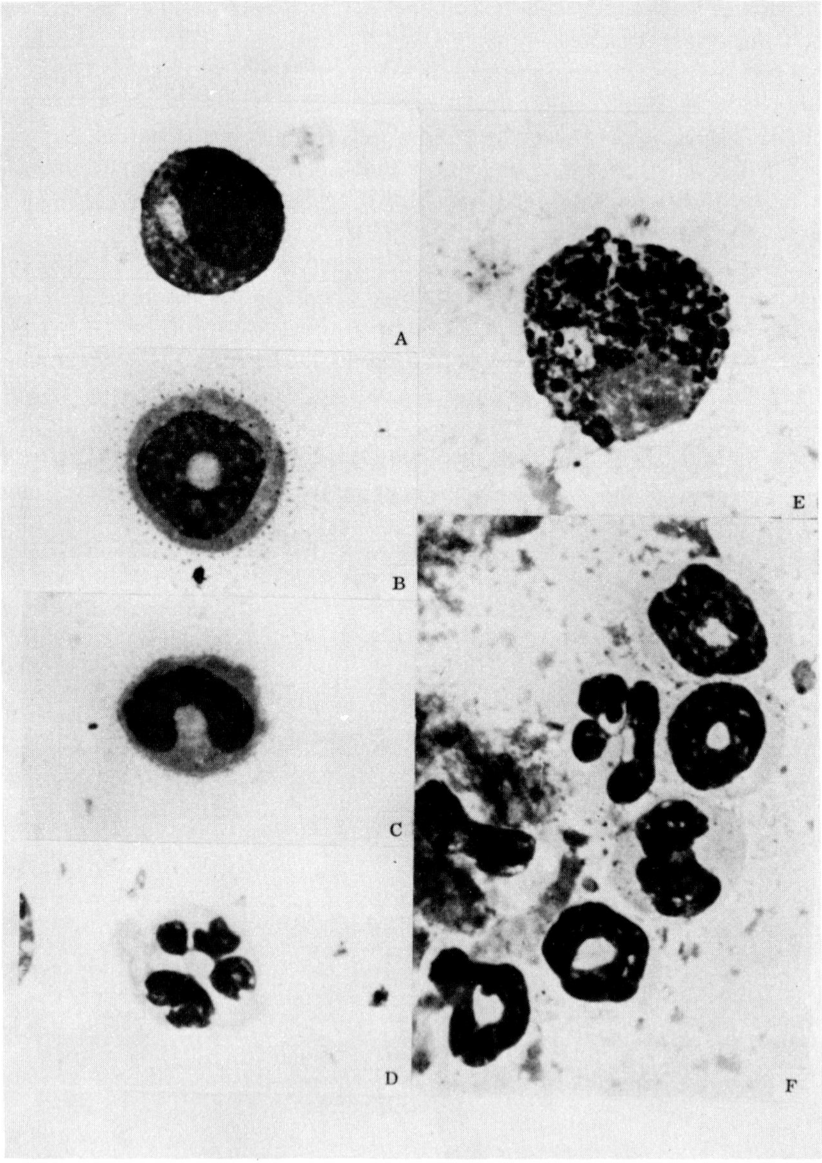

FIG. 4. Differentiation of Fc$^+$C3$^+$D$^+$ cells to mature macrophages and granulocytes by MGI. Undifferentiated blast cell (A), stages in the differentiation to mature granulocytes (B-D), macrophage (E), group of granulocytes in different stages of differentiation (F).

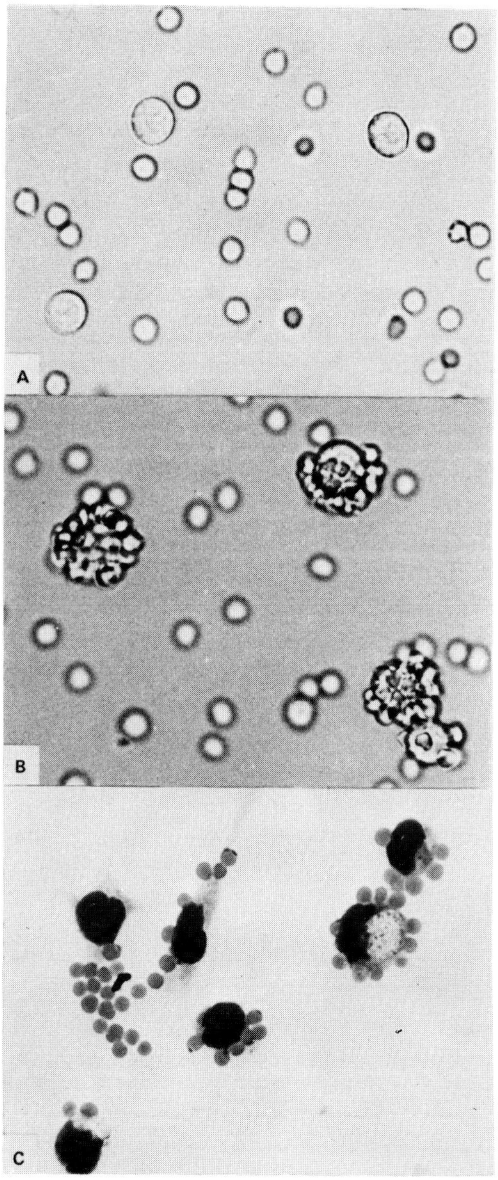

FIG. 5. Induction of C3 rosettes on $Fc^+C3^+D^+$ cells by MGI. Untreated cells (A), rosettes induced by MGI on cells in suspension (B), rosettes induced by MGI on cells attached to a Petri dish (C).

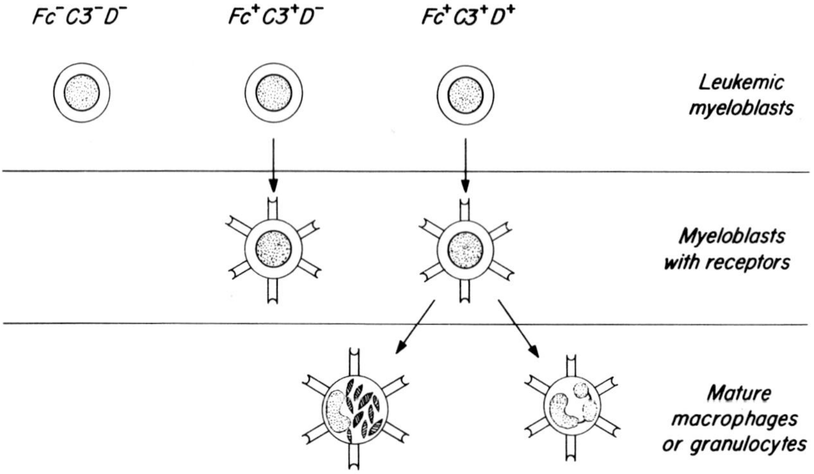

FIG. 6. Different blocks in the differentiation of myeloid leukemic cells. D = differentiation to mature macrophages and granulocytes. Receptors = receptors for Fc and C3 rosettes.

hormones thus induced some, but not all, the changes associated with normal cell differentiation [20]. Incubation with dibutyryl-cyclic AMP or GMP, theophylline, aminophylline, prostaglandin E_1 or E_2, did not induce the formation of Fc or C3 rosettes, macrophages or granulocytes [17].

The results with steroid hormones have also shown differences in the response of the different types of myeloid leukemic cells. $Fc^+C3^+D^-$ cells

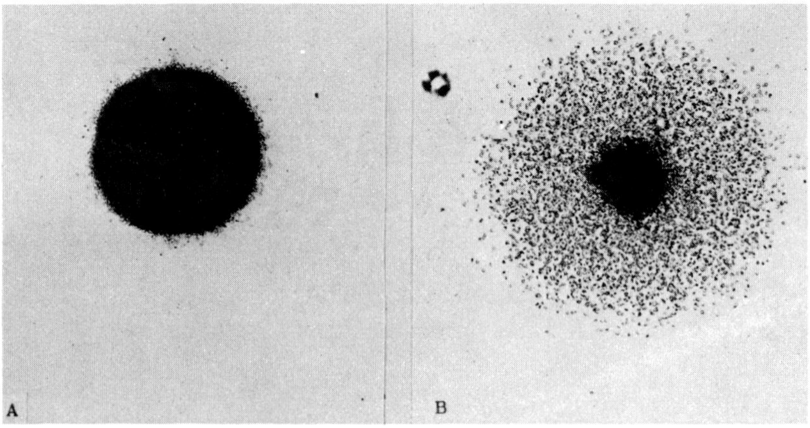

FIG. 7. $Fc^+C3^+D^-$ colony (A) and $Fc^+C3^+D^+$ colony (B) in agar. The cells were grown with MGI.

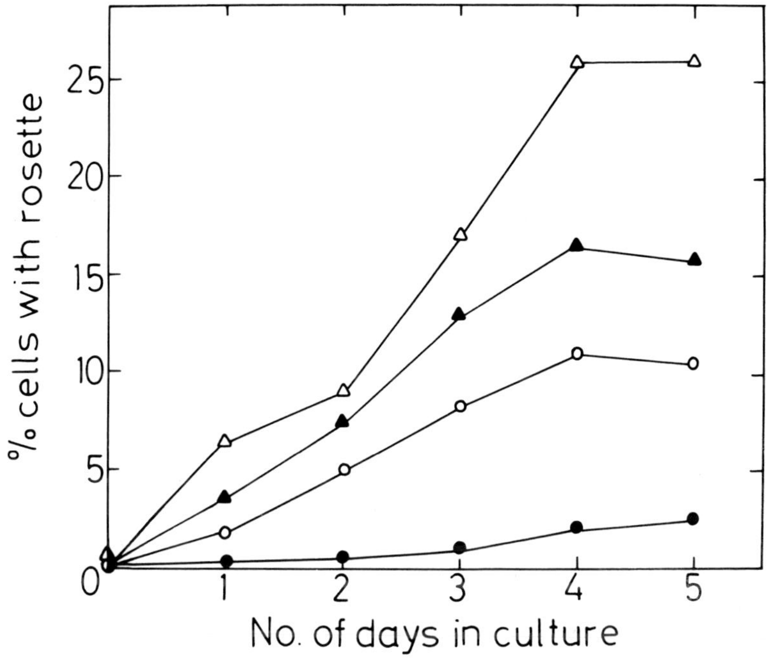

FIG. 8. Induction of C3 rosettes on $Fc^+C3^+D^+$ cells by the steroids prednisolone, dexamethasone and estradiol at different times after incubation with 1 μM steroid. Untreated control (●——●), estradiol (○——○), prednisolone (▲——▲), dexamethasone (△——△). Progesterone, testosterone and cortisone gave results similar to the untreated control.

showed a lower steroid inducibility for C3 rosettes than $Fc^+C3^+D^+$ cells and no induction of macrophages. There was no induction of rosettes or macrophages with $Fc^-C3^-D^-$ cells. Progesterone, testosterone or cortisone did not induce the formation of rosettes or macrophages, but inhibited induction by the inducing hormones. Our results indicate that the steroid induction of surface changes in myeloid leukemia cells involves specific steroid hormone receptors [17,20].

In addition to $Fc^+C3^+D^+$ and $Fc^+C3^+D^-$ clones that can be induced to form Fc and C3 rosettes by MGI and C3 rosettes by a steroid inducer such as dexamethasone, and $Fc^-C3^-D^-$ clones that cannot be induced to form rosettes by either of these inducers, other clones have been isolated that cannot be induced to form rosettes by MGI, but can be induced to form Fc or Fc and C3 rosettes by dexamethasone. The results indicate that there are different cellular sites for MGI and the steroid hormones [17]. Dimethylsulfoxide, which can induce some of the stages of differentiation in erythroid leukemic cells [8,31a], induced C3 rosettes in only one $Fc^+C3^+D^+$ clone but in none of the $Fc^+C3^+D^-$ or $Fc^-C3^-D^-$ clones tested [15].

The inducing hormones prednisolone and dexamethasone can also induce the activity of alkaline phosphatase [35], and tyrosine aminotransferase [31] in some types of mammalian cells, enhance the induction of murine leukemia virus by 5-iododeoxyuridine and induce the synthesis of B-type virus from mammary adenocarcinoma cells [22,25]. The enhancement of induction of murine leukemia virus by these steroid hormones seems to differ from the induction of changes in myeloid leukemic cells. Cordycepin, which inhibits the synthesis of poly A, inhibited virus induction but did not inhibit the induction of changes in the myeloid leukemic cells. Estradiol induced changes in the myeloid leukemic cells and did not enhance the induction of murine leukemia virus by 5-iododeoxyuridine.

Hormones like prednisolone are used in combination chemotherapy for human myeloid leukemia with compounds such as cytosine arabinoside. This compound can induce Fc and C3 rosettes on $Fc^+C3^+D^+$ and $Fc^+C3^+D^-$, but not on $Fc^-C3^-D^-$ cells [19]. The therapeutic effects of prednisolone-like hormones, and compounds such as cytosine arabinoside, may in part be due to their induction of changes associated with differentiation. It will, therefore, be of interest to study the inducibility of Fc and C3 rosettes and other changes associated with differentiation in myeloid leukemic cells from different patients [27] in relation to their response to chemotherapy.

Similar Sequence of Differentiation in Normal and Myeloid Leukemic Cells

In order to compare the sequence of differentiation in normal and myeloid leukemic cells, we first developed a technic for the enrichment of the myeloid precursor cells from normal bone marrow.

An enriched population consisting of early myeloid cells (myeloblasts and promyelocytes) was obtained from normal bone marrow by injection of mice with sodium caseinate and the removal of cells with C3 rosettes by Ficoll-Hypaque density centrifugation. This enriched population had no C3 or Fc rosettes and contained about 90% early myeloid cells as compared with about 15% in the unfractionated bone marrow. Nearly all these early myeloid cells were stained for myeloperoxidase. After seeding with MGI, the enriched population showed a cloning efficiency of about 15%, compared with 0.3% in the unfractionated bone marrow, and both the enriched and the unfractionated cells gave rise to macrophage and granulocyte colonies. The increased cloning efficiency of the enriched population did not seem to be due to the removal of inhibitory cells in the bone marrow [18].

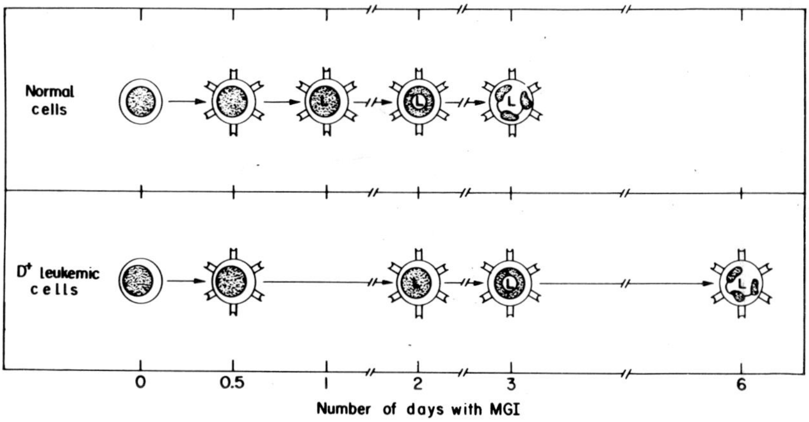

FIG. 9. The sequence and timing of differentiation in normal and $Fc^+C3^+D^+$ leukemic cells after incubation with MGI. Fc and C3 rosettes ( ), lysozyme (L), intermediate stage of granulocyte differentiation ( ), mature granulocyte ( ).

The normal early myeloid cells were induced to differentiate by MGI to mature granulocytes and macrophages. The sequence of granulocyte differentiation was the formation of Fc and C3 rosettes, followed by the synthesis and secretion of lysozyme and then morphologic differentiation to mature cells. The $Fc^+C3^+D^+$ myeloid leukemic cells with no Fc or C3 rosettes before induction had a similar morphology to normal early myeloid cells and showed the same sequence of differentiation. The induction of Fc and C3 rosettes occurred at the same time in both the normal and $Fc^+C3^+D^+$ leukemic cells, but lysozyme synthesis and the formation of mature granulocytes was induced later in these leukemic than in the normal cells (Fig. 9) [18]. The results indicate that normal and $Fc^+C3^+D^+$ myeloid leukemic cells have a similar sequence of differentiation, but that the normal cells had a greater sensitivity for the formation of mature cells by MGI.

Surface Membrane-Cytoskeleton Interactions and the Control of Differentiation

In order to study the association between surface membrane-cytoskeleton interactions and the inducibility of differentiation of myeloid leukemic cells by MGI, we have used as a probe the formation of caps by the lectin concanavalin A (Con A) [34] and the effect of vinblastine on cap formation [3].

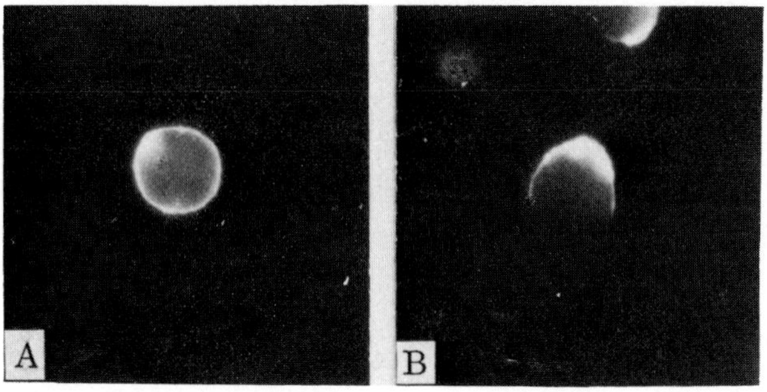

FIG. 10. $Fc^+C3^+D^+$ cells without (A) and with (B) a cap induced by fluorescent Con A.

$Fc^+C3^+D^+$, $Fc^+C3^+D^-$ and $Fc^-C3^-D^-$ cells before induction, showed 50%, 5% and 0% cells with a Con A-induced cap (Fig. 10), respectively [21,34]. Treatment with vinblastine or colchicine, but not with lumicolchicine, increased the frequency of cap formation from 50 to 100% in $Fc^+C3^+D^+$ cells, from 5 to 95% in $Fc^+C3^+D^-$, and from 0 to 50% in $Fc^-C3^-D^-$ cells (Fig. 11). The increased ability to form a cap produced by vinblastine did not change the inducibility of cells for rosettes. Our results indicate that although free surface receptors for Con A and receptors anchored to tubulin can form a cap on myeloid leukemic cells, there are also receptors that may be anchored to structures other than tubulin that did not form a cap. The data suggest that the ability of myeloid leukemic cells to be induced to differentiate by MGI is associated with the frequency of Con A surface receptors that are free or have specific types of anchorage [21]. This association between cap formation and the ability to respond to a differentiation inducer may be useful as an aid in clinical diagnosis in various diseases (Fig. 12) [24].

Chromosome Mapping of the Genes that Control Differentiation and Malignancy in Myeloid Leukemic Cells

The clonal origin and hereditability of the differences in inducibility by MGI suggested that there may be genetic differences between the different cell types and that it may be possible to map the chromosome location of the genes involved. The mouse myeloid leukemic cell clones used, even those that had been in culture for several years, still had a diploid or near diploid modal chromosome number.

An analysis of the chromosome banding pattern (Fig. 13) has shown that none of the cells had a completely normal diploid banding pattern. The $Fc^+C3^+D^+$ cells can be induced to undergo normal cell differentiation even though the cells do not have a completely normal diploid chromosome banding pattern [1,12]. This shows that these malignant cells can be induced by MGI to have a normal differentiation phenotype without a completely normal genotype. There were specific chromosome differences between $Fc^+C3^+D^+$ and $Fc^+C3^+D^-$ cells [12]. In addition, 6 clones of $Fc^-C3^-D^-$ cells derived from 6 independently produced myeloid leukemias showed a loss of a piece of one chromosome 2. This loss was not found in $Fc^+C3^+D^+$, $Fc^+C3^+D^-$ or lymphoid leukemias. Five Fc^+C3^+ mutants derived from an $Fc^-C3^-D^-$ clone with a loss of a piece of one chromosome 2, one normal chromosome 12 and two translocated chromosomes 12, maintained the abnormal chromosome 2 but had lost either the one normal or one of these translocated chromosomes 12. These results indicate that chromosomes 2 and 12 carry genes that control the induction of differentiation of myeloid leukemic cells by MGI. The data suggest that there are suppressor gene(s)

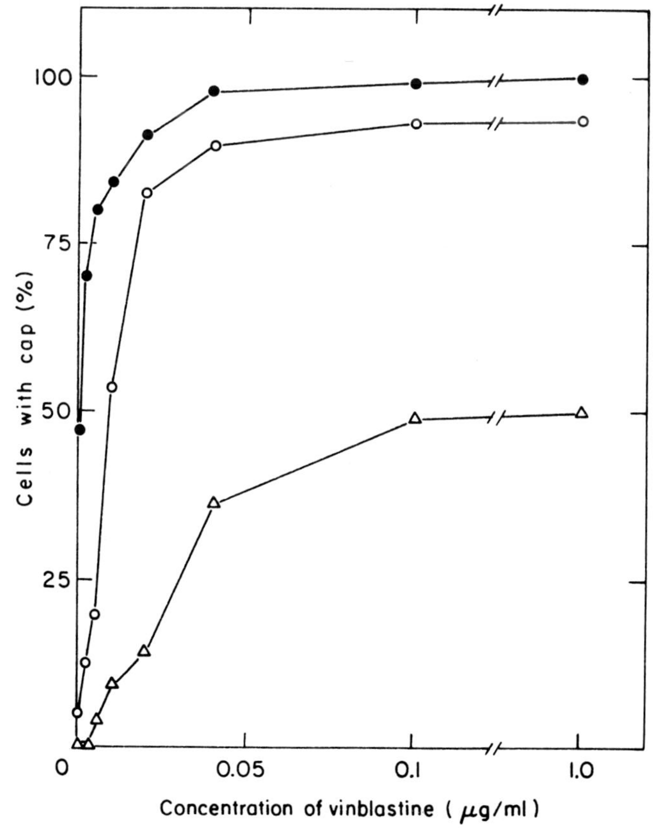

FIG. 11. Frequency Con A induced caps on mouse myeloid leukemic cells in the presence and absence of vinblastine. $Fc^+C3^+D^+$ (●——●), $Fc^+C3^+D^-$ (○——○), $Fc^-C3^-D^-$ (△——△).

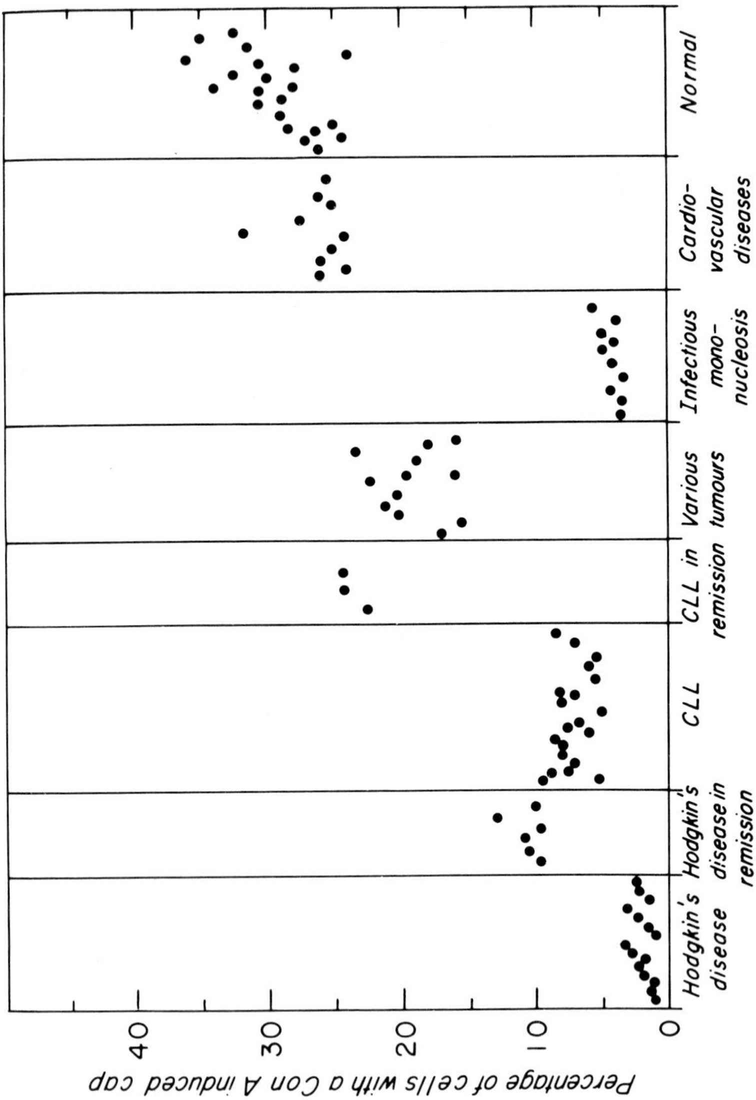

FIG. 12. Frequency of Con A induced caps on human peripheral blood lymphocytes from normal persons, patients with different diseases and patients in remission. CLL = chronic lymphocytic leukemia.

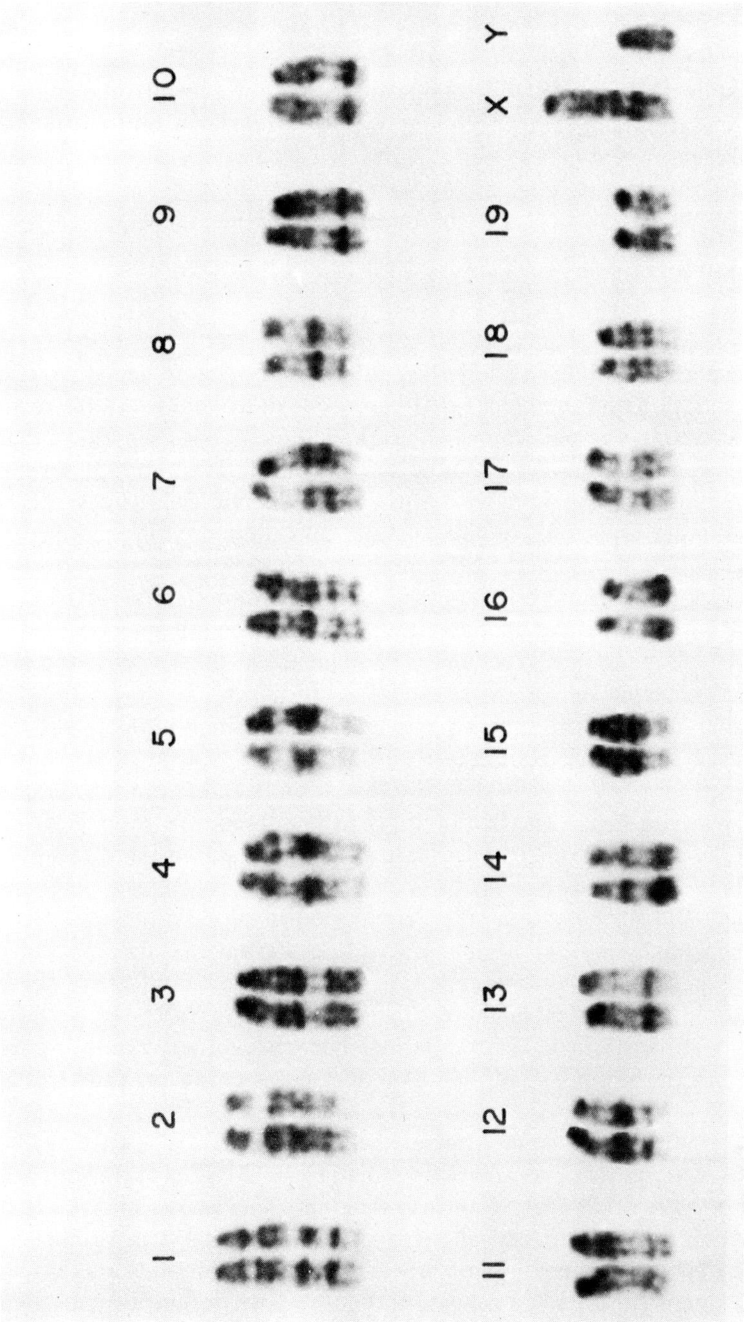

FIG. 13. Chromosome banding pattern of the normal male mouse.

on chromosome 12 that can suppress the inducing gene(s) on chromosome 2. The results also suggest that chromosomes 2 and 12 carry genes that control the malignancy of these myeloid leukemic cells [1]. The data support the suggestion that there is more than one gene that controls malignancy and differentiation and that phenotypic expression depends on the balance between these different genes [1,2,13,34,39].

Conclusions

The development of experimental systems for the culture and cloning of cells has made it possible to study the controls that regulate hematopoietic cell differentiation and the blocks that can occur in leukemia. All the main types of hematopoietic cells can be cloned in culture. It has been shown that the specific macrophage and granulocyte inducer MGI can induce normal cell differentiation in some types of myeloid leukemic cells ($Fc^+C3^+D^+$) and that in other types of myeloid leukemic cells there are different blocks in the process of normal cell differentiation. The ability of the different types of myeloid leukemic cells to be induced to differentiate by MGI, was associated with differences in the membrane-cytoskeleton interactions of specific surface membrane receptors. The sequence of differentiation in the $Fc^+C3^+D^+$ leukemic cells is the same as that in normal myeloid cells, and the leukemic cells studied appear to be less sensitive than normal cells for the induction of mature granulocytes by MGI. The normal cells require MGI for cell viability, growth and differentiation, whereas the $Fc^+C3^+D^+$ leukemic cells are viable and can multiply in the absence of MGI [7]. The change in the cells resulting in a decrease or lack of requirement of this protein for viability and growth, may be one of the causes of myeloid leukemia.

Some but not all the stages of differentiation in myeloid leukemic cells can be induced by steroid hormones, such as dexamethasone and other compounds such as cytosine arabinoside. The therapeutic effects of these compounds in human leukemia may, therefore, be in part due to their induction of some stages of differentiation. There are different cellular sites for MGI and the steroids, and induction does not seem to be mediated by cyclic AMP.

Chromosome banding studies have shown, that $Fc^+C3^+D^+$ cells can be induced to undergo normal cell differentiation by MGI even though the cells do not have a completely normal genotype. The results indicate that differentiation and malignancy of myeloid leukemic cells is controlled by the balance between different genes located on specific chromosomes which have been identified. It can be concluded that the experimental systems used in these studies appear to be particularly favorable for further elucidation of the molecular controls of cell differentiation by specific inducers in normal and malignant cells.

REFERENCES

1. Azumi J, Sachs L: Chromosome mapping of the genes that control differentiation and malignancy in myeloid leukemic cells. Proc Natl Acad Sci USA 74: 253–257, 1977.
2. Bloch-Shtacher N, Sachs L: Chromosome balance and the control of malignancy. J Cell Physiol 87: 89–100, 1976.
3. Edelman GM: Surface modulation in cell recognition and cell growth. Science 192: 218–226, 1976.
4. Fibach E, Gerassi E, Sachs L: Induction of colony formation *in vitro* by human lymphocytes. Nature 259: 127–129, 1976.
5. Fibach E, Hayashi M, Sachs L: Control of normal differentiation of myeloid leukemic cells to macrophages and granulocytes. Proc Natl Acad Sci USA 70: 343–346, 1973.
6. Fibach E, Sachs L: Control of normal differentiation of myeloid leukemic cells. VIII. Induction of differentiation to mature granulocytes in mass culture. J Cell Physiol 86: 221–230, 1975.
7. Fibach E, Sachs L: Control of normal differentiation of myeloid leukemic cells. XI. Induction of a specific requirement for cell viability and growth during the differentiation of myeloid leukemic cells. J Cell Physiol 89: 259–266, 1976.
8. Friend C, Scher W, Holland JG, Sato T: Hemoglobin synthesis in murine virus-infected leukemic cells *in vitro:* stimulation of erythroid differentiation by dimethylsulfoxide. Proc Natl Acad Sci USA 68: 378–382, 1971.
9. Gerassi E, Sachs L: Regulation of the induction of T and B cell colonies *in vitro* by normal human lymphocytes. Proc Natl Acad Sci USA 73: 4545–4550, 1976.
10. Ginsburg H, Sachs L: Destruction of mouse and rat embryo cells in tissue culture by lymph node cells from unsensitized rats. J Cell Comp Physiol 66: 199–220, 1965.
11. Ginsburg H, Sachs L: Formation of pure suspension cells in tissue culture by differentiation of lymphoid cells from the mouse thymus. J Natl Cancer Inst 31: 1–40, 1963.
12. Hayashi M, Fibach E, Sachs L: Control of normal differentiation of myeloid leukemic cells. V. Normal differentiation in aneuploid leukemic cells and the chromosome banding pattern of D^+ and D^- clones. Int J Cancer 14: 40–48, 1974.
13. Hitotsumachi S, Rabinowitz Z, Sachs L: Chromosomal control of reversion in transformed cells. Nature 231: 511–514, 1971.
14. Ichikawa Y, Pluznik DH, Sachs L: *In vitro* control of the development of macrophage and granulocyte colonies. Proc Natl Acad Sci USA 56: 488–495, 1966.
15. Krystosek A, Sachs L: Control of lysozyme induction in the differentiation of myeloid leukemic cells. Cell 9: 675–684, 1976.
16. Landau T, Sachs L: Characterization of the inducer required for the development of macrophage and granulocyte colonies. Proc Natl Acad Sci USA 68: 2540–2544, 1971.
17. Lotem J, Sachs L: Control of Fc and C3 receptors on myeloid leukemic cells. J Immunol 117: 580–586, 1976.
18. Lotem J, Sachs L: Control of normal differentiation of myeloid leukemic cells. XII. Isolation of normal myeloid colony-forming cells from bone marrow and the sequence of differentiation to mature granulocytes in normal and D^+ myeloid leukemic cells. J Cell Physiol, in press, 1976.
19. Lotem J, Sachs L: Different blocks in the differentiation of myeloid leukemic cells. Proc Natl Acad Sci USA 71: 3507–3511, 1974.

20. Lotem J, Sachs L: Induction of specific changes in the surface membrane of myeloid leukemic cells by steroid hormones. Int J Cancer 15: 731–740, 1975.
21. Lotem J, Vlodavsky I, Sachs L: Regulation of cap formation by concanavalin A and the differentiation of myeloid leukemic cells: relationship to free and anchored surface receptors. Exp Cell Res 101: 323–330, 1976.
22. Lowy DR, Rowe WP, Teich N, Harley JW: Murine leukemic virus: high frequency activation *in vitro* by 5-iododeoxyuridine and 5-bromodeoxyuridine. Science 174: 155–156, 1971.
23. Metcalf D: Studies on colony formation *in vitro* by mouse bone marrow cells. I. Continuous cluster formation and relation of clusters to colonies. J Cell Physiol 74: 323–332, 1969.
24. Mintz U, Sachs L: Membrane differences in peripheral blood lymphocytes from patients with chronic lymphocytic leukemia and Hodgkin's disease. Proc Natl Acad Sci USA 72: 2428–2432, 1975.
25. Paran M, Gallo RC, Richardson LS, Wu AM: Adrenal corticosteroids enhance production of type C virus induced by 5-iodo-2-deoxyuridine from cultured mouse fibroblasts. Proc Natl Acad Sci USA 70: 2391–2395, 1975.
26. Paran M, Sachs L: The single cell origin of normal granulocyte colonies *in vitro.* J Cell Physiol 73: 91–92, 1969.
27. Paran M, Sachs L, Barak Y, Resnitzky P: *In vitro* induction of granulocyte differentiation in hematopoietic cells from leukemic and non-leukemic patients. Proc Natl Acad Sci USA 67: 1542–1549, 1970.
28. Pluznik DH, Sachs L: The cloning of normal "mast" cells in tissue culture. J Cell Comp Physiol 66: 319–324, 1965.
29. Pluznik DH, Sachs L: The induction of clones of normal "mast" cells by a substance from conditioned medium. Exp Cell Res 43: 553–563, 1966.
30. Puck TT, Marcus PI: A rapid method for viable cell titration and clone production with HeLa cells in tissue culture: The use of x-irradiated cells to supply conditioning factors. Proc Natl Acad Sci USA 41: 432–437, 1955.
31. Rousseau GG, Baxter JD, Tomkins GM: Glucocorticoid receptors: relations between steroid binding and biological effects. J Mol Biol 67: 99–115, 1972.
31a. Reuben RC, Wife RL, Breslow R et al: A new group of potent inducers of differentiation in murine erythroleukemia cells. Proc Natl Acad Sci USA 73: 862–866, 1976.
32. Sachs L: The analysis of regulatory mechanisms in cell differentiation. *In:* New Perspectives in Biology. Elsevier, Amsterdam, 246–260, 1964.
33. Sachs L: Control of growth and differentiation in normal hematopoietic and leukemic cells. *In:* "Control of Proliferation in Animal Cells." Cold Spring Harbor Lab., New York, 915–925, 1974.
34. Sachs L: Regulation of membrane changes, differentiation and malignancy in carcinogenesis. The Harvey Lectures. Academic Press, New York 68: 1–35, 1974.
35. Sela B, Sachs L: Alkaline phosphatase activity and the regulation of growth in transformed mammalian cells. J Cell Physiol 83: 27–34, 1974.
36. Stephenson JR, Axelrad AA, McLeod DL, Shreeve MM: Induction of colonies of hemoglobin-synthesizing cells by erythropoietin *in vitro*. Proc Natl Acad Sci USA 68: 1542–1546, 1971.
37. Vlodavsky I, Fibach E, Sachs L: Control of normal differentiation of myeloid leukemic cells. X. Glucose utilization, cellular ATP and associated membrane changes in D^+ and D^- cells. J Cell Physiol 87: 167–177, 1976.
38. Worton RGG, McCulloch EA and Till JE; Physical separation of hematopoietic stem cells forming colonies in culture. J Cell Physiol 74: 171–182, 1969.
39. Yamamoto T, Rabinowitz Z, Sachs L: Identification of the chromosomes that control malignancy. Nature New Biology 243: 247–250, 1973.

Neoplastic Stem Cells

G. Barry Pierce, M.D.

No concept of cancer has succeeded in satisfactorily explaining the position of cancer in the flow of biology. Public health concepts, which have been so successful in controlling the great pandemics, have been applied to the study of tumors. It makes good sense that carcinogenic agents should be eliminated from the environment, but whereas the therapy of infectious disease is directed at the etiologic agent, therapy in cancer is directed at the cells of the body which have responded to the etiologic agent. This might seem logical since untreated cancer cells left to their own devices will destroy their host, but nevertheless, it is a significant departure in rationale from therapy of infectious diseases.

In thinking about concepts of cancer, everyone agrees that carcinomas are composed of cells which form tissues that exhibit properties incompatible with the life of the host. Since every other tissue, irrespective of species of origin, arises by the process of differentiation, it seems surprising that these "abnormal" cells and tissues have not been studied from a standpoint of "normal" development. Boveri discounted the idea that tumors might arise by the process of mitosis and differentiation because differentiation was considered to be a rapid process in comparison to carcinogenesis [1]. Furthermore, neoplastic tissues usually contained abnormal mitoses and appeared less differentiated than normal tissue. If the stability and heritability of the neoplastic state could not be attributed to epigenetic mechanisms, the only remaining possibility was structural change in the genome. The most commonly held concept of neoplasia states that carcinomas arise in differentiated tissues from cells which undergo a somatic mutation. The progeny of these mutated cells are believed to undergo a process of dedifferentiation, losing features of the normal phenotype in becoming neoplastic stem cells. This concept of cancer was promulgated before much was known about differentiation, and before anything was known about cell and tissue renewal in either normal or neoplastic tissues. Differentiation yields notoriously stable tissues with heritable properties.

It is the purpose of this paper to describe experiments and clinical observations that lead me to conclude that tumors are a caricature of the

From the Department of Pathology, University of Colorado Medical Center, Denver, Colo.
Supported in part by grant PDT-23R from the American Cancer Society and grants AM-15663 and CA-15823 from the National Institutes of Health.

process of tissue renewal. As such they fit into the mainstream of cellular and developmental biology. It is impossible to pinpoint the exact mechanism by which the neoplastic process is set in motion. Most investigators favor somatic mutation as the inciting event, others believe carcinomas are postembryonic differentiations. Although the data to be cited favor the latter proposition, somatic mutation has not been ruled out.

TISSUE RENEWAL

Growth and development are characterized by controlled mitosis, differentiation and organization of tissues. Cell division and differentiation are continued in the adult and are the processes by which tissues are normally renewed. The reactive cells in this process are called stem cells. They are determined in the sense that although they may not contain the molecules characteristic of a differentiated phenotype, their progeny within very narrow limits can differentiate only into a particular phenotype. In other words, a stem cell of skin does not differentiate into a brain cell or a bone cell. Although determined, the stem cell itself cannot differentiate directly into a differentiated phenotype. Its immediate fate is to divide, and one of its progeny cells through proliferation and differentiation enters into a life cycle that results in a postmitotic functional cell. These functional cells eventually become senescent and die. Senescent cells of seminiferous epithelium, gastrointestinal tract epithelium, upper respiratory tract epithelium, skin, breast, etc., are sloughed from the body. New cells are produced which balance this cellular loss and maintain the status quo, indicating the presence of precise controls.

If a carcinoma is a caricature of tissue renewal, then it too should have stem cells. Makino demonstrated the presence of multiple stem lines in a murine leukemia many years ago, but the developmental significance of this observation was overlooked [9]. The ability of these cells to proliferate and faithfully reproduce themselves was appreciated, but whether or not they could differentiate was not considered.

DIFFERENTIATION IN TUMORS

Teratocarcinomas of mice, first described by Stevens and Little in 1954, were selected for these studies [20]. Teratocarcinoma literally means a malignant tumor of a monster or malformed baby. Definitionally, they contain differentiated elements representing each of the three embryonic germ layers, and they are malignant by virtue of the presence of undifferentiated or "embryonal carcinoma" cells. These bear a close resemblance to embryonic epithelium of the mouse embryo. Occasionally the somatic tissues and embryonal carcinoma are arranged in configurations resembling early mouse embryos, which have been called embryoid bodies. It was reasoned that if differentiation occurred in tumors, the differentiation of embryonal carcinoma cells into skin, teeth, muscle, glands, etc., should be dramatic and easily documented.

Solution of the problem became possible when it was discovered that the injection of minced teratocarcinoma into the intraperitoneal cavity of mice resulted in the formation of free-floating embryoid bodies in the peritoneal fluid [14]. The smallest embryoid body resembled a mouse embryo of about 4 1/2 days of gestation, and was composed of a core of embryonal carcinoma overlain by proximal endoderm. When transplanted subcutaneously in an animal, these small embryoid bodies invariably developed into teratocarcinomas containing 12 or more differentiated tissues as well as embryonal carcinoma. A series of more mature embryoid bodies containing embryonal carcinoma, mesoderm and endoderm were transplanted subcutaneously in animals, and groups were examined by serial section daily to determine the behavior of the individual tissues. The endodermal and mesodermal cells had a limited capacity for proliferation, although they could differentiate into tissues of the respective germ layers. Only the embryonal carcinoma had tumorigenic activity. These cells invaded through the walls of the embryoid bodies and developed into new teratocarcinomas. The smallest tumors contained only embryonal carcinoma and a few primitive neural elements; they developed into tumors containing the 12 to 15 somatic tissues characteristic of teratocarcinomas.

Confirmation for the observations that embryonal carcinoma cells could differentiate into a multiplicity of tissues was obtained by cloning teratocarcinomas from single embryonal carcinoma cells; 42 such tumors were produced and each contained the differentiated tissues and embryonal carcinoma characteristic of teratocarcinomas [8]. The next step was to determine the nature of the tissues derived by differentiation of embryonal carcinoma cells. Were they benign or malignant? For these experiments it was necessary to separate the differentiated tissues from embryonal carcinoma and study their behavior when transplanted into appropriate animals [14]. Transplants of differentiated tissues resulted in tumors equivalent to benign cystic teratomas of the ovary of human beings. They contained well differentiated tissues, representing each of the three germ layers, which persisted in the subcutaneous space of the mouse for as long as six months with no evidence of proliferation or malignancy. The conclusion was reached that stem cells of teratocarcinoma could differentiate into benign somatic tissues. These observations have been extended by Mintz et al. [11]

and Papaioannou et al. [12], and it can now be said that embryonal carcinoma can differentiate into normal cells. The teratocarcinoma is a caricature of the process of embryogenesis.

Confirmatory experiments were undertaken, utilizing a squamous cell carcinoma which had been chemically induced by Snell [17]. The tumor was slowly growing, well differentiated, and composed of large masses of undifferentiated cancer cells in which islands of mature keratinizing epithelium were present. The latter have been referred to as "squamous pearls." Animals bearing such tumors were injected with tritium-labeled thymidine, and tissues were prepared at intervals for autoradiography and electron microscopy. At 2 hours, labeled cells were found only in the undifferentiated areas of the tumors, and these cells proved to be extremely undifferentiated, with no ultrastructural evidence of squamous differentiation. In contrast, by 96 hours many labeled cells were found in pearls. These cells had developed the morphologic features of mature squamous cells. It was concluded that undifferentiated cells migrated into pearls and differentiated. Their benign nature was demonstrated by transplantation studies in which pearls were dissected from the tumors and transplanted subcutaneously into animals. They never developed into tumors, whereas comparable amounts of undifferentiated tumor produced squamous cell carcinomas in a third of the cases. It was concluded that they were malignant stem cells of the squamous cells, which were benign. These tumors were considered caricatures of squamous epithelial renewal [17].

Confirmation for the idea that carcinomas were caricatures of tissue renewal was obtained in experiments using adenocarcinomas of the breast [21]. In addition, for the first time, we obtained evidence to the effect that malignant stem cells were under environmental control. In this regard, Mendelsohn [10] showed that about one-half of the cells in spontaneous adenocarcinomas of the breast of mice were noncycling and presumably arrested in G-0. We showed that some of these cells had differentiated to benign postmitotic states, but in addition, there were significant numbers of undifferentiated cells that were not cycling. These were considered to be stem cells in G-0. Nothing is known about the biology of these cells other than that they were not cycling, presumably as a result of an intimate environmental control.

There is widespread support in the ultrastructural literature for the idea that tumors are caricatures of tissue genesis. Most investigators, impressed by the classical dogma of somatic mutation and dedifferentiation, have interpreted their data to the effect that the various degrees of differentiation in tumors were the result of variable degrees of dedifferentiation. The data in these studies are equally supportive of the notion that these are differentiating systems, and what has been attributed to dedifferentiation is in fact abortive differentiation.

These briefly reviewed studies supporting the idea that tumors are caricatures of tissue renewal have important implications for tumor biology. Many unexplained phenomena begin to make sense. For example, immuno-oncologists have been impressed with the immune response associated with certain tumors. Seminomas and medullary carcinomas of breast and stomach that have an immune response have a better prognosis than corresponding tumors that do not. This more favorable prognosis is usually attributed to the immune response, but an alternative explanation is even more compelling. It has been pointed out that when a normal epithelium evolves senescent cells, these cells are eliminated from the body. Tumors lack this system of elimination and present the reticuloendothelial system with the debris. It is well known that sperm are highly antigenic, and if a seminoma produced differentiated senescent cells these too should be antigenic. Thus, it is quite likely that the good prognosis for this kind of tumor is the result of the ability of the tumor cells to differentiate, and not to the immune response to senescent cells.

Tumors appear to be composed of two compartments. One contains malignant stem cells, some proliferating and some arrested in G-O. Although the latter are not expressing the malignant phenotype at the particular time, it is presumed that when the environment changes, they can commence cycling. The other compartment is the differentiated compartment. The cells of the differentiated compartment, although derived by differentiation from malignant stem cells, are not malignant and play no further role in the neoplastic process. Since these compartments vary in composition between tumors, it is small wonder that biochemists have found wide variations in the enzymes in individual tumors of a particular kind. Cloning eliminates the variability occasioned by the presence of multiple stem lines, but it does not reduce the variability in differentiation within a particular stem line. In normal tissue, the proliferating compartment is small and for a given normal tissue relatively constant in amount. In malignant tumors it is usually large but variable in amount. The dilution occasioned by the presence of benign cells in these tumors is of great significance. Obviously the best control for biochemical experiments of malignant stem cells would be to compare the data with those obtained from normal stem cells. This is easier to say than do, however.

Similarly, the electron microscopist, when making quantitative comparisons of membrane differentiations between normal and malignant cells, must be careful to determine whether the tumor cells under examination are malignant or benign. It is our impression in studies of teratocarcinomas and squamous cell carcinomas that when the first manifestations of differentiation are visible in cells, those cells no longer exhibit malignant characteristics.

ORIGIN OF NEOPLASTIC STEM CELLS

If a tumor develops in skin painted with carcinogen, the tumor will be composed of skin cells and not of brain or bone cells. This tells us that the responding cell is already determined for a particular differentiation. Since tumors develop only in tissues capable of mitosis, it would appear that the target cells in carcinogenesis are the normal stem cells and their partially differentiated progeny.

The cell of origin has been identified in only one spontaneously developing tumor. Stevens has demonstrated that testicular teratocarcinomas of mice are derived from primordial germ cells. These tumors are often present at birth and, by extrapolation, Stevens calculated that the carcinogenic event must have occurred on or about the 12th day of fetal life [18]. He dissected genital ridges from mouse embryos of 12 days gestational age and transplanted them beneath the testicular capsule of adult mice. Eighty per cent of the resultant fetal testes contained foci of embryonal carcinoma cells, recognizable 7 days after transplantation. They quickly developed into teratocarcinomas. When a gene was incorporated into strain 129 animals for the absence of germ cells, the incidence of teratocarcinomas in fetal testes approached 0 [19].

In a correlated light and ultrastructural study of transplanted genital ridges, a close resemblance in terms of differentiation was found between embryonal carcinoma cells and the primordial germ cells from which they originated [16]. Each type was undifferentiated, with a few stunted profiles of rough endoplasmic reticulum and atrophic Golgi complexes. The cytoplasm had numerous free polysomes arranged in configurations compatible with the synthesis of cell cytoplasm.

The observation that the cell of origin for a highly malignant tumor might be an undifferentiated cell raised the possibility that other tumors might also have their origin from undifferentiated cells. Franks and Wilson compared the ultrastructure of normal fibroblasts with ones transformed by oncogenic viruses [4]. They had comparable degrees of differentiation.

It was decided to compare the differentiation of normal stem cells with their respective adenocarcinomas to determine if a pool of undifferentiated cells existed that could serve as targets in carcinogenesis [15]. Normal stem cells of the breast of mice were identified in simple ducts between pregnancies, and on the 8th day of pregnancy, using autoradiography with the electron microscope. They proved to be equivalent in their degrees of differentiation with the least differentiated cells in mammary adenocarcinomas similarly labeled. Each had sparse elements of rough endoplasmic reticulum, an atrophic Golgi complex and numerous polysomes unattached to membranes. Normal and neoplastic stem cells were each able to divide, and some of their differentiated progeny differentiated into lactogenic cells. Lactogenic cells were always postmitotic.

The situation in the colon is much more complex than in the breast [15]. Mucous, columnar and argentaffin cells have each been identified as stem lines in normal colon by Chang and Leblond [2]. Animals bearing chemically induced adenocarcinomas of the colon were injected with tritium-labeled thymidine, and the ultrastructure of the least differentiated columnar, mucous and argentaffin cells was compared with the least differentiated labeled cells in the adenocarcinoma. The vacuolated and mucous cells of normal colon proved to be no more differentiated than their counterparts in the adenocarcinoma. During these studies, an extremely undifferentiated cell type was identified in the colonic tumor. Its epithelial nature was attested to by the desmosomes which connected it to mucous, columnar, or argentaffin cells. Since all other cell types of the tumor were represented in normal tissue, search was made in normal colonic epithelium for this undifferentiated cell. Identical cells were found lying along the basement membrane in normal colonic crypts. These were not labeled in normal colon, although they were not infrequently found to be synthesizing DNA in the tumors. The conclusion was reached that these colonic cells are the counterpart of hemocytoblasts in the marrow. In other words, it would appear that if a tissue has more than one differentiated terminal cell type, each cell type will have its own stem line, which in turn will differentiate from an undifferentiated cytoblast. If this hypothesis is true, then it should be possible to clone pure mucous or columnar or argentaffin cell lines from the adenocarcinoma of the colon. These experiments are in progress, and to date we have cloned two mucous tumor lines. Finally, if one of the undifferentiated colonic cells is cloned, it should give rise to a tumor containing all of the other cell types. The answer to this proposition will be known shortly.

ORIGIN OF BENIGN AND MALIGNANT TUMORS

The relationship of benign and malignant tumors is controversial. Students of human pathology do not believe them to be related, but many experimentalists have observed that application of chemical carcinogen results first in the appearance of benign tumors, followed later by the appearance of malignant ones. These observations are consistent with the idea that benign tumors are a stage in the development of malignant ones, but there is an alternative explanation [14]. Since the same carcinogen can give rise to benign or malignant tumors, it is conceivable that each develops independently. The state of differentiation of the reacting cells probably determines whether or not the resulting neoplasm will be benign or malignant; the environment would be responsible for the appearance of benign tumors before malignant ones. This idea is illustrated in Figure 1. If the oncogenic event involves an almost terminally differentiated cell, that cell would be converted to the stem cell of a benign tumor. This cell would not only closely resemble its normal counterpart morphologically, but it would be so little altered that it would find in the tissue of origin all of the ingredients for phenotypic expression and maximal rate of growth. Thus, it would appear first. If, on the other hand, the same carcinogenic stimulus involved a primitive stem cell, that stem cell should be converted to a malignant stem cell, which again would look very much like the undifferentiated stem cell. These cells are so altered in relationship to their normal counterparts that they could be considered as "foreign" in the tissue

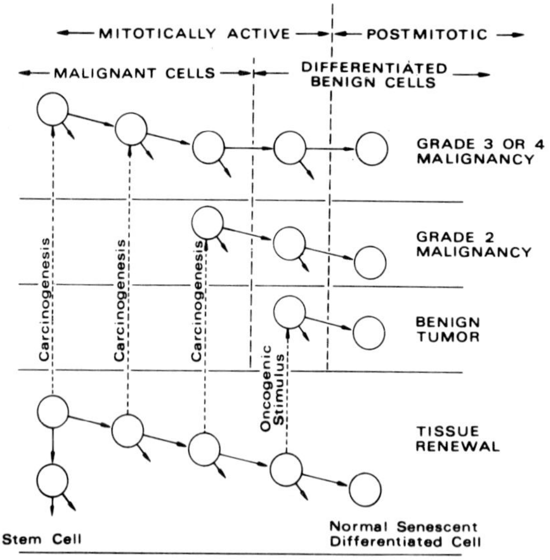

FIG. 1. Scheme of oncogenesis based on tissue renewal as it might occur in gut, breast, skin, upper respiratory tract, etc. Normal tissue renewal is at the bottom of the figure. A variety of cells in various stages of differentiation are usually involved in the carcinogenic event. If a well differentiated cell responds, a benign tumor results. If the stem cell responds, the tumor will be a caricature of normal tissue renewal. If all of the normal cells are transformed, the tumor will be composed of a heterogeneous collection of cells and the characteristics of the tumor will change with time as selection occurs.

of origin. They would reproduce slowly, some of their progeny would differentiate, and a prolonged period of time is required for them to obtain a critical mass and express their phenotype.

This scheme raises several interesting questions. Is there any evidence to support the idea that a critical number of cells is necessary for phenotypic expression? The experiments of Grobstein and Zwilling relate to this problem [7]. They showed in developing systems that a threshold number of cells is required for a differentiation to occur. In addition, Fisher and Fisher have demonstrated that injection of small numbers of Walker carcinosarcoma cells into the portal circulation does not produce rapidly growing tumors in the liver; rather a situation analogous to dormancy prevails [3]. The cells survive in the liver, and do not express their malignant phenotype until nonspecifically stimulated. When so stimulated, a threshold number of cells quickly accumulates and the malignant phenotype is expressed.

Further support for the idea that threshold numbers of cells are required for a differentiation to occur is to be had from the experiments of Mintz et al. [11]. When a few embryonal carcinoma cells were placed in a blastocyst, the cells were successfully incorporated in the developing embryo and an allophenic mouse resulted. On the other hand, when Papaioannou et al. placed larger numbers of embryonal carcinoma cells in the blastocyst, allophenic mice were produced that had nodules of teratocarcinoma [12]. Thus it can be concluded that environment can overcome bias of small numbers of maligant cells, but not large numbers of them. Threshold numbers of cells are required for a differentiation to occur and when that number is achieved, they create their own optimal microenvironment which is less susceptible to modulation.

PROGRESSION AND AUTONOMY

The concept of progression states that tumors gain or lose unit characters independently with time. With repeated transplantation, tumors may lose differentiated features, gain the ability to metastasize and acquire a faster rate of growth. Each of these is an example of gain or loss of a unit character. Greene demonstrated that very small tumors that had not metastasized would not transplant successfully to the anterior chamber of a heterologous host [6]. If allowed to develop until metastasis had occurred, these tumors could be transplanted successfully. The behavior of the tumor was considered to reflect the behavior of individual cells, and it was supposed that sequential changes had occurred in these cells, leading to the autonomous state. This is the point of view accepted by most oncologists, but it is worth pointing out that alternative explanations bear on this point.

If a spectrum of cells, some exceedingly undifferentiated and some well differentiated, are involved in the carcinogenic event, it would be anticipated that the initial tumor would be composed of a disproportionately large number of slowly growing tumor cells in relationship to the fast growing ones. With time, and particularly with repeated transplantation, selection would occur for the fast growing elements, which in turn would change the biologic nature of the tumor.

If this idea were true, it should be possible to disassemble a tumor and find multiple lines, each representing various stages of differentiation. This has been done with melanomas [5]. Multiple lines have been obtained from the same tumor, some exceedingly well differentiated with production of large amounts of melanin and a slow rate of growth, and, at the other end of the spectrum, amelanotic and rapidly growing lines. With repeated selection, as in the experiments in progression, the original pigmented tumors always become amelanotic and fast growing. This indicates that it is possible to segregate the component stem lines from these tumors. On the other hand, in nature, segregation occurs so slowly that the conclusion is reached that the cells of a tumor must exert a marked influence on each other. This is most dramatically seen in adenocarcinomas of the breast in which stem cells arrested in G-0 have been discovered. The principal impact of this discovery has been a concern with chemotherapy. As a biologist, it would appear that the important impact should be the discovery of the environmental factors which preclude these G-0 malignant stem cells from expressing the malignant phenotype. Discovery of these controls could conceivably lead to alternatives to cytotoxic chemotherapy.

SUMMARY

When tumors are viewed from a developmental standpoint, they begin to make sense. The carcinomas studied are all caricatures of the process of cell and tissue renewal. Although proved in only one instance, malignant stem cells appear to take origin in normal, undifferentiated stem cells. Normal stem cells of breast and colon are as undifferentiated as their malignant counterparts, and they could serve as targets in carcinogenesis. This obviates the notion of dedifferentiation to explain the undifferentiated appearance of tumors. What has been considered to be dedifferentiation is really differentiation.

Carcinomas are composed of heterogeneous collections of malignant stem cells and their partially and terminally differentiated progeny. Recognition of this complexity of cellular composition has implications for interpretation of immunologic, biochemical and ultrastructural evidence pertaining to tumors. Only the stem cells of the carcinomas studied are malignant.

Finally, the role of environment in modulating phenotypic expression is stressed. Deficiencies in our knowledge in this area are extremely serious. If we understood the mechanisms controlling malignant proliferation in the case of latent, dormant, or stem cells arrested in G-0, we would have an alternative to cytotoxic chemotherapy.

REFERENCES

1. Boveri T: The Origin of Malignant Tumors. Baltimore: Williams & Wilkins Co, 1929. (English translation of "Zur Frage der Entstehung Maligner Tumoren". Jena, G Fisher, 1914.)
2. Chang G, Leblond CP: Renewal of the epithelium in the descending colon of the mouse. Am J Anat 131: 73–99, 1971.
3. Fisher B, Fisher E: Experimental evidence in support of the dormant tumor cell. Science 130: 918–919, 1959.
4. Franks LM, Wilson PD: "Spontaneous" neoplastic transformation *in vitro:* The ultrastructure of the tissue culture cell. Eur J Cancer 6: 517–523, 1970.
5. Gray JM, Pierce GB: Relationship between growth rate and differentiation of melanoma *in vivo.* J Natl Cancer Inst 32: 1201–1210, 1964.
6. Greene HSN: A conception of tumor autonomy based on transplantation studies. Cancer Res 11: 899–903, 1951.
7. Grobstein C, Zwilling E: Modification of growth and differentiation of chorioallantoic grafts of chick blastoderm pieces after cultivation at a glass-clot interface. J Exp Zool 122: 259–284, 1953.
8. Kleinsmith LJ, Pierce GB: Multipotentiality of single embryonal carcinoma cells. Cancer Res 24: 1544–1551, 1964.
9. Makino S, Makino S: Further evidence favoring the concept of the stem cell in ascites tumors of rats. Ann NY Acad Sci 63: 818–830, 1956.
10. Mendelsohn ML: Autoradiographic analysis of cell proliferation in spontaneous breast cancer of C_3H mouse. J Natl Cancer Inst 28: 1015–1029, 1962.
11. Mintz B, Illmensee K, Gearhart JD: Developmental and experimental potentialities of mouse teratocarcinoma cells from embryoid body cores. *In:* Sherman MI, Solter D (Eds): Teratomas and Differentiation. New York, Academic Press, 1975, pp 59–82.
12. Papaioannou VE, McBurney MW, Gardner RL: Fate of teratocarcinoma cells injected into early mouse embryos. Nature 258: 70–73, 1975.
13. Pierce GB: Neoplasms, differentiation and mutations. Am J Path 77: 103–118, 1974.
14. Pierce GB, Dixon FJ, Verney EL: Teratocarcinogenic and tissue forming potentials of the cell types comprising neoplastic embryoid bodies. Lab Invest 9: 583–602, 1960.
15. Pierce GB, Nakane PK, Martinez-Hernandez A et al: Ultrastructural comparison of differentiation of stem cells of adenocarcinomas of colon and breast with their normal counterparts. J Natl Cancer Inst. Submitted 1976.
16. Pierce GB, Stevens LC, Nakane PK: Ultrastructural analysis of the early development of teratocarcinomas. J Natl Cancer Inst 39: 755–773, 1967.
17. Pierce GB, Wallace C: Differentiation of malignant to benign cells. Cancer Res 31: 127–134, 1971.

18. Stevens LC: Testicular teratomas in fetal mice. J Natl Cancer Inst 28: 247–256, 1962.
19. Stevens LC: Origin of testicular teratomas from primordial germ cells in mice. J Natl Cancer Inst 38: 549–552, 1967.
20. Stevens LC, Little CC: Spontaneous testicular teratomas in an inbred strain of mice. Proc Natl Acad Sci USA 40: 1080–1087, 1954.
21. Wylie CV, Nakane PK, Pierce GB: Degrees of differentiation in nonproliferating cells of mammary carcinoma. Differentiation 1: 11–20, 1973.

Malignancy vs. Normal Differentiation of Stem Cells as Analyzed in Genetically Mosaic Animals

Beatrice Mintz, Ph.D.

There has been increasingly expressed, in recent years, the view that neoplasia may be essentially a disease or derangement of cell differentiation and not merely of cell growth. This suspicion has been fed by a variety of clues: Some tumors express proteins ordinarily found only in fetal stages of normal tissues; tumors of certain tissues may display a "confusion" of tissue identity and may form products usually associated with other tissues; and, occasionally, there are tumors whose histologic composition gradually shifts to a more benign or differentiated state.

While this recent emphasis on developmental aspects of tumorigenesis has led to new insights into the problem, it has also unnecessarily engendered a false dichotomy, i.e., the notion that a genetic (or mutational) and an epigenetic (or nonmutational) hypothesis of neoplastic conversion must inevitably be juxtaposed. There appears to be good evidence, of a statistical nature, that some cancers, studied in the human population, originate as a direct result of mutational events [4], and there is also evidence that chemical carcinogens often act as mutagens [5]. On the other hand, there is now an unequivocal demonstration that at least one cancer, a malignant mouse teratocarcinoma, originates without mutation [9,10], as a result of cells finding themselves in an abnormal or inappropriate environment; and one might reasonably hypothesize that at least some, though not all, other cancers might originate comparably. In *both* the initially mutational and the initially nonmutational neoplasms, aberrations of differentiation appear to be critically implicated.

Laboratory animals—particularly mice—in which genetically dissimilar cells have been made experimentally to coexist throughout development [8] have provided novel ways in which various questions concerning the developmental origins of malignancy have been analyzed *in vivo*. Many fea-

From the Institute for Cancer Research, Fox Chase, Philadelphia, Pa.

These investigations were supported by United States Public Health Service Grants Nos. HD-01646, CA-06927, and RR-05539, and by an appropriation from the Commonwealth of Pennsylvania.

tures of tumorigenesis *in vivo* cannot be simulated *in vitro,* and the framework of the entire organism is indispensable for a full understanding of malignancy.

Studies of mammary tumor formation in such allophenic mice demonstrate that genes responsible for hereditary susceptibility to these tumors are expressed only, or chiefly, in the potentially tumorous cells of the mammary gland: When the animals possess both susceptible-strain and nonsusceptible-strain cells closely intermingled throughout development, the tumors develop autonomously in the susceptible-strain cells [7]. Even if the immune system in a particular cellular mosaic animal happens to be of the genetically nonsusceptible strain, it does not provide protection against mammary tumor formation, hence does not exercise an effective surveillance role.

In the early stages of mammary tumorigenesis, premalignant hyperplastic nodules are found which, in allophenic mice, often contain both cells of the susceptible and the nonsusceptible genotypes. However, transplantation of such nodules to a recipient of the nonsusceptible strain results in selective destruction of the (histoincompatible) susceptible-strain cells and a "liberation" of the previously entrapped normal cells, which are then able to resume completely normal growth [6]. Further examination of this phenomenon has led to the conclusion that the companionship of normal, untransformed cells may, in the early stages of tumorigenesis, be indispensable for the survival and growth of the transformed ones.

The hypothesis of mutational origin of a tumor carries with it the expectation that a tumor should generally arise clonally from a single mutated cell. Women heterozygous for electrophoretically distinguishable forms of the X-linked enzyme glucose-6-phosphate dehydrogenase have only one enzymatic type expressed in any somatic cell. Therefore, tumors from heterozygous patients have been used to examine this proposition. From such studies, many kinds of tumors appear to be monoclonal in origin [2]. However, when a more sensitive method was used in allophenic mice in which tumorous liver parenchyma cells were individually distinguishable histochemically by means of a strain-specific quantitative difference in β-glucuronidase activitity, it became apparent that some hepatomas contained cells of two different genetic strains [1]. Thus, the possibility remains open, at least for this kind of tumor, that more than one cell may have been transformed and that mutational change may not have been initially involved. While numerous studies of tumors have shown chromosome changes and aneuploidy, it remains unclear to what extent these changes were primary or causal, as opposed to secondary, in tumor progression.

While malignancy *in vivo* has often been referred to as "dedifferentiation," there appears to be no sound basis for that assertion, which would leave unexplained the orderly hierarchy of intermediate levels of differentiation in some tumors or the presence of normally differentiated nondividing cells. A more reasonable view, consistent with the facts, is that neoplasia is a failure in the normal progression of differentiation, due either to a mutational event in an initially normal stem cell, or to a nonmutational stem-cell change in gene expression as a consequence of a change in the cellular microenvironment [11]. According to this hypothesis, a chemical carcinogen, or an oncogenic virus, might cause a mutational change in certain specific instances or, in others, might change the milieu so that cellular gene expression is modified. Pierce [12] has earlier pointed to the normal stem-cell-renewal system as an appropriate model in which the normal stem cell is the target cell that becomes transformed in carcinogenesis, though he has generally favored a non-mutational mechanism of inception of malignancy.

As diagrammed elsewhere [11], the stem cell in a normally differentiating clone has a "proliferative option" as well as a "differentiative option"; in a transformed clone, the proliferative option increases and the differentiative option decreases. The extent to which differentiation is reduced will determine whether the tumor will ultimately be malignant (relatively undifferentiated or anaplastic, with dividing stem cells) or benign (well-differentiated, with many terminal nondividing cells). In tumor progression, increasing selection for more rapidly proliferating cells may promote increasing malignancy. In tumor regression, while the same stem cell may initially have been the target, the transformed stem cells may be lost and replaced by their differentiated nondividing mitotic progeny. We know very little about normal stem cells, and they have been recognized in only a few tissues. Future progress in understanding normal differentiation as well as malignancy will require more intensive studies of stem cells in all tissues.

A genetic change leading to malignancy would usually be irreversible; malignancy resulting from a nongenetic alteration might be reversible to normalcy, provided cascades of further changes had not supervened. The mouse teratocarcinomas experimentally produced by Stevens [14] seemed to be possible nonmutational tumors of the most developmentally primitive kind of stem cell because they generated many (but not all) kinds of tissues in the chaotically arranged solid tumors. In addition, we found our tumor of choice to be still euploid, despite eight years' elapsed time during which the tumor had been transplanted. Reversibility to normalcy was tested by introducing the embryonal carcinoma stem cells into recipient blastocysts bearing many genetic markers. In this way, virtually all the tissues of any mouse arising from such an experiment could be analyzed for their blastocyst vs. tumor-cell origin [3,10]. The early embryo environment was chosen for potential "normalization" of the tumor stem cells because it is only here that a normal embryo cell is able to embark on full-spectrum differentiation.

The results of this experiment were dramatic and decisive. Normal healthy mice were obtained in which tumor-strain contributions were present to varying degrees. In the most successful cases, there were significant contributions in all somatic tissues, including some tissues (e.g., liver, thymus, kidney, lung) never seen in the solid tumors that form in transplant hosts. The tissues apparently functioned quite normally and synthesized their specific products (e.g., immunoglobulins, hemoglobin, liver proteins) coded for by strain-type alleles at known loci. In addition, a tumor-contributed color gene, *steel,* not previously known to be present in the carcinoma cells, was detected from the coat phenotype. Cells derived from this carcinoma, which is of X/Y sex chromosome type, also gave rise to normal functional sperms, some of which transmitted the *steel* gene to the progeny [10]. These diverse tissues were obtained even when only a single tumor cell was injected into each blastocyst [3]. In only a small minority of cases did the animals have teratomas; these were not re-reversals, but instances of nonintegration of the injected cells, which continued to grow as a tumor in a culture chamber. Thus, after almost 200 transplant generations as a highly malignant tumor, these teratocarcinoma stem cells were totipotent and reversible to complete normalcy. They were even able to express, in an orderly sequence, many genes previously silent in the tumor of origin.

These results furnish the first unambiguous example in animals of a nonmutational basis for malignancy and of reversal to normalcy. The question is now open as to whether some malignancies of more specialized tissue stem cells might also be nonmutational in origin. This remains to be tested, by introducing the stem cells (if still fairly euploid) into appropriate embryonic hosts.

In planning to undertake the experiments involving teratocarcinoma cell normalization, I had in mind an additional purpose [11]. For some time, it had seemed to me that mammalian developmental genetics was handicapped by the limited numbers of available mutant genes whose products or effects had been biochemically defined. Such genes are potentially the most powerful tools for unraveling the secrets of regulation of differentiation in higher organisms, as they were for probing the regulatory systems of bacteria. In the bacterial work, mutagenized cultures were subjected to known selective screens. Comparable procedures have been used with nondifferentiating mammalian cultured cell lines [13]. In work at the more complex levels, mutant mice or embryos are generally recognized only by gross morphologic, behavioral, or other deviations, and the biochemical lesion has yet to be defined in most cases. It would therefore be highly desirable to be able to mutagenize totipotent mammalian cells *in vitro,* to select for clones with specific biochemical alterations, and then, in effect, to turn these totipotent cells into mice. By introducing the mutagenized cells into blastocysts, the resultant cellular mosaic situation would even permit many lethal mutants to be rescued. Somatic tissues of various kinds and germ cells would differentiate *in vivo*. The mutation could be mapped by recombination in the germ line during meiosis, and mammalian gene complexes and genetic systems increasingly defined. Mutations affecting enzymes, antigens, histocompatibility or immunoglobulin molecules, surface properties, and many other phenotypes may be expected. Another, particularly rewarding, prospect is that mouse models of human genetic diseases may now become feasible for many kinds of heritable defects. Genetic diseases in man often comprise a complex clinical syndromy made all the more difficult to analyze by virtue of the succession of indirect effects generated during development. With experimentally accessible mouse models, "tailor-made" for specific diseases, the prospect of deciphering the complexities, and even of devising treatments for these ailments, becomes more realistic [11]. The proof that mouse teratocarcinoma stem cells are completely reversible to normal full-scale differentiation now opens the door to this kind of mutational approach to the analysis of differentiation and disease in higher organisms.

REFERENCES

1. Condamine H, Custer RP, Mintz B: Pure-strain and genetically mosaic liver tumors histochemically identified with the β-glucuronidase marker in allophenic mice. Proc Natl Acad Sci USA 68: 2032–2036, 1971.
2. Friedman JM, Fialkow PJ: Cell marker studies of human tumorigenesis. Transplant Rev 28: 17–33, 1976.
3. Illmensee K, Mintz B: Topipotency and normal differentiation of single teratocarcinoma cells cloned by injection into blastocysts. Proc Natl Acad Sci USA 73: 549–553, 1976.
4. Knudson A: Mutation and human cancer. Adv Cancer Res 17: 317–352, 1973.
5. McCann J, Ames BN: Detection of carcinogens as mutagens in the *Salmonella*/ microsome test: Assay of 300 chemicals: Discussion. Proc Natl Acad Sci USA 73: 950–954, 1976.
6. Mintz B, Slemmer G: Gene control of neoplasia. I. Genotypic mosaicism in normal and preneoplastic mammary glands of allophenic mice. J Natl Cancer Inst 43: 87–95, 1969.
7. Mintz B: Neoplasia and gene activity in allophenic mice. *In:* Genetic Concepts and Neoplasia, 23rd Annual Symp on Fundamental Cancer Res, M. D. Anderson Hospital and Tumor Institute. Baltimore, Williams & Wilkins, 1970, pp 477–517.
8. Mintz B: Allophenic mice of multi-embryo orign. *In:* Daniel J Jr (Ed): Methods in Mammalian Embryology. San Francisco, W. H. Freeman, 1971, 186–214.
9. Mintz B, Illmensee K, Gearhart JD: Developmental and experimental potentialities of mouse teratocarcinoma cells from embryoid body cores. *In:* Sherman MI, Solter D (Eds): Teratomas and Differentiation. New York, Academic Press, 1975,pp 59–82.
10. Mintz B, Illmensee K: Normal genetically mosaic mice produced from malignant teratocarcinoma cells. Proc Natl Acad Sci USA 72: 3585–3589, 1975.
11. Mintz B: (1976). Gene expression in neoplasia and differentiation. Harvey Society Lectures. In press.
12. Pierce GB Jr: Teratocarcinoma: Model for a developmental concept of cancer. Curr Top Develop Biol 2: 223–246, 1967.
13. Puck TT: Biochemical and genetic studies on mammalian cells. In Vitro 7: 115–119, 1971.
14. Stevens LC: The biology of teratomas. Adv. Morphog 6: 1–31, 1967.

Theory of Mass Transport and Ligand Binding for Macromolecular Interactions Induced by Small Molecules

John R. Cann, Ph.D.

Our interest in the mass transport of interacting macromolecules stemmed from observations made already two decades ago that a variety of highly purified proteins display bimodal, moving-boundary electrophoretic patterns in acidic media containing acetate or other carboxylic acid buffers. At first we though that the two peaks in the electrophoretic pattern might correspond to different isomeric forms of the protein; and a theory of electrophoresis was formulated for kinetically controlled macromolecular isomerization,

$$A \underset{k_2}{\overset{k_1}{\rightleftharpoons}} B \qquad \text{(I)}$$

where k_1 and k_2 are the specific rates of interconversion between two isomers with different electrophoretic mobilities. These early calculations [7,14] predicted that the electrophoretic, sedimentation and chromatographic patterns of such a system can show one, two or three peaks, depending upon the half-times of reaction relative to the duration of the transport experiment. Slow interconversion gives two peaks corresponding to the separated isomers; rapid interconversion gives one peak with weight average mobility; while intermediate rates can give three peaks. This is so for both moving-boundary and zonal modes of operation. Concurrent with this theoretical work, however, experimental evidence was obtained [2] to support interpretation of the patterns shown in acetate buffer in terms of rapid and reversible complexing of the protein with undissociated buffer acid, with concomitant increase in the net change on the protein molecule without significant change in its frictional coefficient. Subsequently, it was shown that the change in net positive charge on the protein is most probably

From the Department of Biophysics and Genetics, University of Colorado Medical Center, Denver, Colo.

This work was supported in part by Research Grant 5R01 HL13909-25 from the National Heart and Lung Institute, National Institutes of Health, United States Public Health Service. This publication is No. 681 from the Department of Biophysics and Genetics, University of Colorado Medical Center, Denver, Colo.

due to binding of the undissociated buffer acid to side-chain carboxyl groups with accompanying increase in their pK [3]. It was also found that under appropriate conditions of pH and buffer concentration, interaction of aliphatic acids with ribonuclease A is a cooperative phenomenon associated with a change in tertiary structure without substantial alteration in secondary structure [4], but that binding of undissociated acid to a protein need not necessarily be cooperative for resolution of bimodal reaction boundaries or zones [3]. Resolution of the electrophoretic pattern into two peaks was unexpected by many, since it had been intuitively held that such an interaction would give a single peak. Thus, extension of the theory of electrophoresis to include this new class of macromolecule-small molecule interactions became crucial for umambiguous interpretation, not only of our own experimental observations, but also of the results of more conventional applications of electrophoresis to biologic problems.

At this juncture Ted Puck introduced me to Walter Goad and a long and fruitful collaboration on the theory of electrophoresis and sedimentation of interacting macromolecules was launched. This is but one of many instances illustrating how Ted's scientific acumen, breadth of interest and intellectual generosity created the stimulating environment enjoyed by the Department of Biophysics and Genetics.

Collaboration with Walter Goad soon provided [8] a theoretical basis for interpretation of the electrophoretic patterns. A theory of electrophoresis was formulated for reversibly interacting systems of the type

$$P + nHA \rightleftharpoons P(HA)n \qquad \text{(II)}$$

where P represents a protein molecule or other macromolecular ion in solution and P(HA)n its complex formed by binding n moles of a small, uncharged constituent, HA, of the solvent medium, e.g., undissociated buffer acid. It is assumed that P and P(HA)n possess different electrophoretic mobilities and that equilibrium is established instantaneously. These calculations account for the essential features of the moving-boundary electrophoretic behavior of proteins in acidic media containing varying concentration of carboxylic acid buffer.

Upon realization that a single macromolecule can give moving-boundary patterns showing two peaks, the computations were extended to zone electrophoresis. As evidenced by Fig. 4 of reference [9], zone electrophoretic patterns can also show two well resolved and stable peaks arising from interaction of a single macromolecule with a small neutral molecule in accordance with Reaction II. This, despite instantaneous reestablishment of equilibrium during differential transport of P and P(HA)n. Resolution occurs because of changes in the concentration of HA accompanying reequilibration and maintenance of the resulting concentration gradients of the electrically neutral molecule along the electrophoresis column, the peaks in the pattern corresponding to different equilibrium compositions and not simply to P and P(HA)n. Although the protein concentration eventually becomes very small between the peaks, it never vanishes. In other words, the patterns are bimodal reaction zones, but in practical terms it may be said that a single macromolecule can give two zones due to reversible and rapid interaction with the buffer medium.

This important prediction was verified experimentally [1], and it now appears that multiplicity of zones due to interaction is of rather frequent occurrence. However, with due precaution [5], fractionation can distinguish between interaction and heterogeneity. The protein eluted from each unstained zone is subjected to zone electrophoresis in the same buffer that was used for the original separation. For interactions, the fractions will behave like the unfractionated material and show multiple zones, while for heterogeneity a single zone will be obtained. In the many conventional analytical and preparative applications of zone electrophoresis it is obviously desirable to avoid conditions conducive to the generation of transport patterns which do not faithfully reflect the inherent state of homogeneity of the protein. Multizoning due to interaction with constituents of the solvent medium may often be avoided by judicious choice of buffer [5].

The above electrophoretic considerations also apply to ultracentrifugation. Thus, it is conceivable that binding of small molecules or ions by a protein might cause a significant change in its sedimentation coefficient due to a change in macromolecular conformation. Given appropriate conditions, such a system can, in principle, show a bimodal reaction zone upon sedimentation through a density gradient containing the ligand (Fig. 5 in reference [6]). A more interesting possibility is that binding of a small molecule mediates association or dissociation of the protein, and we have developed the theory of sedimentation for ligand-mediated association-dissociation reactions [2,6,10,11,13,17].

While analytical sedimentation patterns have been calculated for both dimerization and tetramerization, the results for dimerization are, in certain respects, the more revealing with regard to fundamental principles. Two different mechanisms of reversible, ligand-mediated dimerization have been examined:

$$2M + nX \rightleftharpoons M_2X_n \qquad (n = 1 \text{ or } 6) \tag{III}$$

and

$$M + X \rightleftharpoons MX, \qquad K_1 \tag{IVa}$$

$$M + MX \rightleftharpoons M_2X, \qquad K_a \tag{IVb}$$

in which M is the monomeric form of the macromolecule, and X is the small ligand molecule or ion. Both mechanisms can give sedimentation patterns showing two well resolved peaks of large molecule despite instantaneous establishment of equilibrium (Figs. 83–86 in reference [2] and Fig. 3A in reference [13]). As in the case of electrophoresis, resolution of the reaction boundary into two peaks is dependent upon generation of stable concentration gradients of unbound ligand across the sedimenting boundary by reequilibration during differential transport of monomer and dimer. In general, the areas under the two peaks do not faithfully reflect the initial equilibrium composition. Nor do the peaks migrate with velocities per unit centrifugal field commensurate with the sedimentation coefficients of monomer and dimer. Moreover, these quantities cannot be determined by extrapolation of the velocities per unit field to infinite dilution of macromolecule at constant ligand concentration. The conditions for such sedimentation behavior require a sufficiently strong interaction of ligand with macromolecule to generate the gradients of ligand upon which resolution depends. These conditions are satisfied by two systems which provide experimental verification of the theoretical predictions; namely, the reversible dimerization of hexameric New England lobster hemocyanin to a dodecamer mediated by the binding of Ca^{++} and H^+ [19,21] and the dimerization of tubulin induced by vinblastine (compare Fig. la-e in reference [18] with Fig. 2 in reference [11]; and Fig. 2A in reference [18] with Fig. 84c in reference [2]).

For weak interactions which require a large excess of ligand over macromolecule for dimerization, the concentration of unbound ligand along the centrifuge cell cannot be significantly perturbed by reequilibration during the sedimentation process; and the system effectively collapses to the nonmediated dimerization reaction

$$2M \rightleftharpoons M_2 \qquad \text{(V)}$$

In this limit, the sedimentation pattern shows a single peak as predicted for simple dimerization by the Gilbert theory [15,16].

As already noted, the described theory is for very rapid chemical equilibration; but calculations have also been made for kinetic formulations of Reaction Set IV and Reaction V. In the case of Reaction Set IV, it is assumed that ligand-binding per se (Reaction IVa) equilibrates rapidly, so that the monomer-dimer interconversion (Reaction IVb) becomes rate-limiting. This treatment revealed that conclusions concerning the sedimentation behavior of ligand-mediated interactions in the limit of instantaneous establishment of equilibrium are also valid for kinetically-controlled reactions characterized by half-times as long as 20–60 sec.

Because zone sedimentation is so extensively employed in biologic research, we extended our calculations to encompass the zonal behavior of a wide variety of reversible, rapidly equilibrating ligand-mediated interactions [6,10]. These include macromolecular association and dissociation involving dimer, trimer and/or tetramer in sequential as well as single-step reaction schemes which are either noncooperative or cooperative in ligand and/or macromolecule. Aside from a particular tetramerization reaction highly cooperative with respect to both ligand and macromolecule, each of these interactions can, under appropriate conditions, give zone patterns showing two peaks (e.g., Fig. 4 in reference [6] and Fig. 2 in reference [10]). Thus, with only a single exception, the generalization seems justified that ligand-mediated interactions in general have the potentiality for showing bimodal reaction zones and boundaries irrespective of reaction mechanism. Accordingly, it is imperative that subsequent to fractionation experiments designed to distinguish between interaction and heterogeneity [12], appeal be made to the combined application of at least two independent biophysical methods in order to elucidate the mechanism of the interaction. The same applies to gel-permeation chromatography (see Fig. 6 in reference [6]).

The foregoing discussion has focused on ligand-mediated macromolecular interactions. In the ligand-mediated mechanism of dimerization, obligatory binding of ligand to the macromonomer must precede formation of the dimer, as illustrated by Reaction Set IV and the reaction scheme

$$M + X \rightleftharpoons MX, \qquad K_1 \qquad \text{(VIa)}$$

$$2MX \rightleftharpoons M_2X_2, \qquad K_a \qquad \text{(VIb)}$$

Actually, macromolecular interactions induced by small molecules may in principle proceed by either the ligand-mediated mechanism or another extreme mechanism, termed the ligand-facilitated mechanism. In the ligand-facilitated mechanism of dimerization, two monomeric molecules associate to a dimer, which is then stabilized by the binding of ligand as illustrated by the scheme

$$2M \rightleftharpoons M_2, \qquad K_a \qquad \text{(VIIa)}$$

$$M_2 + X \rightleftharpoons M_2X, \qquad 2\,k_o \qquad \text{(VIIb)}$$

$$M_2X + X \rightleftharpoons M_2X_2, \qquad 1/2\,k_o \qquad \text{(VIIc)}$$

Recently, comparative transport calculations were made for Reaction Sets VI and VII, the equilibrium constant of the over-all reaction ($2M + 2X \rightleftharpoons M_2X_2$) being the same for both. The results indicate that the ligand-mediated and ligand-facilitated mechanisms cannot be distinguished by either analytical or zone sedimentation. Both mechanisms can give bimodal sedimentation patterns which show only small quantitative differences. Nor are the shapes of the ligand concentration profiles distinctive. Moreover, the changes in shape of the sedimentation patterns with decreasing ligand concentration at constant macromolecule concentration and vice versa show no distinguishing qualitative features. Examination of the concentration profiles of all species present indicates that the transport process is dominated by the liganded dimer, and it makes little difference how it came to be liganded.

In contrast to sedimentation, relaxation kinetic measurements are capable of making some distinctions between mechanisms. Tai and Kegeles [21] have applied relaxation kinetics to the Ca^{++}-hemocyanin system mentioned earlier. Their results are in accord with the predictions for Ca^{++}-mediated dimerization of hexameric hemocyanin, and rule out both Ca^{++}-facilitated dimerization and a "compromise" mechanism in which binding of Ca^{++} to the hexameric molecule is obligatory for its dimerization, while binding to the dodecamer is required for stabilization. These are elegant but technically demanding experiments within the reach of a relatively few laboratories, and current developments in molecular and cellular biology point to the need for a routine method. During the course of our comparative transport calculations on ligand-mediated and ligand-facilitated dimerization it was noted that these mechanisms are distinguishable by the distinctive features of their Scatchard plots for binding of ligand. This observation is being pursued, and a progress report is given below.

Ligand-binding data for several dimerization mechanisms have been generated by computer simulation. The results are expressed as the mean number of moles of ligand bound per mole of constituent macromolecule, ν, at the equilibrium concentration of unbound ligand, [X]. The constituent concentration, $\bar{C}$, is the total concentration of macromolecule in all its forms; e.g., for Reaction Set IV, $\bar{C} = [M] + [MX] + 2\,[M_2X]$ and $\nu = ([MX] + [M_2X])/\bar{C}$. The results are displayed in Figures 1A, 2 and 3 as Scatchard plots [20] of $\nu/[X]$ vs. ν.

It can be shown analytically that the intercept of the Scatchard plot with the ordinate for the individual model systems is: (a) Reaction Set IV, $K_{app} = K_1 + K_1K_a\,\bar{C}$, where the apparent binding constant, K_{app}, of ligand to macromolecule extrapolates to the intrinsic binding constant, K_1, at infinite dilution of macromolecule; (b) Reaction Set VI, K_1; and (c) Reaction Set VII, $K_{app} = k_o\,[(1 + 8\bar{C}K_a)^{1/2} - 1]/\{2 + [(1 + 8\bar{C}K_a)^{1/2} - 1]\}$, which extrapolates to zero at infinite dilution of macromolecule. It can also be shown analytically that for Reaction Set IV, the Scatchard plots are stationary at their midpoint ($\nu = 0.5$) where $1/[X] = K_1$, independent of K_a and $\bar{C}$.

Let us first consider the results for the ligand-mediated mechanism. The Scatchard plots for Reaction Set IV (Fig. 1A) are concave toward the abscissa. This shape is reminiscent of nonassociating systems characterized by heterogeneity of binding sites with respect to their intrinsic affinity for ligand. In contrast to heterogeneity, however, the extent of binding is dependent upon macromolecule concentration, the plots for difference $\bar{C}$ showing the analytically predicted stationary point. These features can be understood as follows: At ligand concentrations less than the midpoint

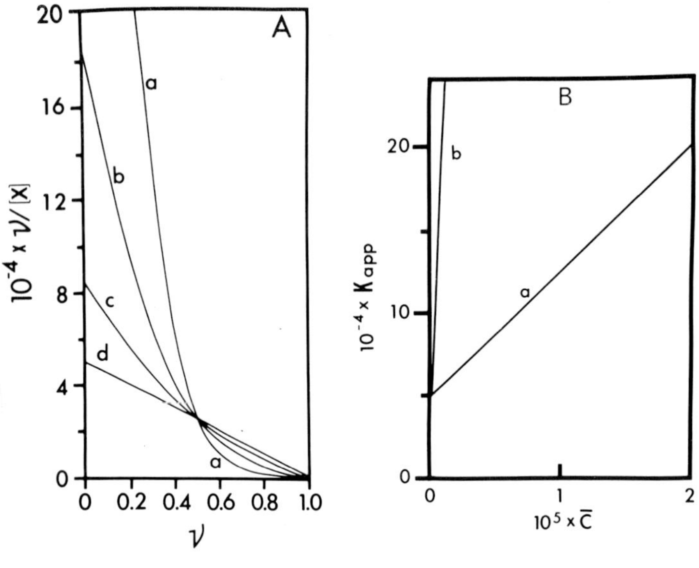

FIG. 1. Theoretical ligand-binding data calculated for the ligand-mediated mechanism of dimerization schematized by Reaction Set IV. A—Scatchard plots for different macromolecule concentrations, $K_1 = 5 \times 10^4 M^{-1}$ and $K_a = 1.5 \times 10^5 M^{-1}$: a, $\bar{C} = 7.27 \times 10^{-5}$M; b, 1.80×10^{-5}; c, 4.55×10^{-6}M; and d, limit of infinite dilution of macromolecule. B—Extrapolation of apparent binding constant given by the intercept of the Scatchard plot with the ordinate to infinite dilution of macromolecule: a, $K_1 = 5 \times 10^4 M^{-1}$ and $K_a = 1.5 \times 10^5 M^{-1}$; b, $K_1 = 5 \times 10^4 M^{-1}$ and $K_a = 4.5 \times 10^6 M^{-1}$.

concentration, dimerization (Reaction IVb) enhances ligand-binding. Thus, for given $\nu < 0.05$, [X] decreases (i.e., ν/[X] increases) with increasing $\bar{C}$. On the other hand, at higher ligand concentrations dimerization actually inhibits binding by removing unliganded monomer from the reaction arena. Moreover, when the strength of binding is the same order of magnitude or weaker than dimerization, very high concentrations of ligand are required to reverse the dimerization reaction via binding to the released unliganded monomer, thereby driving the binding reaction (IVa) to completion. Thus, for given $\nu > 0.5$, ν/[X] decreases with increasing $\bar{C}$. As predicted analytically, the apparent binding constant, as given by the intercept with the oridinate, increases with increasing $\bar{C}$; but extrapolation to infinite dilution of macromolecule (Fig. 1B) erases the role of dimerization in determining the extent of ligand-binding at finite $\bar{C}$ and yields and intrinsic binding constant, K_1.

In contrast to Reaction Set IV, Reaction Set VI is cooperative in the sense that dimerization enhances ligand-binding at all ligand concentrations. This cooperativity is reflected in Scatchard plots which are open downward in shape (Fig. 2). Here, too, the plots are dependent upon macromolecule concentration. For given $\nu < 1$, [X] decreases with increasing $\bar{C}$, so that the plots deviate ever more markedly from that expected for simple binding in the absence of dimerization. The role of dimerization in enhancing ligand-binding is seen with particular clarity in the dependence of the Scatchard plots upon the equilibrium constant for dimerization, K_a, at constant $\bar{C}$. As illustrated in Figure 2B, the greater K_a, the greater is the cooperativity. In agreement with analytical prediction for this reaction scheme, the intercept of the Scatchard plots with the ordinate gives K_1 irrespective of $\bar{C}$ and K_a.

Turning our attention to the ligand-facilitated mechanism, the open-downward shape of the Scatchard plots displayed in Figure 3 for Reaction Set VII reflects the cooperativity conferred by stabilization of the dimer via the binding reaction VIIb. The increase in cooperativity with increasing macromolecule concentration derives from a mass action effect on Reaction VIIa; the higher the concentration of unliganded dimer, the greater the extent of ligand-binding by the system. The most intriguing feature of this system is that, in agreement with the analytical prediction, the apparent binding constant decreases with decreasing macromolecule concentration

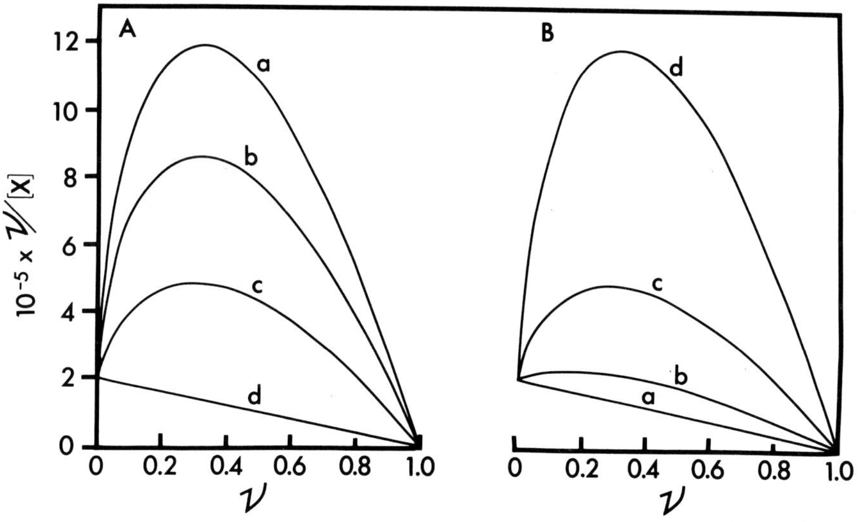

FIG. 2. Theoretical ligand-binding data calculated for the ligand-mediated mechanism of dimerization schematized by Reaction Set VI. A—Scatchard plots for different macromolecule concentrations, $K_1 = 2 \times 10^5 M^{-1}$ and $K_a = 7.5 \times 10^5 M^{-1}$: a, $\bar{C} = 1.4 \times 10^{-4}$M; b, 7.0×10^{-5}M; c, 1.8×10^{-5}M; and d, limit of infinite dilution of macromolecule. B—Scatchard plots for different values of K_a at fixed value of $K_1 = 2 \times 10^5 M^{-1}$ and $\bar{C} = 1.4 \times 10^{-4}$M: a, $K_a = 1 \times 10^2 M^{-1}$; b, $1 \times 10^4 M^{-1}$; c, $1 \times 10^5 M^{-1}$; d, $7.5 \times 10^5 M^{-1}$.

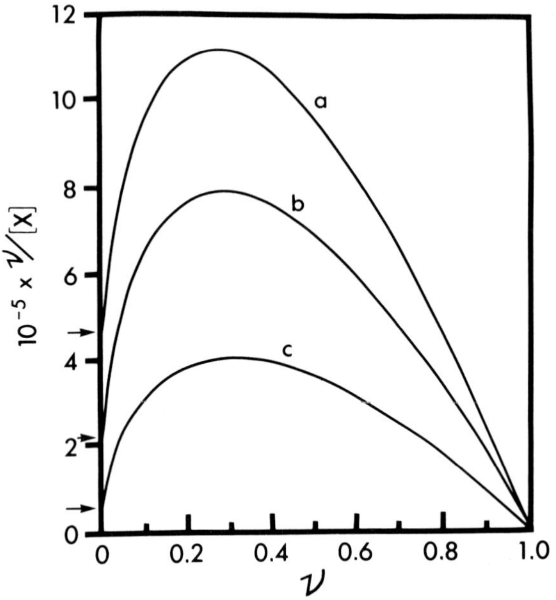

FIG. 3. Theoretical ligand-binding data calculated for the ligand-facilitated mechanism of dimerization schematized by Reaction Set VII. Scatchard plots for different macromolecule concentration, $K_a = 1 \times 10^2 M^{-1}$ and $k_o = 1.732 \times 10^7 M^{-1}$: a, $\bar{C} = 1.4 \times 10^{-4}M$; b, $7.0 \times 10^{-5}M$; c, $1.8 \times 10^{-5}M$. Arrows point to intercepts on the ordinate. Note that the equilibrium constant for the over-all reaction ($2M + 2X \rightleftharpoons M_2X_2$) is the same as in Fig. 2A. A similar picture was obtained for $K_1 = 1 \times 10^2 M^{-1}$ and $k_o = 2 \times 10^5 M^{-1}$.

and extrapolates to zero at infinite dilution of macromolecules. This is because the extrapolation erases the dimerization upon which ligand-binding is dependent at finite $\bar{C}$, which poses the question as to how k_o can be determined. In principle it can be obtained by extrapolation of the apparent binding constant to infinite concentration of macromolecule, but in practice this may be difficult. A more attractive procedure would be nonlinear least-squares fitting of the experimental binding data to the appropriate theoretical equation.

To summarize, the analysis of binding data by means of Scatchard plots can distinguish between the two extreme mechanisms by which ligand-binding may induce dimerization of a macromolecule. Thus, the open-downward shape of the plots for the ligand-facilitated mechanism schematized by Reaction Set VII serves to distinguish it from the ligand-mediated Reaction Set IV; while the dependence of the apparent binding constant upon macromolecule concentration distinguishes it from the ligand-mediated Reaction Set VI. Likewise, the two ligand-mediated schemes are readily distinguishable one from another. It must be pointed out, however, that this treatment ignores possible extraneous binding of ligand. The complications introduced by binding to sites not specifically involved in the dimerization process per se are being examined.

Finally, the new insights provided by the theory of mass transport for macromolecular interactions induced by small molecules have important implications for interpretation of the many analytical and preparative applications of electrophoresis, sedimentation and chromatography in biochemistry and molecular biology. It is to be expected that these advances will also find application in fundamental studies on a variety of biochemical reactions such as protein-drug interactions and the interaction of enzymes with cofactors and allosteric affectors. Likewise, the calculations on ligand-binding by associating macromolecules should prove useful in the analysis of such diverse biologic systems as the interaction of vinblastine and other vinca alkaloids with tubulin and the binding of lectins to cells.

REFERENCES

1. Cann JR: Multiple electrophoretic zones arising from protein-buffer interaction. Biochemistry 5: 1108–1112, 1966.
2. Cann JR: Numerical solution of exact conservation equations. *In:* Cann JR: Interacting Macromolecules. The Theory and Practice of Their Electrophoresis, Ultracentrifugation, and Chromatography. New York, Academic Press, 1970.
3. Cann JR: Interaction of acetic acid with poly-L-glutamic acid and serum albumin. Biochemistry 10: 3707–3712, 1971.
4. Cann JR: Cooperative interaction of aliphatic acids with ribonuclease A. Biochemistry 10: 3713–3722, 1971.
5. Cann JR: Significance of interactions in electrophoretic and chromatographic methods. Methods in Enzymology 25: 157–178, 1972.
6. Cann JR: Theory of zone sedimentation for non-cooperative ligand-mediated interactions. Biophys Chem 1: 1–10, 1973.
7. Cann JR, Bailey HR: Theory of isomerization equlibrium in electrophoresis. II. Arch Biochem Biophys 93: 576–579, 1961.
8. Cann JR, Goad WB: Theory of moving boundary electrophoresis of reversibly interacting systems. J Biol Chem 240: 148–155, 1965.
9. Cann, JR, Goad WB: Theory of zone electrophoresis of reversibly interacting systems. Two zones from a single macromolecule. J Biol Chem 240: 1162–1164, 1965.
10. Cann JR, Goad WB: Bimodal sedimenting zones due to ligand-mediated interactions. Science 170: 441–445, 1970.
11. Cann JR, Goad WB: Theory of sedimentation for ligand-mediated dimerization. Arch Biochem Biophys 153: 603–609, 1972.
12. Cann, JR, Goad WB: Measurements of protein interactions mediated by small molecules using sedimentation velocity. Methods in Enzymology 27: 296–306, 1973.
13. Cann JR, Kegeles G: Theory of sedimentation for kinetically controlled dimerization reactions. Biochemistry 13: 1868–1874, 1974.
14. Cann JR, Kirkwood JG, Brown RA: Theory of isomerization equilibrium in electrophoresis. I. Arch Biochem Biophys 72: 37–41, 1957.
15. Gilbert GA: Discussion. Faraday Soc. 20: 68–71, 1955.

16. Gilbert GA: Sedimentation and electrophoresis of interacting substances. I. Idealized boundary shape for a single substance aggregating reversibly. Proc Roy Soc (London) Ser A 250: 377–388, 1959.
17. Goad WB: Numerical methods. *In:* Cann JR: Interacting Macromolecules. The Theory and Practice of Their Electrophoresis, Ultracentrifugation, and Chromatography. New York, Academic Press, 1970.
18. Lee JC, Harrison D, Timasheff SN: Interaction of vinblastine with calf brain microtubule protein. J Biol Chem 250: 9276–9282, 1975.
19. Morimoto K, Kegeles G: Subunit interactions of lobster hemocyanin I. Ultracentrifuge studies. Arch Biochem Biophys 142: 247–257, 1971.
20. Scatchard G: The attractions of proteins for small molecules and ions. Ann NY Acad Sci 51: 660–672, 1949.
21. Tai MS, Kegeles G: Mechanism of the hexamer-dodecamer reaction of lobster hemocyanin. Biophys Chem 3: 307–315, 1975.

Allosteric Mechanics

Walter B. Goad, Ph.D.

In 1970–1971 I was lucky enough to hear at first hand Max Perutz' developing synthesis of his lifelong work on the structure of hemoglobin, culminating in his stirring and luminous picture of how an allosteric protein actually works [12]—stirring not least because allostery provides one of the keys to freeing the logic of biologic systems from, in Monod's phrase [10], chemical necessity as it governs the nonliving world. I should like to present a formulation of the molecular mechanics and energetics of allosteric linkage that responds directly to Perutz' synthesis. I present it here, to Ted Puck, because it also seems to me to respond to what he has taught so many of us by the force of innumerable examples: to knead and interconnect our scientific understanding until it has the texture and feel of direct experience. Ted does it from a range of reference that extends over almost all of science and that emphasizes science's relation to general human needs; no wonder, then, that he so often reminds one of an Old Testament prophet, speaking for the integrity and wholeness of an entire culture, in this case the culture of science.

INTRODUCTION

In each subunit of hemoglobin, the bond to an O_2 ligand is provided by extensive rearrangement of the electronic structure centered on the iron atom; this includes the bonds that coordinate iron to the heme group and, through an amino acid side chain (the imidazole of a histidine residue), also coordinate it to a fold of the polypeptide chain many residues away from the folds which anchor the heme group directly. Thus this system of bonds has a purchase on two quite distinct regions of the polypeptide chain. Hoard [5] first suggested that in the rearrangement, this electronic system becomes

From the Theoretical Division, University of California, Los Alamos Scientific Laboatory, Los Alamos, N.M.

This work was done under the auspices of the U.S.E.R.D.A.

The author is very grateful to Dr. Max Perutz and to Dr. Francis Crick for the hospitality of the MRC Laboratory of Molecular Biology at Cambridge during 1970–71. He also thanks Mrs. Mary Plehn for expert assistance with the manuscript and Mr. Benjamin Atencio for the drawings.

more compact: four electrons from the iron and two from the ligand, that before binding had parallel spins, are now paired with zero total spin, and although a sufficiently accurate and detailed quantum-mechanical analysis is yet to be made, it is plausible that this reflects primarily the change in Pauli exclusion, which permits low spin systems to be more compact than those of high spin; in atomic spectra this is well understood and gives rise to Hund's rule [8]. In any case, the shortening of the iron bonds necessitates some change in the folding of the protein. The continuing researches of Perutz and his colleagues [13] make it clear that there are other interactions of ligand and globin that may also be important, and I will come back to these. I want first to identify the mechanisms by which the link through the iron system modulates ligand affinity; at the end it will be seen that other interactions are readily incorporated.

My strategy is to extend the thermodynamic distinction between work and heat to the molecular level and to identify the work available from rearrangement of the iron-bond system and its relation to mass-action equilibrium constants. With these tools, plus equilibrium ligand-binding data, I will interpret Perutz' stereochemical mechanism in detail. The key point is that some of the work available is required to adapt the protein structure to new bond lengths and the remainder is available to extract a ligand from solution; the latter determines ligand affinity and, since it depends on the ease of adapting protein structure, this couples ligand affinity at one site with changes elsewhere in the protein—in particular with those brought about by the binding of other ligands.

MOLECULAR THERMODYNAMICS

To a good approximation (the Born-Oppenheimer separation [1]), the electronic system as a whole is at every instant in its quantum ground state as appropriate to the current nuclear configuration. And (according to the Hellmann-Feynman theorem [4]) as the nuclei move about, although changing the electronic ground state and with it both the electron density distribution and the electron kinetic energy, these changes are brought about solely by the coulomb forces between electrons and nuclei; or, put the other way round, the forces on each nucleus are entirely given by the coulomb forces from all other charges in the system. In these terms, attractive Van der Waals forces are owing to a slight increase in the electronic charge density between two nuclei when they are well separated and the quantum dynamics of the electrons is dominated by stronger influences [9]; a covalent bond represents a striking concentration of electron density between two nuclei as a result of these stronger influences; and the strong Fermi repulsion at the close approach of two atoms (the repulsive part of the Van der Waals force) is owing to an emptying of the electron density between them as the electrons seek more space, going into regions such that coulomb forces are now pulling the two nuclei apart. This description in terms of electron densities and coulomb forces is general and accurate so far as it goes, and makes the nuance and complexity of quantum dynamics implicit rather than explicit; it is needed in what follows.

The ligand affinity—that is, the equilibrium binding constant K_a—is a number such that if there are N ligands in the volume V with protein and solvent, the probability P that a binding site is occupied is

$$P = K_a \frac{N}{V}(1\text{-}P) \qquad \text{(a)}$$

To find a molecular expression for this, let us think of a single protein molecule within V, and suppose that the quantum states of the entire system—protein, ligand, and solvent—are enumerated by an index *i*. Statistical mechanics tell us that in equilibrium at temperature T, as the system jostles and is jostled by its environment—a bath that maintains the temperature—it visits each of the states *i* in proportion to $\exp(-E_i/KT)$, where E_i is the state's total energy. In our formulation of the molecular physics, the total energy (but for the kinetic energy of the nuclei), $E(R_1, R_2, \ldots)$, was defined, a function of the positions of all the nuclei. In the approximation that the nuclei move according to Newtonian dynamics, their momenta and positions are independent variables, and the relative frequency with which they occupy any particular array of positions is then

$$f(R_1, R_2, \ldots) = \exp[-E(R_1,R_2, \ldots)/KT] \qquad \text{(b)}$$

independently of their velocities. It is readily shown that quantum effects replace Eq. (b) by a complicated average of its values over a region extending of the order of $(h^2/MKT)^{1/2}$ in each nuclear coordinate from the point $R_1,R_2, \ldots$. As this is c.a. .1A° for carbon, Eq. (b) itself is a reasonably good approximation. Then to get P we integrate Eq. (b) over all positions of the nuclei for which there is a ligand at the binding site and divide by the integral over all possible configurations.

We will have to look at the carpentry of these integrals, even though in the end most of the complication will be circumvented by calculating the work done in binding a ligand.

Only those regions of the nuclear coordinate space contribute appreciably that are near minima of E; the contribution from any such region can always be written $\delta S \exp(-E_{min}/KT)$, where E_{min} is the local energy minimum and δS is a volume—a precise number if the integral is evaluated accurately—whose magnitude is of the order of the volume over which, as we leave the local minimum, the energy remains within nKT of the minimum, where n is the dimensionality of δS. In particular, for all covalently bonded species, only regions corresponding to small variations about the covalent structures contribute, and so the first step is to transform the variables of integration to center-of-mass and relative coordinates for the covalent species.

Assuming that ligand is sufficiently dilute that there is very little of the 3N dimensional space of its center-of-mass coordinates for which two ligand molecules are close to one another, essentially all positions within V are equivalent for unbound ligands and the integral over each from ligand position gives a factor V. However, as all free ligands are identical in the quantum statistical sense, every contribution to the integral that just corresponds to a permutation of the ligand molecules is to be omitted; therefore, if there are N free ligands, this part of the integration gives a factor $V^N/N!$, which is approximately $(V/N)^N$ for large N. There are also factors from integrals over internal coordinates of the ligand (corresponding to rotation and vibration as perturbed by the presence of solvent), and a factor for protein, which depends on whether or not a ligand is bound to it. Let us just write these with an obvious notation to obtain the integral corresponding to N free ligands and an unliganded protein

$$\left(\int fdR_1dr_2\ldots\right)_u = \left(\frac{V}{N}\right)^N \left(\Delta Se^{-E_m/KT^N}\right)^N_{\text{free ligand}} \left(\Delta Se^{-E_m/KT}\right)_{\text{Protein, u}} \cdot \qquad \text{(c)}$$

We have omitted a factor for the solvent, taking into the other factors any difference in it as between the liganded and unliganded configurations. Any common factor will cancel when P is calculated. For the corresponding integral when the protein is liganded, let us preserve the dimensionality and significance of the factors by making a (somewhat arbitrary) separation between the energy E_m of the protein and that of the ligand at the minimum energy configuration of both, which of course is determined by their interaction. The integral over the ligand's center-of-mass while bound we will denote by V_0.

$$\left(\int fdR_1dR_2\ldots\right)_L \qquad \text{(c')}$$

$$= \left(\frac{V}{N}\right)^{N-1} \left(\Delta Se^{-E_m/KT}\right)^{N-1}_{\substack{\text{free}\\\text{ligand}}} V_0\left(\Delta Se^{-E_m/KT}\right)_{\substack{\text{bound}\\\text{ligand}}} \left(\Delta Se^{-E_m/KT}\right)_{\text{Protein, L}}$$

Calculating P as

$$P = \int_L fdR_1\ldots \Big/ \left(\int_L fdR_1\ldots + \int_u fdR_1\ldots\right)$$

and comparing with Eq. (a), we get an expression for K_a;

$$K_a = V_0 \frac{\left(\Delta Se^{-E_m/KT}\right)_{\substack{\text{bound}\\\text{ligand}}}}{\left(\Delta Se^{-E_m/KT}\right)_{\substack{\text{free}\\\text{ligand}}}} \frac{\left(\Delta Se^{-E_m/KT}\right)_{\text{Protein, L}}}{\left(\Delta Se^{-E_m/KT}\right)_{\text{Protein, u}}} \qquad \text{(d)}$$

Note that V_0 is a volume of molecular dimension—in fact something like the volume to which the bound ligand is confined. Comparing with the thermodynamic form for K_a,

$$K_a = \frac{e^{-\Delta F/KT}}{(N/V)_{\text{Standard State}}} \tag{e}$$

we see that $1/V_0$ gives a kind of absolute definition of the standard state for which ΔF, the logarithm of the other two factors in Eq. (d), has a straightforward molecular interpretation involving only shifts in the internal states of the protein and ligand.

I now come to the point of this section, to identify ΔF in Eq. (e) as work done by coulomb forces in binding a ligand.

Let us note that mass-action equilibrium constants are conventionally connected with thermodynamic quantities by analyzing thought-experiments utilizing, typically, a set of membranes, each permeable to only one constituent of the reaction, and pistons to register external work. This is essential to over-all accounting for energy and free energy [3]. The position of a piston is an external parameter of which the energies of possible states of the system, E_i, are functions; the piston does work on the system by changing the E_i. In fact, if the piston moves from position X_1 to X_2, the work done is exactly the change in the energy levels, averaged over their occupation [14]:

$$W = \int_{X_1}^{X_2} dx \left\{ \sum_i \frac{\partial E_i(x)}{\partial x} e^{-E_i(x)/KT} \Big/ \sum_i e^{-E_i(x)/KT} \right\} \tag{f}$$

Note that the equilibrium distribution of energy level occupation is maintained throughout—this is the statistical meaning of reversibility—and therefore, in general, heat is exchanged with the bath as the system occupies its changed levels in proportion to $\exp(-E_i/KT)$.

Just so, let us visualize a process in which the binding of a ligand is accomplished continuously, under external control—at least in our imagination. Suppose a ligand with its center-of-mass out in the solvent, at R , say, is moved slowly to the binding site, its center-of-mass being placed at the minimum energy position, R_m. According to Eq. (f), the work done on the agent that moves the ligand is the net ΔE for the process, averaged over the equilibrium distribution of all the other nuclear coordinates $R_1, R_2, \ldots$ as the process is carried out. Letting R_c be the position of the ligand's center of mass,

$$W_b = \int_{R_s}^{R_m} dR_c \left\{ \frac{\int dR_1 dR_2 \ldots \dfrac{\partial E(R_c, R_1, R_2 \ldots)}{\partial R_c} \exp[-E(R_c, R_1, R_2, \ldots)/KT]}{\int dR_1 dR_2 \ldots \exp[-E(R_c, R_1, R_2 \ldots)/KT]} \right\} \tag{g}$$

$$= -KT \ln \left\{ \int dR_1 dR_2 \ldots \exp[-E(R_c, R_1, R_2 \ldots)/KT] \right\}_{R_c = R_s}^{R_c = R_m}.$$

If we reexamine the arguments that led up to Eq. (d), we see that Eq. (g) is closely related, and that

$$K_a = V_0 e^{-W_b/KT}. \tag{h}$$

This connects the ligand affinity with two quantities that have direct physical significance: V_0 is the effective volume to which the center-of-mass of the ligand is confined when bound (if there were no forces acting on it there, the probability that it finds itself in V_0 would, per ligand, be V_0/V, as Eqs. (a) and (h) predict). The negative of W_b is the reversible work that we could extract as the ligand is pulled into the binding site, were we able to attach a rod of negligible size to its center-of-mass. This work is done, as we have remarked, purely by coulomb forces acting on the nuclei as the electron density distribution changes to effect binding; the electron density itself is of course responding to the exigencies of quantum dynamics as the configuration of the nuclei changes. So, if we were hanging on to that rod, we would register the work that is done as the energy levels of the system are lowered by coulomb forces, as Eq. (f) states; just as in the thought experiments of classical thermodynamics, this gives us an accounting for the balance of free energy. We shall say that W_b is the work available to extract a ligand from solution; by Eq. (h) it determines the ligand affinity.

APPLICATION TO HEMOGLOBIN

In reference [1], Perutz describes the structural changes that follow ligand binding and interprets shortening of the iron bonds as their "trigger." To see the thermodynamics of this mechanism, consider binding at a particular subunit as the reversible process formulated in the last section: As the ligand approaches and the electronic system centered on the iron atom combines with that of the ligand, shifting to provide a net attractive force on the ligand nuclei, it apparently also shifts to pull the iron atom closer to the plane of the heme group; its purchase on another arm of the polypeptide chain (via coulomb forces acting on one or more nuclei of the histidine imidazole) means that this arm must be pulled closer to the heme. Remembering that at every point along the binding path, we consider the system to visit its states—which in our approximation are the configurations of all the nuclei—in proportion to $\exp(-E(R_1,R_2,\ldots)/KT)$, let us consider how work is taken up by the protein as we proceed along the pathway to binding. We must note that E can take on very many local minima by the polypeptide chain's adjustment of myriad nonbonded interactions among its side chains and with solvent, as limited by the relative rigidity of bond lengths in its network of covalent bonds, but as permitted by very many almost-free rotations about single bonds (the energy "landscape" in the space of the protein's nuclear coordinates has many peaks and pockets). Many bonds are undoubtedly strained—that is, moved away (or perhaps farther away) from their individual energy minima, as suggested by Hopfield [6]. But it must also be so that at one or more points along the pathway, relocating some atoms, or groups of atoms, relative to each other relaxes many strains. Certainly at the end, with ligand bound, according to Perutz, there have been extensive relocations extending all the way out to the subunit's contacts with neighboring subunits. According to Eisenberger et al. [2], fine-structure x-ray absorption spectroscopy shows that the iron bonds are not appreciably strained. The work done on the protein structure thus may well move it through a series of local minima—pockets of the energy landscape—and probably at the end leaves it in or near one. (The work done may be of either sign; for simplicity, we have spoken as if it were positive.)

Changes in entropy are implicit in these considerations, being represented in the total configurational volume that contributes appreciably to integrals over the frequency of occupation, as in equation (g)—that is, in the size of the regions of the energy landscape over which the energy is not too much above its local minimum. Thermally, as already remarked, entropy changes consist in the heat that must flow in or out of the system to have it visit its changed energy levels in proportion to exp(-E/KT), while the change in free energy—the work—is given entirely by the average change in the energy levels themselves.

In any case, the structure is brought to a new conformation and the work available from the changes in electronic configuration at the iron atom is divided between the two functions of adapting the protein structure and extracting a ligand from solution. Since the latter, by equation (h), determines ligand affinity, the ligand affinity is modulated by the work required to adapt the protein to liganding. Perhaps it should be noted that according to the arguments advanced above, these changes in ligand affinity are unrelated to whether or not there are prominent—or indeed any—bond strains: The rates with which the system transits to and fro between any pair of states depends on the contours of the energy landscape in between, but the ratio of forward and backward rates which gives an equilibrium constant depends only on the states' energies.

In Perutz' mechanism, some of the work to adapt the protein is done against the liganded subunit's contacts with its neighbors. Thus, when one of the neighbors is liganded, the work required to adapt it is modified, the principal effect in the models of Koshland, et al. [7]. Apparently when several subunits have been liganded, the intersubunit contacts are sufficiently modified that a new relative orientation of all four subunits becomes energetically optimal, as foreseen by Monod et al. [11]. The contraption in Figure 1 depicts the mechanical logic of this situation. Stretching of one of the large springs attached to an Fe atom represents the work of adapting a

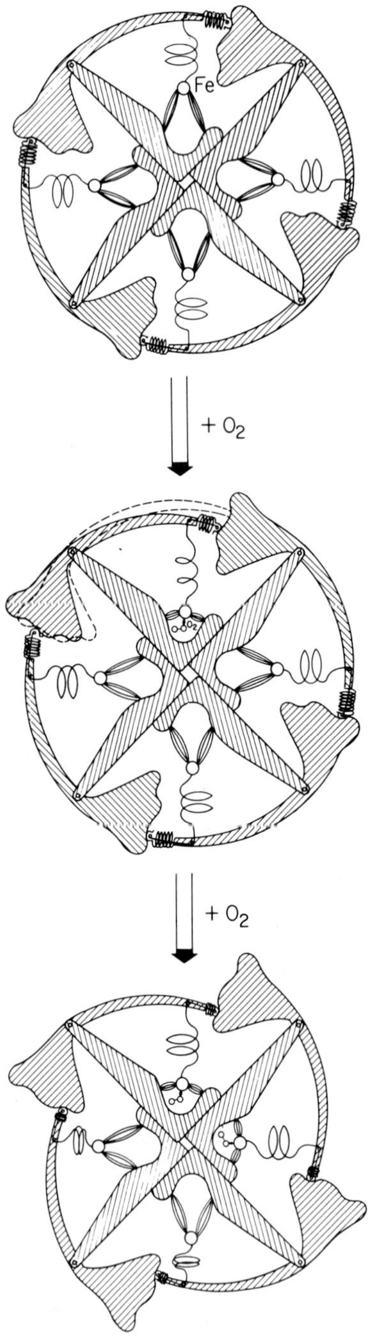

FIG. 1. A contraption depicting the mechanical logic of allosteric linkage via the quatenary transition. The two bonds attaching Fe to the interior of each subunit represent those coordinating the iron atom to heme, which must shrink upon ligand binding. The larger springs and the rocker arms represent the deformable part of tertiary structures; the smaller springs provide deformable inter-subunit contacts. Ligand affinity is modulated by the work required to deform the structure.

subunit's tertiary structure, while compression of the springs on the rocker arm represents work done against intersubunit contacts. The figure is drawn as if the quatenary transition occurs on adding the second ligand.

Linkage with the binding of other ligands (for example, organic phosphates) may be interpreted in the same way—if in binding the ligand, the forces that accomplish that also do work on the protein structure and this modifies the work required to adapt the structure to the binding of a different ligand, then their affinities are linked.

The last paragraph brings forward the point that although for ease in exposition we have throughout spoken as if the Fe-bond system were the only link between ligand and protein, the formulation is in fact independent of details of the electronic mechanisms. If the ligand's interactions with other moieties in the binding pocket are also important, these electronic systems also, in general, divide the work available from their rearrangement between helping (or interfering) with extracting a ligand from solution and adapting the protein. The net cooperative linkage depends on the over-all balance of these terms as produced by the total interaction.

ANALYSIS OF BINDING ISOTHERMS

It is interesting to see actual values for the work terms in hemoglobin. Tyuma et al. have reported extensive measurements of hemoglobin-O_2-DPG equilibria [15]. By straightforward generalization of Eqs. (a) and (h) we find the degree of saturation (fraction of all O_2 binding sites occupied) at given O_2 and DPG concentrations:

$$Y = \frac{1}{4}\,\frac{\sum_{n=1}^{4} nQ_n\left(e^{-W_n/KT} + C_D V_D e^{-W_n^D/KT}\right)(C_0V_0)^n}{\sum_{n=0}^{4} nQ_n\left(e^{-W_n/KT} + C_D V_D e^{-W_n^D/KT}\right)(C_0V_0)^n}\,. \qquad \text{(i)}$$

Q_n are statistical weights: 1, 4, 6, 4, 1 for n = 0 to 4. For oxygen and DPG respectively, C_0 and C_D are (molar) concentrations; V_0 and V_D are the volumes defining affinity, Eq. (h); and W_n and W_n^D are the work of binding exactly n O_2 ligands when a DPG molecule is or is not bound, respectively. Since the α and β subunits are not equivalent, each W_n represents an average over the various permutations with which subunits can be liganded: thus, for n = 2

$$e^{-W_2/KT} = \frac{1}{6}\left(e^{-W_{\alpha_1\alpha_2}/KT} + 2e^{-W_{\alpha_1\beta_1}/KT} + 2e^{-W_{\alpha_1\beta_2}/KT} + e^{-W_{\beta_1\beta_2}/KT}\right).$$

Although Tyuma et al. themselves extract the equivalent of our parameters by fitting their data, they do so by fitting the saturation curves for each

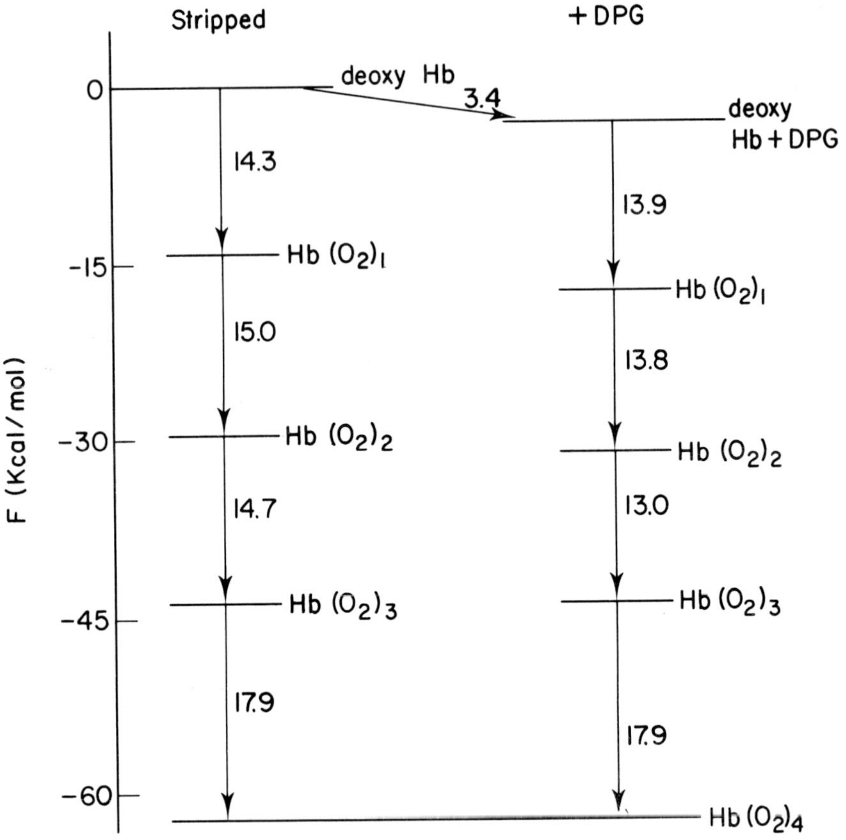

FIG. 2. Free energy-level diagram showing the allosteric linkage for O_2 and DPG as hemoglobin ligands. The numbers represent the work available to extract each ligand from solution. Calculated by fitting the data of Tyuma. et al. [15] for human Hb A, pH 7.4, 25°C, 0.1 M Na Cl and 0.05 M bisTris buffer. DPG = 2,3 diphosphoglycerate.

DPG concentration, one at a time. We have found that more consistent results are obtained by fitting all of the data simultaneously (using, of course, a computer). The result is shown in Figure 2 as a free-energy-level diagram, a mode adopted by Weber [16]. The work available to bind the n'th ligand is written between the appropriate pair of levels.

In the left-hand column one sees that for hemoglobin without bound DPG, the work done on the protein by binding of its first O_2 ligand reduces the work needed to adapt it to binding the second, as Koshland et al. suggest, though the difference (between 14.3 and 15.0 Kcal/mol) is barely significant, given the experimentalists' assessment of their experimental errors and the great sensitivity of the parameters to small changes in the data. The most significant change is clearly due to the quatenary transition, which appears to occur on the binding of the third O_2 ligand. This interpretation is supported by comparing the left- and right-hand columns, the latter made up from the W_n^D; that is, it describes hemoglobin with a molecule of DPG bound. One sees that as each of the first two ligands is bound, of the order of 1 Kcal/mol of work is done against the DPG-protein contacts, and, given the uncertainties, as much as 2 Kcal/mol when the third O_2 is bound; thus DPG is successively loosened, and freed as the third O_2 is bound, which almost certainly locates the quatenary transition there.

CONCLUSION

I have tried to organize and express an approach to thinking quantitatively about allostery around the way in which the work available from the forces that bind a ligand is divided between deforming the protein and the work of extracting the ligand from solution. While much remains to be learned about the details of the forces—especially the subtle and complex quantum dynamics of the electronic systems whose charge densities provide the forces—it seems to me that at the next level up, Perutz has determined the mechanism of heme-heme cooperation; the interaction is transmitted through the protein structure, by its movement, precisely as work is conveyed by rods and springs between macroscopic devices. And we see in considerable and concrete detail how a protein can capture some of the free energy available from chemical rearrangement at one site, transmit it to another site, and use it there to assist chemical rearrangement; which suggests the possibility that the hemoglobin mechanism may be a paradigm not only for allosteric linkage, but for energy transduction as well. In hemoglobin, only one or two kcal/mol of the 15 kcal/mol or so available is transduced, but this is presumably the result of evolution to satisfy a particular function. An exciting challenge is to discover and understand the theoretical limits of this efficiency.

REFERENCES

1. Born M, Oppenheimer JR: Zur Quantenthorie der Molekein. Ann Physik 84: 457–484, 1927.
2. Eisenberger P, Shulman RG, Brown GS, Ogawa S: Structure-function relations in hemoglobin as determined by x-ray absorption spectroscopy. Proc Natl Acad Sci USA 73: 491–495, 1976.
3. Fermi E: Thermodynamics. New York, Dover, 1956, pp 98–138.
4. Feynman RP: Forces in molecules. Phys Rev 56: 340–343, 1939.

5. Hoard JL: Stereochemistry of prophyrin. *In:* Chance B, Estabrook RW, Yonetami T (Eds). New York, Academic Press, 1966, p 9.
6. Hopfield JJ: Relation between structure, co-operativity and spectra in a model of hemoglobin action. J Mol Biol 77: 207–227, 1973.
7. Koshland DE, Nemethy G, Filmer D: Comparison of experimental binding data and theoretical models in proteins containing subunits. Biochemistry 5: 365–385, 1966.
8. Kuhn HG: Atomic Spectra. New York, Academic Press, 1969, pp 267–272.
9. London F: On centers of Van der Waals attraction. J Phys Chem 46: 305–316, 1942.
10. Monod J: Chance and Necessity. New York, Knopf, 1971.
11. Monod J, Wyman J, Changeux J: On the nature of allosteric transitions: a plausible model. J Mol Biol 12: 88–118, 1965.
12. Perutz MF: Stereo chemistry of cooperative effects in haemoglobin. Nature 228: 726–739, 1970.
13. Perutz MF, Kilmartin JV, Nagai K et al: Influence of globin structures on the state of the heme. Biochemistry 15: 378–387, 1976.
14. Schrodinger E: Statistical Thermodynamics. Cambridge, Cambridge University Press, 1948, pp 10–12.
15. Tyuma I, Imai K, Katsuhiko S: Analysis of oxygen equilibrium of hemoglobin and control mechanism of organic phosphates. Biochemistry 12: 1491–1498, 1973.
16. Weber G: Ligand binding and internal equilibria in proteins. Biochemistry 11: 864–878, 1972.

Control of Cell Shape by Adenosine 3′:5′-Phosphate in Chinese Hamster Ovary Cells: Studies of Cyclic Nucleotide Analogue Action, Protein Kinase Activation, and Microtubule Organization

Abraham W. Hsie, Ph.D., J. Patrick O'Neill, Ph.D., Albert P. Li, B.S., Linda S. Borman, B.A., Claus H. Schröder, Ph.D. and Kohtaro Kawashima, Ph.D.

Since the establishment of the Chinese hamster ovary (CHO) cell line in culture, these cells have been utilized for various aspects of genetic, cellular, and biochemical studies. In the past seven years, we have found that CHO cells are valuable for studies on the role of adenosine 3′:5′-phosphate (cAMP) in the regulation of mammalian cell physiology. These studies began with the observation that treatment of CHO cells with an analogous of cAMP, N^6,$O^{2'}$-dibutyryl adenosine 3′:5′-phosphate (dibutyryl cAMP) results in the conversion of the cells from a normally epithelial-like morphology to an elongated fibroblast-like form [3,5]. This morphologic conversion has been shown to result in only minor changes in the growth properties of the cells. The major changes are in membrane structure and activity and in the microtubular system of the cells [1,2,13,14]. The morphologic conversion can be seen within 1 hour; it shows the specificity

From the Biology Division, Oak Ridge National Laboratory (operated by Union Carbide Corporation for the U.S. Energy Research and Development Administration), and University of Tennessee-Oak Ridge Graduate School of Biomedical Sciences, Oak Ridge, Tenn.

The senior author, AWH, wishes to dedicate this article to Dr. Theodore T. Puck on his 60th birthday.

JP O'N, Postdoctoral Investigator, supported by Carcinogenesis Training Grant CA 05296 from the National Cancer Institute; APL, University of Tennessee Predoctoral Fellow; LSB, Predoctoral Fellow, supported by Carcinogenesis Training Grant CA 05296 from the National Cancer Institute; CHS, Postdoctoral Investigator (present address: Institut für Virologie an Deutschen Krebsforschungzentrum, Heidelberg, Germany); KK, Postdoctoral Investigator (present address: Department of Physiological Chemistry, University of Tokyo, Japan).

Abbreviations used in this article: AMP, adenosine 5′-phosphate; cAMP, adenosine 3′:5′-phosphate; cGMP, guanine 3′:5′-phosphate; CHO, Chinese hamster ovary.

expected of an authentic cAMP-mediated response, in that elongation is not observed with adenine, adenosine, adenosine 5′-phosphate (AMP), butyric acid, or cAMP itself. Several other analogues of cAMP are active, in addition to the commonly used dibutyryl cAMP. Treatment of cells which results in morphologic conversion causes the activation of the cAMP-dependent protein kinase, implicating this enzyme in the process which leads to cell elongation.

Here we will summarize our studies of the cAMP-mediated morphologic changes in CHO cells, particularly in terms of the mechanisms through which cAMP analogues act and the changes which occur in microtubule structure.

The properties of the CHO cell clone, CHO-K_1, used in these studies and the conditions of growth have been described previously [3,5]. The procedures employed in these studies have been published in detail previously and are referenced when mentioned.

Properties of the Morphologically Converted Cells

When CHO cells are incubated with dibutyryl cAMP, the cells are converted from a compact, epithelial-like morphology to an elongated, fibroblast-like form. This effect is seen both in colonies of cells grown in the presence of this cAMP analogue (Fig. 1) and in single cells treated for short time periods. Untreated cells grow randomly, without obvious orientation among themselves, while treated cells are oriented along their long axis in a "contact-inhibited" or growth-controlled fashion. The morphologic conversion is observable within 1 hour, is readily reversible upon removal of the cAMP analogues, and causes only minor changes in the growth properties of the cells. Treated and untreated cells have similar plating efficiencies (90–100%) and population doubling times (untreated, 12–14 hours; treated, 13–15 hours).

Associated with the morphologic conversion are changes in properties of the cell membrane [3,14,17,18]. Untreated cells have a highly knobbed surface and the cell membrane is very active, showing rapid, oscillating movements. Treated cells have a smooth surface, with only occasional knobs at the poles of the elongated cells, and movement of the elongated cell membrane is limited to slow ruffling. Another property of the cells which is presumably associated with the cell membrane is cell adhesion, which is also altered. Untreated cells are readily dissociated from the substratum by trypsin and appear to detach spontaneously during growth, leading to the formation of satellite colonies. Treated cells appear to be more firmly adherent to the substratum and to each other, since they are less readily dissociated by trypsin and show little tendency to form satellite colonies. A

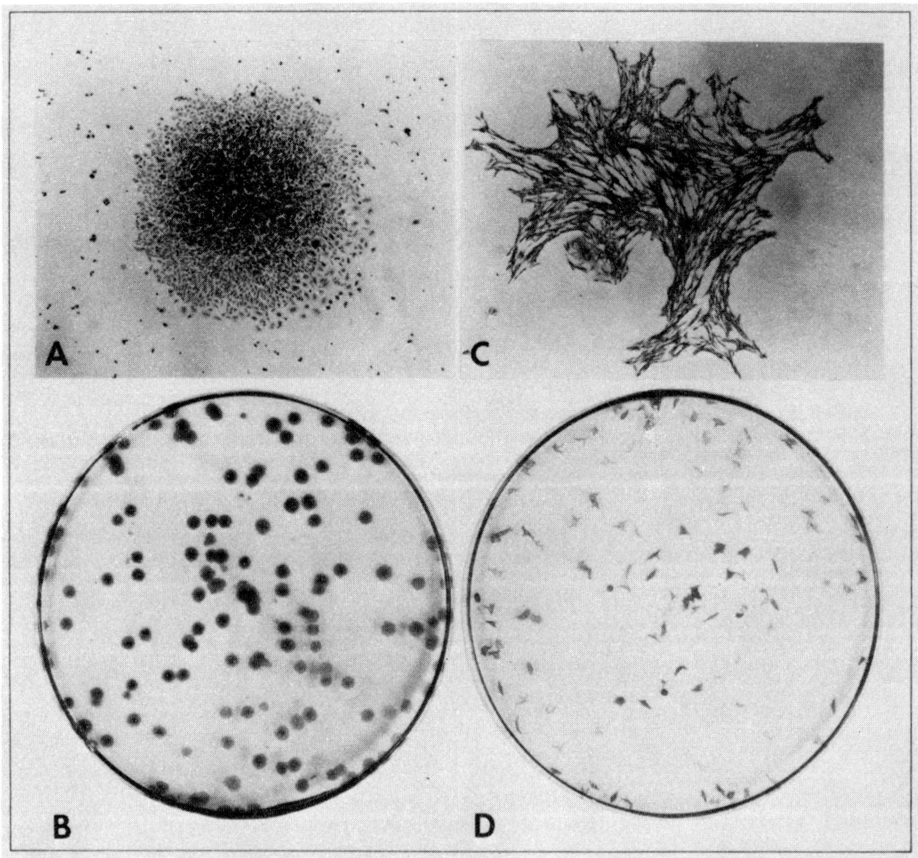

FIG. 1. Effects of dibutyryl cAMP on the morphology of CHO cells. (*A,B*) Colonies grown from single cells on standard medium for 7 days; × 50 and × 2, respectively. (*C,D*) Colonies grown in the presence of 1 m*M* dibutyryl cAMP for 7 days; × 50 and × 2, respectively. (Adapted from [5].)

third property of the cells which is attributed to the cell membrane is also altered—the agglutinability by plant lectins such as concanovalin A and wheat germ agglutinin. Untreated CHO cells are agglutinated by small amounts of concanavalin A (75% agglutination in 15 minutes with 50 μg/ml), while treated cells are refractory to agglutination (75% agglutination in 15 minutes requires more than 750 μg/ml). If both cell types are first treated with trypsin (0.05%, 5 minutes), they show identical agglutination properties. These observations of membrane alterations in cells treated with dibutyryl cAMP are consistent with the idea that such treatment causes the usually fluid cell membrane to become more constrained. Both cell-to-cell interaction and cell-to-substratum association are affected.

The involvement of microtubules and microfilaments in the process of cell elongation was deduced from the observation that colchicine, vinblastine, and cytochalasin B prevent the dibutyryl cAMP-mediated morphologic conversion [5]. Electron microscopic studies revealed that treated, elongated cells contain long, parallel arrays of microtubules which are not seen in untreated cells [1,13]. These microtubules might serve as the internal "skeleton" which allows the cells to assume the elongated shape. To explore this phenomenon further, electron microscopic studies were performed as follows: Cells were fixed *in situ*, embedded, and sectioned, always parallel to the substratum. A series of sections was obtained, proceeding from the bottom to the top of the cell. In this way changes in microtubule structure in the bottom, middle, and top of both treated and untreated cells could be compared. Relatively isolated cells were studied to avoid the complications of cell contact [2]. The long, parallel arrays of microtubules are best seen in more oblique sections of treated cells and do not occur in untreated cells. Microtubules appear to be evenly distributed in the top, middle, and bottom of untreated cells. However, treated cells show a two- to three-fold increase in the number of microtubules in the middle and bottom of the cell, and an increase in orientation parallel to the substratum surface and the new long axis of the cell. Thus, it appears that the elongation of the cell is due to both increased polymerization of tubulin into longer or more microtubules and reorientation. A role for the cell membrane is suspected in this reorientation, perhaps in terms of microtubular attachment.

A mutant line, derived from the parental CHO cell line, has been isolated, and these cells do not elongate when treated with dibutyryl cAMP [1]. Electron microscopic studies have shown that treatment of these cells does result in increased polymerization of microtubules but no reorientation [2]. This shows that microtubule polymerization and orientation can be dissociated and, further, that the orientation is ultimately responsible for the cell elongation. The mutant cells do show the change in agglutinability seen with wild-type cells, demonstrating that at least this aspect of the membrane change is unaffected. Continued study of the microtubule system in this mutant should yield insight into the control of microtuble organization in CHO cells.

Additional Aspects of the Morphologic Conversion

The ability of the treated cells to elongate is not affected by inhibition of RNA or protein synthesis [3,5,12]. When cytochalasin B was used to enucleate CHO cells, it was found that the enucleated cells elongate in a manner identical to nucleated cells [16]. It thus appears that the process of cell elongation per se does not involve the induction of gene transcription or translation but is accomplished by changes in preexisting cell products, for instance, the stimulation of microtubule subunit polymerization. Possible mechanisms are described below.

A cell-cycle phase specificity was deduced from the observation that, in asynchronous cultures at confluence, cells are more responsive to dibutyryl cAMP. Using the mitotic shake-off method of cell synchrony, it was found that only cells in the G_1 phase of the cell cycle initiate elongation when treated [11]. These cells elongate within 1 and 2 hours. Cells treated with dibutyryl cAMP during other phases of the cell cycle show very little elongation until the next G_1 phase is reached. However, once elongated, cells retain this morphology throughout all phases of the cell cycle. In this way, the morphologic conversion of CHO cells shows an important aspect of a differentiation function—the involvement of the G_1 phase of the cell cycle. Compartmentalization of microtubule subunits, which are needed for cell division as well as the morphologic conversion, might explain this G_1 phase specificity.

Agents Which Affect the Morphologic Conversion

Both dibutyryl cAMP and N^6-monobutyryl cAMP cause cell elongation. Adenosine, adenine, AMP, butyric acid, and cAMP itself are inactive. In synergism with low amounts of dibutyryl cAMP, prostaglandins and androgens (testosterone and δ_1-testolactone) are also active [3,5,15]. Estrogens (17β-estradiol and diethylstilbestrol) antagonize the action of dibutyryl cAMP (unpublished observations of N. Nozawa and A. W. Hsie).

Other analogues of cAMP have been tested for cell-elongating action: $O^{2'}$-monobutyryl cAMP, guanine 3′:5′-phosphate (cGMP), 8-bromo cGMP and N^2,$O^{2'}$-dibutyryl cGMP are inactive, while 8-benzylthio cAMP and 8-bromo cAMP are active in this system.

Molecular Aspects of the Morphologic Conversion

Since dibutyryl cAMP and N^6-monobutyryl cAMP are active in inducing the cell elongation while unsubstituted cAMP is inactive, we began a study comparing the properties of these three compounds. Using cell-free extracts, we found that both cAMP analogues inhibit the cAMP phosphodiesterase enzymes and that the inhibition is competitive with respect to cAMP [4]. The N^6-monobutyryl derivative is the better inhibitor. We also studied the uptake of radiolabeled cAMP and dibutyryl cAMP into CHO cells and their intracellular metabolism. With cAMP, we found uptake of radiolabel and accumulation of AMP, ADP, ATP, and a small amount of adenosine. However, no radiolabeled cAMP was found in the cells. With dibutyryl cAMP, we found stable intracellular accumulation both of dibutyryl cAMP and N^6-monobutyryl cAMP at concentrations of 75 and 90 μM, respectively, and of a compound identified as N^6-monobutyryl-AMP. However, no radiolabeled cAMP was found within the cells, although the meaured amount of cAMP increased 20-fold in treated cells [4]. Additional studies of the degradation of the butyryl derivatives of cAMP by CHO cell extracts [10] showed that the butyryl derivatives of cAMP are degraded by cAMP phosphodiesterase and by both N^6- and $O^{2'}$-deacylating activities. From these types of studies we concluded that the inability of cAMP to induce cell elongation is due to the fact that there is no increase in intracellular cAMP level, probably because of rapid degradation. The results of incubation of cells with dibutyryl cAMP are envisioned as follows: (a) partial hydrolysis of dibutyryl to both N^6- and $O^{2'}$-monobutyryl cAMP in the medium; (b) uptake of all three butyryl derivatives, presumably at different rates; (c) rapid hydrolysis of phosphodiester bonds as well as rapid deacylation of intracellular $O^{2'}$-monobutyryl cAMP; (d) slow hydrolysis of phosphodiester bonds and slow deacylation of intracellular N^6-monobutyryl cAMP; (e) deacylation of intracellularly accumulated dibutyryl cAMP, primarily to N^6-monobutyryl cAMP; (f) metabolism of the AMP and adenosine formed as a result of the various types of degradation; (g) accumulation of N^6-monobutyryl-AMP which the cell does not appear to metabolize further; (h) accumulation of N^6-monobutyryl cAMP, which inhibits cAMP phosphodiesterase and results in the increased intracellular level of endogenous cAMP.

At this point, we turned to investigate the effect of this increase in intracellular cAMP on the physiology of the cell. An obvious candidate was the cAMP-dependent protein kinase which has been shown to be activated by cAMP in many cell and tissue systems. It is generally believed that activation of protein kinase(s) is the major mechamism by which cAMP exerts its physiologic function as a second messenger in response to various hormone actions in eukaryotic cells.

We found that the protein kinase activity in CHO-cell crude extracts is cAMP-dependent and is stimulated three- to four-fold by 1 μM cAMP. This stimulation by cAMP occurs with the dissociation of the holoenzyme into regulatory and catalytic subunits, which bind cAMP and catalyze phosphorylation, respectively. The enzyme from untreated CHO cells is largely the high-molecular-weight form of holoenzyme which is cAMP-dependent. However, treatment of cells with dibutyryl cAMP converts the activity to the lower-molecular-weight form which is cAMP-independent [8]. This shows that incubation of CHO cells with dibutyryl cAMP, which results in an increase in the intracellular level of cAMP, causes the activation of the intracellular cAMP-dependent protein kinase, and offers the

TABLE 1. Distribution of cAMP-Binding and Protein Kinase Activities in Subcellular Fractions of CHO Cells[a]

					Protein kinase activity									
					− Histone					+ Histone				
	Protein		cAMP-binding		−cAMP		+cAMP			−cAMP		+cAMP		
Fraction	mg	% distr.	Spec. act.	% distr.	Spec. act.	% distr.	Spec. act.	% distr.	Activity ratio (+cAMP/−cAMP)	Spec. act.	% distr.	Spec. act.	% distr.	Activity ratio (+cAMP −cAMP)
Nuclei (480-g pellet)	46.2	32.0	0.44	3.7	5.2	30.9	6.2	33.2	1.2	55.8	20.6	110.1	5.4	2.0
Crude mitochondria (5,000-g pellet)	18.9	13.1	0.87	3.0	3.4	8.3	2.3	5.0	0.7	21.0	3.2	145.0	2.9	6.9
Plasma membrane (7,000-g pellet)	2.8	1.9	1.36	0.6	3.5	1.3	3.7	1.2	1.1	138.7	3.1	490.8	1.5	3.5
Microsome (27,000-g pellet)	2.6	1.8	1.00	0.4	7.8	2.6	7.2	2.2	0.9	167.7	1.4	400.7	1.1	6.1
Microsome-ribosome (100,000-g pellet)	8.8	6.1	1.74	2.8	6.8	7.7	6.4	6.5	1.0	227.3	16.0	742.3	6.9	3.3
Cytosol (100,000-g supernatant)	65.0	45.1	7.64	89.6	5.9	49.3	6.9	51.9	1.2	107.7	55.9	1190.3	82.2	11.1

[a] Specific cAMP-binding activity is expressed as pmoles of [^{3}H]cAMP bound per mg of protein in each fraction assayed. Protein kinase activity was assayed in the presence and absence of 1 μM cAMP, and the specific activity is expressed as pmoles of [^{32}P]phosphate incorporated per min per mg of protein.

obvious possibility that protein phosphorylation is the mechanism through which tubulin is stimulated to polymerize into microtubules, orient through membrane interaction, and cause the cells to elongate.

We have pursued characterization of the protein kinase in CHO cells in an effort to explore the role of phosphorylation of cell components in the morphologic transformation. Studies on the subcellular distribution of the protein kinase are presented in Table 1 [6]. Specific cAMP binding and protein kinase activity with and without added substrate have been measured. Over 80% of both total cAMP-dependent protein kinase (using histone as substrate) and cAMP-binding activity is present in the cytosol fraction. The protein kinase activity in this fraction is also the most cAMP-dependent; it is stimulated 11-fold by 1 μM cAMP. The other fractions shows small amounts of the protein kinase activity. When the fractions are assayed in the absence of added histone, a low level of activity representing phosphorylation of endogenous substrate(s) is seen. The cAMP-dependence of this activity is difficult to assess because of the limiting amount of substrate present. However, the presence of both protein kinase activity and substrate for phosphorylation allows the possibility that many cell components might be phosphorylated when the intracellular level of cAMP is increased and thus modify various physiologic characteristics.

Since over 80% of the protein kinase activity is found in the cytosol, the property of the cytosol enzyme was further characterized [7]. There appear to be two different protein kinases in this fraction; their properties are presented in Table 2. The two enzymes can be separated by DEAE-cellulose

TABLE 2. Properties of Protein Kinase I and Protein Kinase II in the Cytosol Fraction

Property	Protein kinase I	Protein kinase II
Elution in DEAE-cellulose column	50 mM NaCl	150 mM NaCl
Sedimentation coefficient		
Holoenzyme	6.4S	4.8S
Catalytic subunit	3.0S	2.9S
Molecular weight		
Holoenzyme	105,000	68,000
Catalytic subunit	34,000	32,000
K_m, ATP		
−cAMP	1.25×10^{-4} M	1.67×10^{-4} M
+cAMP	0.76×10^{-4} M	0.88×10^{-4} M
K_a, cAMP	5.56×10^{-9} M	4.54×10^{-9} M
Activation by cAMP (1 μM)	10X	2X
Substrate specificity	histone > protamine > phosvitin	phosvitin > histone > protamine

TABLE 3. Properties of cAMP and Its Analogues

Compound	Cell elongation activity[a]	cAMP phosphodiesterase kinetics[b] (M)	cAMP increase *in vivo*[c] (pmoles/10^6 cells)	Protein kinase activity ratio *in vivo*[d]	Protein kinase activation *in vitro*,[e] $K_a(M)$
cAMP	−	1.2×10^{-6}, 8.9×10^{-6}	0	0.40	1.0×10^{-7}
dibutyryl cAMP	+	2.3×10^{-4}, 7.1×10^{-4}	26.4	0.95	2.0×10^{-5}
$O^{2'}$-monobutyryl cAMP	−	1.1×10^{-5}, 2.5×10^{-4}	0	0.33	7.0×10^{-6}
N^6-monobutyryl cAMP	+	3.5×10^{-5}, 1.5×10^{-4}	28.9	1.01	6.0×10^{-7}
8-bromo cAMP	+	8.3×10^{-5}, 4.2×10^{-4}	0	0.93	1.3×10^{-7}
8-benzylthio cAMP	+	4.0×10^{-5}, 2.1×10^{-4}	0	0.92	1.6×10^{-7}
cGMP	−	1.7×10^{-4}	0	0.26	5×10^{-6}
N^2,$O^{2'}$-dibutyryl cGMP	−	1.9×10^{-4}	0	0.32	$\sim 10^{-5}$
8-bromo cGMP	−	2.0×10^{-3}	0	0.33	$>10^{-5}$

[a] Ability of compound to elongate cells when present in medium at 1 mM for 6 hours. +, cells elongate; −, cells do not elongate and remain similar to untreated cells.

[b] Effect of compound on cAMP phosphodiesterase of CHO cell extracts assayed as described previously [4]. For cAMP, apparent K_m is given; for the remainder, the apparent K_i of competitive inhibition is given.

[c] Increase in intracellular cAMP levels over untreated control when cells are incubated with compound at 1 mM for 6 hr. In all cases controls for washing of the plates have been performed and are subtracted. cAMP was measured as described previously [4].

[d] Cells were treated with the compound at 1 mM for 6 hr, cell extracts were made, and cAMP-dependent protein kinase was assayed as described previously [8]. The ratio of activity assayed in the absence of 1 μM cAMP to that found in the presence of cAMP is given. In untreated cultures the kinase activity is stimulated three- to four-fold by cAMP (activity ratio, 0.25–0.30). Totally activated protein kinase would have an activity ratio of 1.0.

[e] The cAMP-dependent protein kinase from untreated cell extracts was assayed in the presence of different amounts of the various compounds. The amount needed to give 50% of the maximally activated enzyme activity is given as the apparent K_a.

column chromatography and also differ in molecular weight. Kinetically they are similar in their affinity for both ATP and cAMP, but they differ in degree of activation by cAMP as well as in substrate preference.

Recently we have extended our studies of the effects of cyclic nucleotides in CHO cells to other analogoues of cAMP [9]. In addition to dibutyryl cAMP and N^6-monobutyryl cAMP, only 8-bromo cAMP and 8-benzylthio cAMP have been found to induce morphologic transformation (Table 3). As found previously, $O^{2'}$-monobutyryl cAMP is ineffective, probably due to its rapid degradation within the cell [9]. Only dibutyryl cAMP and N^6-monobutyryl cAMP cause a net increase in the intracellular level of cAMP, presumably due to phosphodiesterase inhibition. The four analogues which cause cell elongation also result in the activation of the cAMP-dependent protein kinase in the treated cells. Since treatment with 8-bromo cAMP or 8-benzylthio cAMP does not cause any measurable increase in the intracellular level of cAMP, we propose that they activate the protein kinase directly. Treatment with dibutyryl cAMP and N^6-monobutyryl cAMP appears to result in kinase activation due to the increase in intracellular cAMP. The *in vitro* activation of the protein kinase in cell extracts is consistent with this proposal, since both 8-bromo cAMP and 8-benzylthio cAMP are similar to cAMP in their ability to activate (Table 3). N^6-monobutyryl cAMP is a less effective activator of the protein kinase.

SUMMARY AND CONCLUSIONS

Treatment of CHO cells with analogues of cAMP causes a morphologic conversion, attended by reappearance of several "normal" characteristics of cultured cells, which involves the polymerization and reorientation of microtubules. Unsubstituted cAMP is ineffective due to its rapid degradation by the cellular phosphodiesterase. One effect of the treatment with cAMP analogues is the intracellular activation of cAMP-dependent protein kinase(s). The cellular components which serve as substrates for phosphorylation are not known at present.

There appear to be two mechanisms by which cAMP analogues act. Those such as dibutyryl cAMP and N^6-monobutyryl cAMP cause a net increase in the intracellular level of cAMP, presumably through inhibition of cAMP phosphodiesterase. The increased level of cAMP activates the protein kinase. Other analogues, such as 8-bromo cAMP and 8-benzylthio cAMP, do not cause an increase in intracellular cAMP but instead appear to activate the protein kinase directly.

REFERENCES

1. Borman LS, Dumont JN, Hsie AW: Relationship between cyclic AMP, microtubule organization and mammalian cell shape: Studies on Chinese hamster ovary cells and their variants. Exp Cell Res 91: 422–428, 1975.
2. Borman LS, Hsie AW: Modifications of the microtubule system of Chinese hamster ovary cells by dibutyryl cyclic AMP. In Vitro 12: 285–286, 1976.
3. Hsie AW, Jones C, Puck TT: Further changes in differentiation state accompanying the conversion of Chinese hamster cells to fibroblastic form by dibutyryl adenosine cyclic 3′:5′-monophosphate and hormones. Proc Natl Acad Sci USA 68: 1648–1652, 1971.
4. Hsie AW, Kawashima K, O'Neill JP, Schröder CH: Possible role of cyclic AMP phosphodiesterase in the morphological transformation of Chinese hamster ovary cells by dibutyryl cyclic AMP. J Biol Chem 250: 984–989, 1975.
5. Hsie AW, Puck TT: Morphological transformation of Chinese hamster ovary cells by dibutyryl adenosine cyclic 3′:5′-monophosphate and testosterone. Proc Natl Acad Sci USA 68: 358–361, 1971.
6. Li AP, Hsie AW: Subcellular distribution and the possible roles of cyclic AMP dependent protein kinase in Chinese hamster ovary cells. Fed Proc 35: 1713, 1976.
7. Li AP: Studies on adenosine 3′:5′-phosphate-dependent protein kinase in Chinese hamster ovary cells. Ph.D. thesis, University of Tennessee, 1976.
8. Li AP, Kawashima K, Hsie AW: *In vivo* activation of cyclic AMP dependent protein kinase in Chinese hamster ovary cells treated with dibutyryl cyclic AMP. Biochem Biophys Res Commun 64: 507–513, 1975.
9. Hsie AW: Actions of adenosine 3′:5′-phosphate in Chinese hamster ovary cells. In Vitro 12: 284–285, 1976.
10. O'Neill JP, Schröder CH, Hsie AW: Hydrolysis of butyryl derivatives of cyclic AMP by Chinese hamster ovary cell extract and characterization of the products. J Biol Chem 250: 990–995, 1975.
11. O'Neill JP, Schröder CH, Riddle JC, Hsie AW: The cell cycle specificity of the morphological conversion of Chinese hamster ovary cells by dibutyryl cyclic AMP. Exp Cell Res 97: 213–217, 1976.
12. Patterson D, Waldren CA: The effects of inhibitors of RNA and protein synthesis on dibutyryl cyclic AMP mediated morphological transformation of Chinese hamster ovary cells *in vitro*. Biochem Biophys Res Commun 50: 566–573, 1973.
13. Porter KR, Puck TT, Hsie AW, Kelley D: An electron microscope study of the effects of dibutyryl cyclic AMP on Chinese hamster ovary cells. Cell 2: 145–162, 1975.
14. Puck TT, Waldren CA, Hsie AW: Membrane dynamics in the action of dibutyryl adenosine cyclic 3′:5′-monophosphate and testosterone on mammalian cells. Proc Natl Acad Sci USA 69: 1943–1947, 1972.
15. Puck TT, Wenger L: Substitution of testolactone for testosterone in the "reverse transformation" of Chinese hamster cells. IRCS Med Sci (73-6) 1-8-2, 1973.
16. Schröder CH, Hsie AW: Morphological transformation of enucleated Chinese hamster ovary cells by dibutyryl adenosine 3′:5′-monophosphate and hormones. Nature [New Biol] 246: 58–60, 1973.
17. Storrie B: Antagonism by dibutyryl adenosine cyclic 3′:5′-monophosphate and testosterone of cell rounding reactions. J Cell Biol 59: 471–479, 1973.
18. Storrie B: Effect of dibutyryl adenosine cyclic 3′:5′-monophosphate and testolactone on concanavalin A binding and cell killing. J Cell Biol 61: 247–252, 1974.

Cell Killing by Viruses—Single-Cell Survival Procedure for Detecting Viral Functions Required for Cell Killing

Philip I. Marcus, Ph.D.

The demise of a host through virus infection is the ultimate consequence of an initial encounter of a single cell with a lethal virus. In cell culture, virus-cell interaction can produce a comparable effect—cell killing, an event preceded by a cytopathology specific for each virus. In spite of the detailed *catalogue raisonné* of events which characterize death of a host or host cell, the molecular reactions which result in cell killing by animal viruses, or its prevention through viral interference, still pose an enigma for the molecular virologist. The absence of cell killing by some of these same viruses, as in the development of persistent infection or an oncogenic state, compounds the enigma but stokes our zetetic zeal.

Quantification of an event invariably precedes its definition at the molecular level; but the quantitative approach to cell killing by animal viruses lay dormant, though simmering, when I joined T. T. Puck in 1954 as his second graduate student. Although Dulbecco two years earlier had solved the problem of quantitating animal virus infectivity by plaque assay [7], cell killing experiments based on the classic studies of Luria and Delbrück [18] with bacterial viruses and cells were still beyond the reach of animal virologists for lack of comparable quantitation of the mammalian host cell. With our development of single-cell plating procedures for animal cells in 1955 [23,35,36] it became possible to detect and quantitate the lethal action of animal viruses on cells. Through single-cell survival curve analysis we were able to obtain a precise measure of a virus's capacity to kill cells, i.e., prevent colony (clone) formation, thereby defining virions in terms of cell killing (CK) particle activity [21,22,25,39].

In contrast to the measurement of infectious virus which requires that all viral gene functions be expressed and that sufficient new virus be produced

From the Microbiology Section, University of Connecticut, Storrs, Conn.

These studies were supported by grants from the Damon Runyon-Walter Winchell Cancer Fund (DRG 1284), and the National Cancer Institute (CA 20882) and benefited from use of the Cell Culture Facility supported by National Cancer Institute grant CA 14733.

The author thanks Margaret J. Sekellick for a critical reading of the manuscript and her collaboration on much of the work reported here.

so that multiple cycles of infection may ensue and an amplified response be generated, the detection of CK particles requires only that the initial infecting virion prevent cell reproduction, namely the formation of a macroscopically visible and countable colony. New infectious virus need not be produced. Clearly, then, virions which kill cells can be infectious—in which case, CK particles are equivalent to plaque-forming (PF) or infectious particles, or they can be defective in replication, but still act to kill cells. In this latter case, such particles may be intrinsically toxic, i.e., contain a sufficient quantity of a cell killing factor to bring about cell death, or they may require the expression of some viral gene functions, short of the total required for infectivity. Finally, cellular events triggered by less than full viral replication, such as viral transformation, must occur in the absence of cell killing—reflecting either restricted replication of the virus, or a viral or host-modified control over the expression of a cell-killing factor, as in persistent infection or viral interference. A review of all these aspects of cell killing by viruses is beyond the scope of this article and will be treated in depth in a subsequent communication.

We will limit the content of this communication to studies designed to determine whether all, some, or none of the proteins which constitute a virion, or are expressed through its genome, are required for cell killing. In particular, we will consider cell killing by a virus with well defined genetics—vesicular stomatitis (VS) virus [40] (see below, *The Virus*). Large numbers of temperature-sensitive (ts) virus mutants representing the 5 complementation groups reported for the Indiana serotype of this virus are available, and assignment of the biochemical functions of specific polypeptides to each complementation group is almost complete. In addition, sub-virion sized structures which contain all the polypeptides of the complete virus are available in the form of nonreplicating defective-interfering (DI) particles [11]. Thus, we are in an excellent position to determine which viral protein, or proteins, are required for the expression of the cell-killing particle activity of this virus. The experiments described below represent our attempts to define this requirement for cell killing by VS virus, with the hope that such knowledge will provide new insights into how these molecules act to express cell killing and how cell killing by viruses can be prevented.

I dedicate this resumé of our quantitative studies on cell killing by viruses to my mentor, Theodore H. Puck, on his 60th birthday year. They stand as a tribute to his wisdom and foresight—for certainly he is the least suprised of the two of us that the single-cell plating procedures he has championed these past 22 years would quietly (perhaps not so quietly) await the maturation of animal virus genetics to again serve as a formidable tool in unraveling an enigma in molecular biology, this time in a virus-infected cell.

THE VIRUS

Vesicular stomatitis (VS) virus, the prototype rhabdovirus, is a bullet-shaped virion which consists of a ribonucleocapsid core surrounded by a membrane of lipoprotein. The membrane contains about 1000 molecules of maxtrix (M) protein buried within the lipid complex which derives from the host cell, and about 200 molecules of a glycoprotein (G) in the peplomers (spikes) that constitute the receptor molecules which stud the surface of the virion. Beneath the lipoprotein membrane lies a transcribing ribonucleoprotein complex which consists of one genome molecule of [−] single stranded (ss) RNA (3.6×10^6 daltons), contained within about 2000 molecules of viral specific nucleoprotein (N), associated with about 40 molecules of a "nonstructural" (NS) protein and some 50 molecules of a large molecular weight (190,000 daltons) L protein—the transcriptase of the virion [40]. Three proteins, N, NS, and L, along with the RNA genome as template, are required for primary transcription, i.e., mRNA synthesis from the parental strand [8]. Transcription is readily demonstrable *in vitro* [2] and *in vivo* [24]. The single negative-strand genome of RNA contains the genes for these 5 polypeptides of the VS virus, Indiana serotype; all are found both in the virion and infected cell [40]. The gene order of transcription has been established as initiating at a single site at or near the 3′ terminus and occurs according to the sequence 3′-N-NS-M-G-L-5′ [1]. The ts-mutants of the Indiana serotype comprise 5 complementation groups: mutations affecting the L protein (transcriptase) are localized in group *I* [14,37], the cistrons coding for proteins M and G are defined by groups *III* and *V*, respectively [15], and proteins N and NS may be provisionally assigned to the cistrons represented by groups *IV* and *II*, respectively [40,42].

VS virus characteristically produces noninfectious subvirions, shorter in length than the full genome, but capable of interfering with the replication of complete virus [11]. These so-called defective interfering (DI) or truncated (T) particles, although smaller in length than complete infectious particles, possess two unique attributes of interest to our studies: (1) they contain the same 5 polypeptides as the complete infectious virus used as helper in their replication cycle, yet they are by themselves unable to replicate, and (2) they contain ssRNA of a [−] stranded configuration—like that of the genome, or as recently discovered [16], ssRNA of a [±] configuration as covalently linked [−](genome)-[+](messenger) RNA. This latter form, upon deproteinization self-anneals ("snaps-back") to form an RNAase-resistant duplex (dsRNA). These unique conditions permit a direct test of the role of virion-associated polypeptides and RNA (ssRNA or potentially dsRNA) in cell killing by this virus—in the absence of viral replication.

THE CELL

African green monkey kidney (GMK) cells of the Vero line [6] were used in our most recent studies. Procedures for their maintenance and growth in single-cell plating experiments have been described [26]. GMK-Vero cells plate with high efficiency (70–100%), and as host cells produce a one-step growth curve with high yields of infectious virions typical of VS virus. These cells are sensitive to DI-particle-induced homotypic interference [26], and possess the unusual attribute of being totally nonresponsive to inducers of interferon [6]. This last named characteristic provides a means of studying cell killing by viruses in the absence of the interferon-mediated interference which often accompanies virus infection. However, although these cells are genotypically nonresponsive to interferon inducers, they nonetheless express interferon-mediated interference when interferon is added exogenously—providing a controlled means of studying cell killing and its prevention by the interferon system.

VIRUS-CELL INTERACTION: THE SINGLE-CELL SURVIVAL TEST

1. Cell-killing particle assay. Populations of VS virus were tested for their capacity to kill GMK-Vero cells by the single-cell survival procedure as first described by Marcus and Puck [21,22,25] and modified by Marcus and Sekellick to permit tests of cell killing by ts-mutants at permissive and nonpermissive temperatures [26,27]. The general procedure is outlined schematically in Figure 1. Briefly, monolayers of cells are infected with various dilutions of virus in order to achieve a range of multiplicities. Following virus attachment and entry, infected cell monolayers are treated with a trypsin:EDTA solution to produce a monodisperse suspension consisting of single uninfected cells and virus-cell complexes. These are plated in duplicate sets, with one set held at permissive temperature (30° in this case) for 20 hours. We have established that this "pulse-infection" at 30° suffices to score as cell-killing particles all virions which may require a full replication cycle [27]. Following the 20 hour "pulse" at 30°, this set of plates is moved to 40° for the development of colonies from surviving cells. The second set of plates is incubated directly at the nonpermissive temperature (40°, for these ts-mutants of VS virus) to determine whether the viral function which is defective at the restrictive temperature is required for the expression of cell killing. Control (uninfected) cells are carried through these same temperature manipulations, which we have found have no adverse effect on the plating efficiency of GMK-Vero cells [27]. Colonies

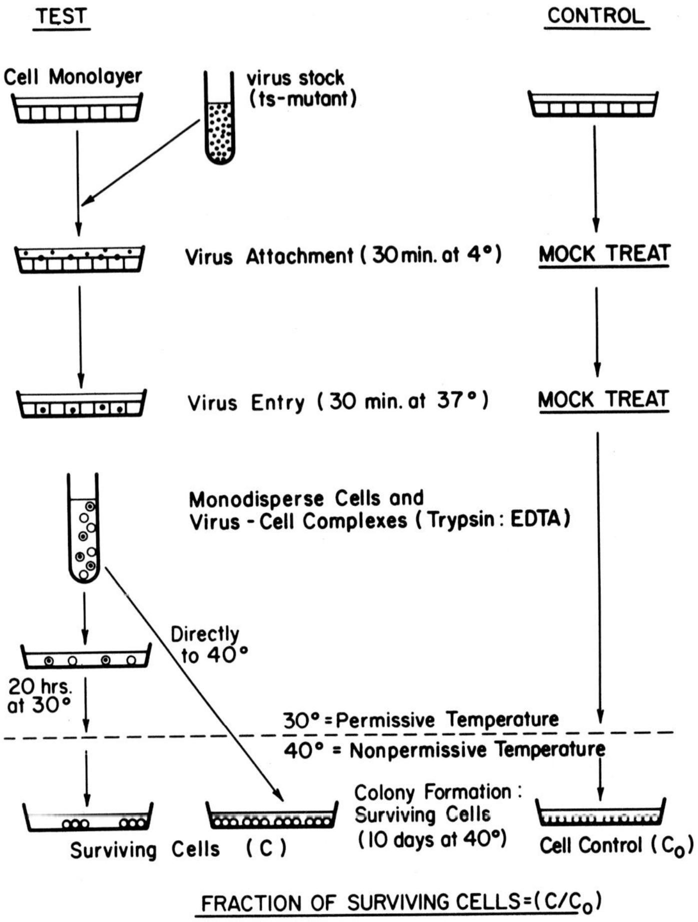

FIG. 1. Cell-Killing Particle Assay: Single Cell Survival Procedure for ts-Virus Mutants.

from surviving cells are fixed, stained and counted after 10 days' incubation at 40°. The fraction of surviving colonies obtained from a typical experiment are illustrated in Figure 2.

The "one-hit" nature of the survival curve for cell killing by viruses is illustrated in Figure 3, and represents an experiment from earlier studies carried out with Newcastle disease virus on HeLa cells by Marcus and Puck 18 years ago [21,22,25]. These results are typical of cell-killing particle assays of VS virus now being carried out on GMK-Vero cells [26,27,28,32].

2. *Cell killing by ultraviolet (UV) and by heat (50°) inactivated wild type vesicular stomatitis virus.* If the expression of cell killing by VS virus requires all viral gene functions, as does the expression of infectivity, the

survival curves for the inactivation of these two functions by UV radiation should be identical. Should the lethal action of VS virus require fewer than all 5 gene functions, or if the virions were intrinsically toxic, then cell-killing particle activity would be more resistant to UV radiation than infectivity. The following experiment was performed: wild type VS virus was exposed to different doses of UV radiation and the fraction of surviving activity of both infectivity (plaque-forming particle activity) and cell-killing particle activity was determined as a function of UV-dose. The results presented in Figure 4 reveal that the capacity of VS virus to kill cells is about 5 times more resistant to UV radiation than is infectivity. It follows from these results that UV radiation can convert a significant fraction of a VS virus preparation to defective (noninfectious) CK particles—providing evidence that cell killing can be expressed in the absence of complete viral replication. Kinetic studies such as those represented by the UV-surviving activity curves cannot be interpreted rigorously, but suggest that about one-fifth of the viral genome must remain undamaged in order to express cell killing. These data do not permit a distinction between the need for transcription of a specific sequence of the genome, or one chosen at random. Data obtained from cell-killing experiments with ts-mutants of VS virus as described below indicate that the former situation prevails. The absence of any multi-hit component to the UV-survival curves for CK particle activity tends to preclude a model in which the VS virus genome

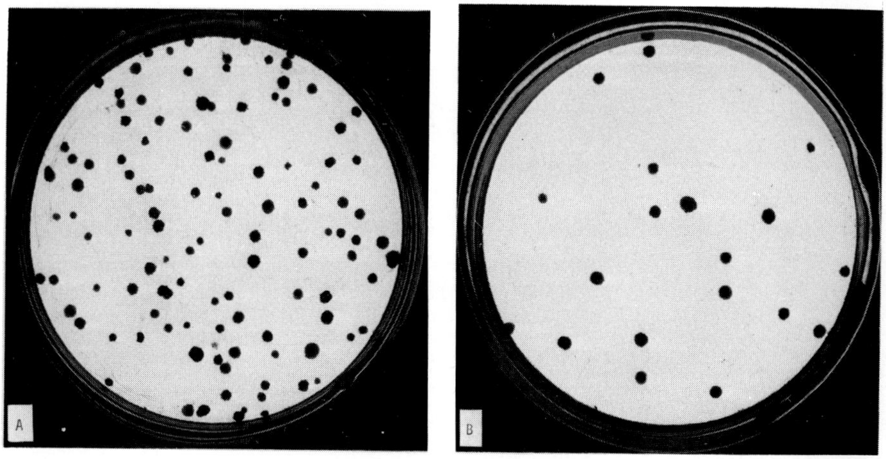

FIG. 2. Colony Formation by HeLA S3-9 Cells Exposed to Newcastle Disease Virus, Strain Vaccine. Plate A received an inoculum of 100 uninfected cells. Plate B received an inoculum of 100 cells which had attached on the average 1.5 cell-killing particles (m_{CKP} = 1.5). The photographs show colonies formed from surviving cells on a 50 mm dish after incubation for 11 days at 37°. (Experimental details in reference [21].)

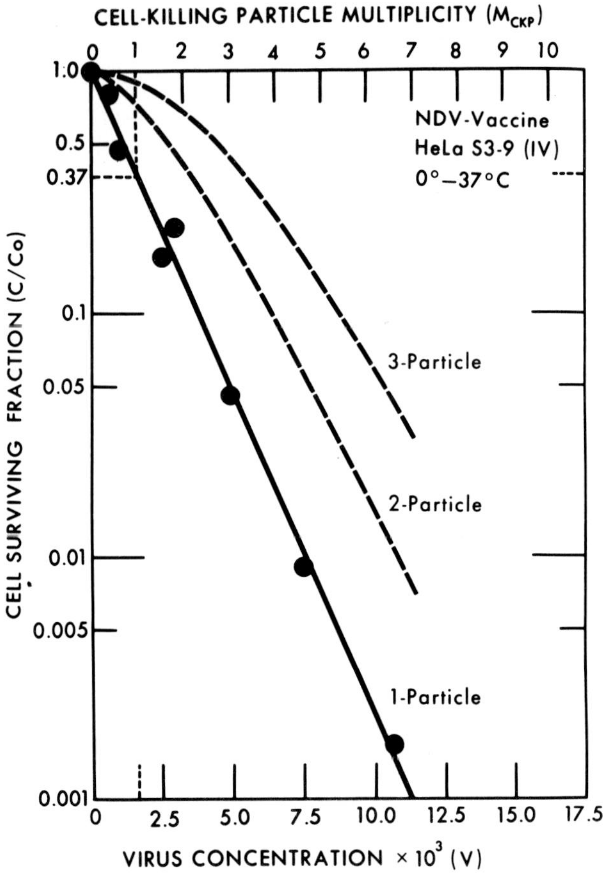

FIG. 3. Survival Curve of HeLa Cells Exposed to Newcastle Disease Virus. Suspensions of monodisperse cells were mixed with different amounts of virus, plated in the standard manner, and the fraction of cell survivors was determined from colony counts and plotted as a function of the virus concentration (cf. Figs. 1 and 2). Theoretical plots for 1-particle-, 2-particles- and 3-particles-to-kill mechanisms are indicated. The three curves have been derived from the Poisson distribution

$$P(r) = \frac{m^r e^{-m}}{r!}$$

where P(r) is the probability of the cells in a given population attaching *r* virus particles when the average multiplicity of infection is *m* (cf. [17,22]). The cell-killing particle multiplicity (m_{CKP}) on the top abscissa was calculated from $C/C_O = e^{-m_{CKP}}$. When the cell surviving fraction $(C/C_O) = 0.37$, then $m_{CKP} = 1.0$.

contains more than a single sequence of information that can be transcribed independently to produce a putative cell-killing factor. We also note that following UV radiation, the capacity of VS virus to accumulate primary transcripts *in vivo* [24] is lost at the same rate as cell-killing particle activity. Because of the relatively low doses of UV radiation used in these experiments, we favor a subtle damage to the VS virus genome, most likely in the form of uracil dimer formation [33].

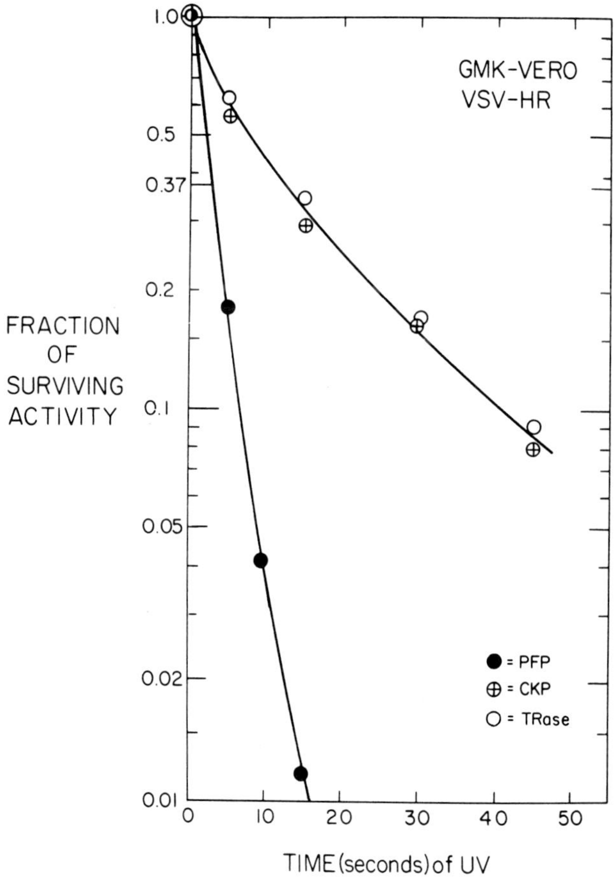

FIG. 4. Surviving Activity Curves for GMK-Vero Cells Infected With Ultraviolet-Irradiated Wild Type Vesicular Stomatitis Virus. Survival curves for plaque-forming (PFP) and cell-killing (CKP) particles and for *in vivo* virion-associated transcriptase (TRase) activity were determined as previously described [27] and were used to calculate the surviving activity curve for each function at different doses of UV radiation. For plaque-forming particles, D^{o} (1/e dose) = 52.3 ergs mm^{-2}. For cell-killing particles, and *in vivo* transcriptase, D_{o} = 262 ergs mm^{-2}.

These experiments provided the first clue that virion-derived primary transcription might be a required, though not necessarily sufficient, reaction for the expression of cell killing by VS virus. Were this the case, and primary transcription the rate-limiting step in cell killing, inactivation of virion-associated transcriptase (L-protein) by heat should lead to the concomitant loss of infectivity, cell-killing particle activity and transcript accumulation. In Figure 5, the upper curve shows that for wild type virus these 3 activities indeed are lost at equivalent rates as a function of time exposed to heat (50°C). Were the activity of the transcriptase molecules rate limiting for cell killing (and in this case for infectivity), then heat inactivation of a temperature-sensitive (ts)-mutant with a labile virion-associated transcriptase [37] should generate surviving activity curves in which these 3 functions are lost at an equal but significantly faster rate than with wild type virus. VS virus tsG114(*I*) is such a mutant: the virion-associated transcriptase of this mutant is so sensitive to heat that its genetic-based biochemical defect is attributable to failure of this enzyme to function at

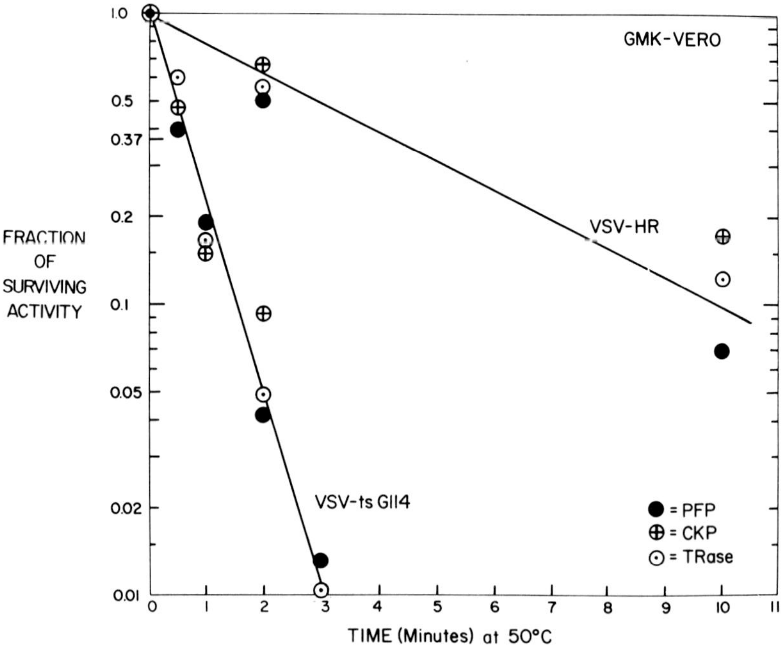

FIG. 5. Surviving Activity Curves for GMK-Vero Cells Infected with Heat (50°)-Treated Wild Type (upper curve) or ts G114 (lower curve) Vesicular Stomatitis Virus. Survival curves for PFP, CKP and *in vivo* virion-associated transcriptase (TRase) were carried out as previously described [27], and were used to calculate surviving activity for each function after exposure of virions to 50° for various periods of time.

nonpermissive temperature [14]. The lower curve in Figure 5 demonstrates that these 3 functions are lost at a rate 6.5 times faster than that observed for wild type virus (the upper curve).

These results with UV- and heat-inactivated virus provide evidence that primary transcription is a prerequisite, though not necessarily sufficient, event for cell killing. Predictably, then, virions or subvirions of VS virus deficient in transcribing capacity would not kill cells. This possibility was tested experimentally as described below.

3. Nontranscribing Defective Interfering Particles Do Not Kill Cells. Defective interfering (DI) particles do not produce detectable primary transcripts [12,34] even though these subvirions contain the same 5 polypeptides present in the complete virion [41]—albeit, proportionally fewer molecules per DI particle. In Figure 6, the upper curve demonstrates that preparations of thrice-gradient purified DI particles do not kill cells as tested up to a multiplicity of 70, a result consistent with our hypothesis that transcription is a requisite event for cell killing. These results also demonstrate that DI particles are not intrinsically toxic, although they contain the same 5 polypeptides present in the complete infectious virion. Figure 6 also illustrates another important observation, namely, that although DI particles block totally the replication of complete virus, they do not block the expression of cell killing by a coinfecting complete virion. Thus, the survival curves for VS virus are the same in the presence or absence of DI particle-induced homotypic interference. It is important to note that although replication of complete virus is aborted totally under these conditions [3,26], virion-derived transcription by the helper (complete) virus is extant [12,34], satisfying the conditions hypothesized for a requirement of primary transcription in cell killing.

4. Cell-killing by Mutant Ts-G114(I). The defect of mutant ts-G114(*I*) has been established [14]: The virions of this mutant contain a transcriptase which functions at 30°, but not at 40°, and hence, primary transcripts do not accumulate at the nonpermissive temperature [14,27,37]. This mutant provides an opportunity to test directly the need for primary transcription in cell killing by VS virus. After having first established that wild type VS virus kills cells equally well at 30° and 40° (27), we tested mutant ts-G114. Figure 7 shows that although all infectious (plaque-forming) particles were being scored as CK particles at 30° (the PF and CK particle titers were equivalent), at 40° this mutant did not kill cells, providing convincing evidence that primary transcription was required for the expression of cell-killing.

5. Time Course for Acquiring a Lethal Dose of Cell-Killing Factor. The lack of cell killing by ts-G114 at 40° permits a determination of the length of time required at permissive temperature to produce and/or accumulate a lethal dose of what may be termed cell-killing factor. GMK-Vero cells

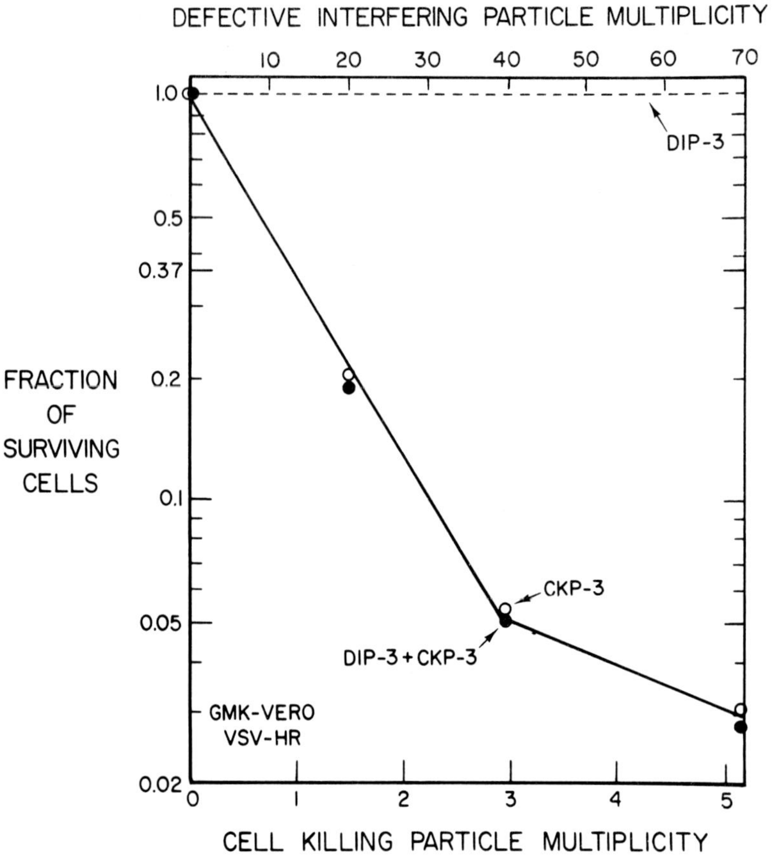

FIG. 6. Survival Curves of GMK-Vero Cells Exposed to Thrice-Gradient Purified Defective-Interfering and Cell-Killing Particles. Survival curves were generated by exposing Vero cells to thrice-gradient purified [26] cell-killing particles (CKP-3 at m.o.i.$_{CKP\text{-}3}$ = 3), defective-interfering particles (DIP-3 at m.o.i.$_{DIP\text{-}3}$ = 100), and a mixture of the two (m.o.i.$_{DIP\text{-}3}$ = 100 + m.o.i.$_{CKP\text{-}3}$ = 3). The bottom abscissa applies to the two survival curves generated when CKP are present. The top abscissa relates to the survival curve produced when only DIP-3 is added to the cell population.

infected with ts-G114 at m_{CKP} = 3 were incubated for various time intervals at 30° prior to shift-up to 40° for colony formation from surviving cells (cf. Fig. 1). Figure 8 (solid circles) demonstrates that 18–20 hours at 30° are required to express the maximum cell-killing particle activity of this mutant virus. The lethal action of the cell-killing particles is expressed in one-half of the cells by 6 hours at 30°. We calculate from the Poisson distribution [17,22] that at m_{CKP} = 3, about one-half of the cells are infected with ≥3 CK particles. A direct comparison between the rate of development of cell-killing factor and the release of infectious virus from the surface of the cell was established by generating a standard virus growth curve for mutant ts-G114 under these same shift-up conditions (Fig. 8, open circles). Clearly, the release of new virus as plaque-forming particles lags considerably behind the appearance of cell-killing factor, i.e., the expression of cell-killing particles. Since only a brief interval (≃1 minute) separates the appearance of newly synthesized virus and its release from the cell surface [10], the disparate rates for the appearance of cell-killing factor and cell-free plaque-forming particles tends to minimize virus budding per se as an important element in the process of cell killing by VS virus.

6. Cell Killing by Mutant ts-W10(IV). While the single-cell survival curves for ts-G114 are representative of ts-mutants which do not kill cells at 40° (Fig. 7), Figure 9, which illustrates data from the use of ts-W10(*IV*), is typical of those mutants which do express cell killing at the nonpermissive temperature. Clearly, the survival curves in Figure 9 demonstrate that this mutant expresses cell killing equally well over the entire range of temperatures encompassing permissive and nonpermissive conditions. However, not

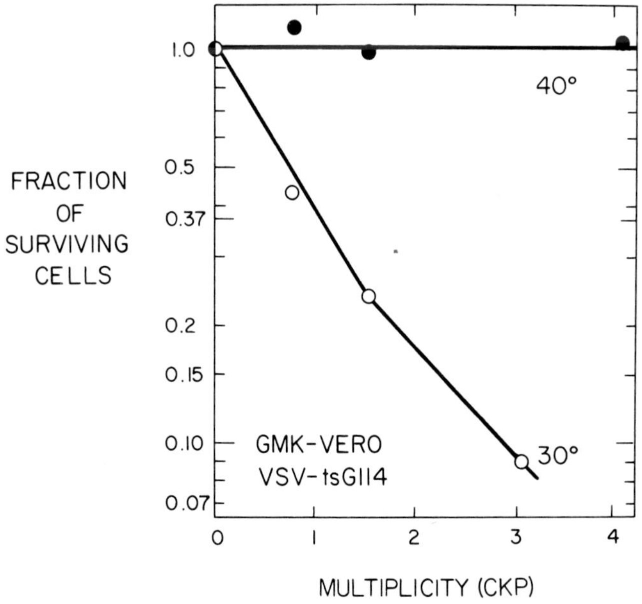

FIG. 7. Cell-Killing Survival Curve of GMK-Vero Cells Infected with Vesicular Stomatitis Virus Mutant ts G114 (I). Virus-infected cells were held for 20 hours at permissive (30°) or nonpermissive (40°) temperatures before incubation at 40° for colony formation from surviving cells—according to the scheme illustrated in Figure 1 (cf. [27]). The abscissa represents CKP multiplicities calculated from the 37% (1/*e*) survival level. For mutant ts G114(I), the CKP and PFP titers were equivalent when measured at 30°.

all mutants in complementation group *IV* express cell-killing at 40°. For example, mutant ts-G41(*IV*) does not kill cells at nonpermissive temperature. Both mutants ts-W10 and ts-G41 are reported to possess an active virion-associated transcriptase at nonpermissive temperature, but are presumably defective in a function required for RNA synthesis subsequent to primary transcription, probably a replicative event [38,42]. We interpret these contrasting results with mutants ts-G41 and ts-W10 to mean that the presumed defect in tertiary structure of ts-W10 protein (N protein), though sufficient to prevent its functioning in normal virus replication, is not so extreme as to hinder its function in producing cell-killing factor. Presumably, the N protein of ts-G41 does not possess even this minimal function. These results indicate that quality rather than quantity of the transcripts synthesized is an important determinant in cell killing, and demonstrate that CK particle assays can resolve genotypic differences within a single complementation group!

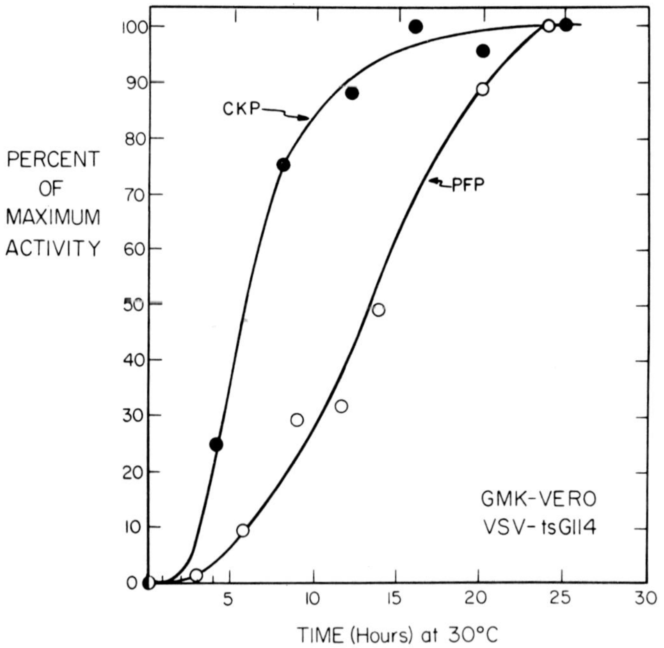

FIG. 8. Time Course for the Expression of CKP and the Release of PFP by Mutant ts G114. GMK-Vero cells were infected at m.o.i. CKP = 3 and incubated at 30°. At various time intervals, samples of medium were removed and assayed at 30° to establish the PFP titer of released virus—producing, in effect, a growth curve. The "growth curve" representing the expression of CKP was generated by holding virus-cell complexes at 30° for the time intervals indicated by the experimental points before shifting to 40° for colony formation from surviving cells (cf. [27]).

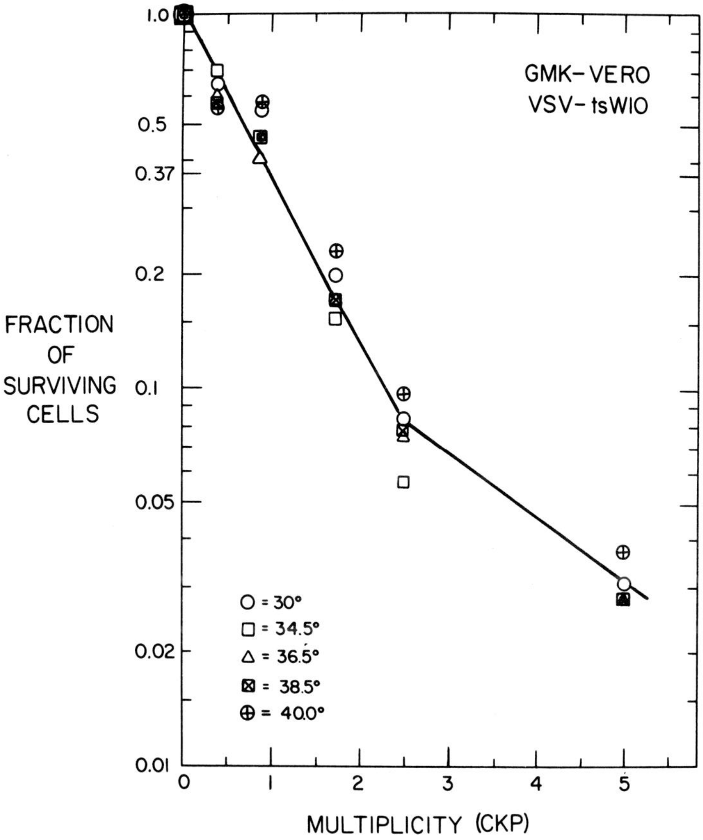

FIG. 9. Cell Killing Survival Curve of GMK-Vero Cells Infected with Vesicular Stomatitis Virus Mutant ts W10(IV). Virus-infected cells were held for 20 hours at the different temperatures noted in the figure prior to incubation at 40° for colony formation. Conditions otherwise as in Figure 7 (cf. [27]).

7. *Cell Killing by ts-Mutants of VS Virus: A Summary.* Table 1 summarizes our results to date and reveals that minimally functioning proteins in 3 complementation groups (*I, II, IV*) appear to be required for the expression of cell killing by VS virus. Cell-killing particle assays for mutants in complementation groups *III* and *V* indicate that newly synthesized and functional membrane proteins M and G, respectively, are not required for cell killing. In a recent study and one now in progress it appears that the capacity of these ts-mutants to turn off host cell protein synthesis at nonpermissive temperature correlates exactly with their capacity to kill cells [19,32]. For example, at 40°, mutant ts-G41 does not kill cells nor does it turn off cell protein synthesis. In contrast, mutant ts-

TABLE 1. Cell Killing by ts-Mutants of Vesicular Stomatitis Virus (Indiana)

Complementation Group	Mutant Designation	Cell Killing Activity at 40.0°
Wild type	$W^{+}HR_{C}$	+*
I_{ckp^-}	tsG114	−
	tsG11	−
	tsT1026	−
I_{ckp^+}	ts05	+
II	tsG22	−
III	tsG31	+
	tsG33	+
	tsT54	+
IV_{ckp^-}	tsG41	−
IV_{ckp^+}	tsW10	+
	ts0100	+
	tsW16B	+
V	ts044	+

* The "+" and "−" designations denote cell killing, or the lack thereof, respectively, for each ts-mutant and the wild type virus at 40.0°, the nonpermissive temperature. All mutants killed cells at the permissive temperature (30°). Studies now in progress indicate that all ts-mutants which kill cells at 40° also turn off cellular protein synthesis within a few hours after infection at 40°; mutants not lethal at 40° do not turn off cellular protein synthesis [32]—a finding in accord with a recent study of protein synthesis inhibition by ts-mutants of complementation groups *I,II* and *IV* [19].

W10 does kill cells at 40° and does turn off protein synthesis [32]. It is important to note that under conditions where mutant ts-G41 does not kill cells, there still is significant primary transcription [30,38]. Thus, although primary transcription is a requisite event for cell killing by VS virus, it is not a sufficient event, and it appears likely that the transcripts coding for viral proteins N (group *IV*) and NS (group *II*) need to be translated into minimally functional polypeptides in order to produce putative cell-killing factor and express cell killing.

Group *I* (L protein = transcriptase) ts-mutants pose a special problem: The UV-radiation data (Fig. 4) reveal a target size for the expression of CK particle activity of one-fifth of the genome. These data are sufficiently accurate and reproducible for us to conclude that the one-half of the genome required to code for the large L protein need not be intact for the virus to express cell killing. This conclusion is strengthened further by the recent assignment of the L protein gene to the most distal point from the mandatory single initiation site at the 3′ terminus of the VS virus genome [1]. This gene position would result in an apparent target size for the L protein cistron equivalent to that of the entire genome, and produce the same rate of inactivation for both plaque-forming and CK particle activity—a result incompatible with the observations reported in Figure 4. Consequently, we interpret these data to reflect a need for minimally functional *virion-associated,* but not newly synthesized, L protein to produce transcripts coding for N and NS proteins. The failure of ts-G114(*I*)

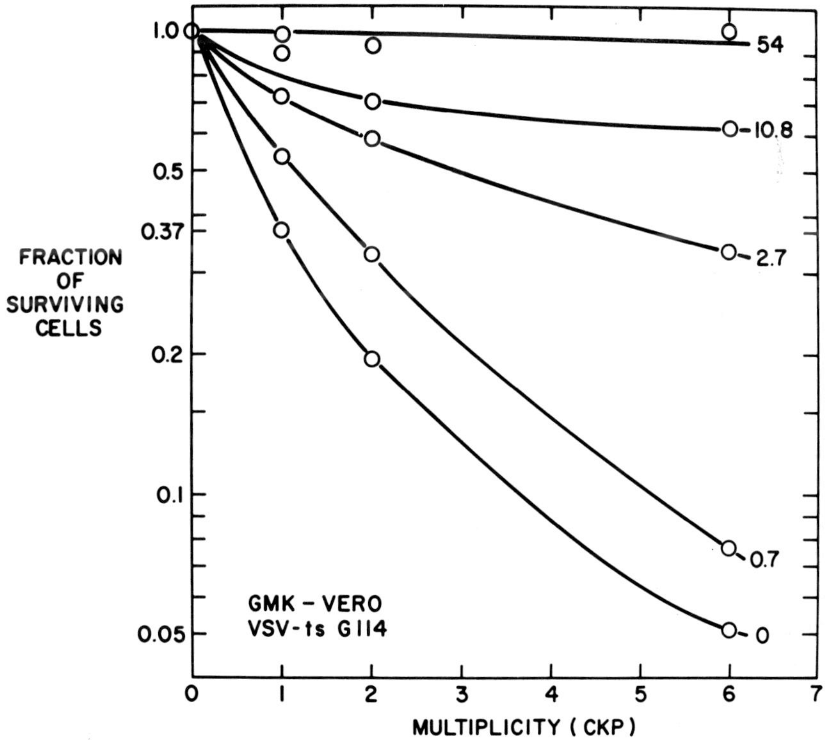

FIG. 10. Survival Curves of GMK-Vero Cells Exposed to Different Concentrations of Human Interferon for 24 hours at 37.5° and Challenged with Varying Multiplicities of VSV ts-G114(I). Following virus attachment and entry, infected and control cell monolayers were dispersed with trypsin, and the cells were counted and plated in growth medium containing VSV antiserum. All plates were held at 30.0° for 24 hours (pulse-infection) before shifting to 40.0° for colony formation from surviving cells. The abscissa represents the CKP multiplicities calculated from the 37% survival level in mock-treated cells ("O" interferon). The concentrations of interferon, in PR_{50} (VSV) units/ml, are indicated to the right of each survival curve. Individual points represent the averages of duplicate plates and agreed within ± 20% of the mean value (cf. [28]).

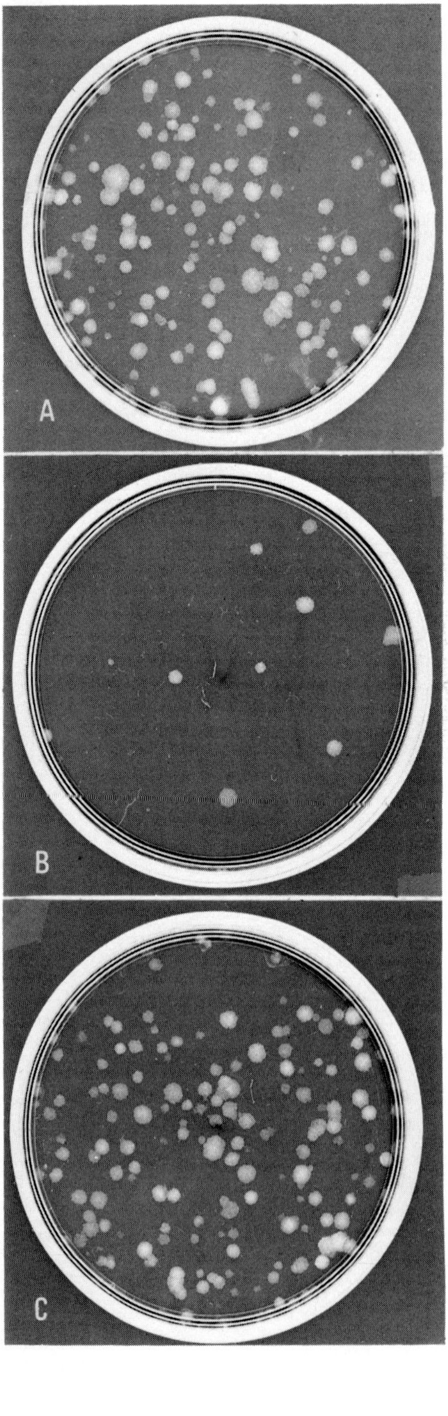

FIG. 11. Colony Formation by GMK-Vero Cells Treated With Human Interferon and Challenged with VSV ts-G114(I). Photographs show colonies from surviving cells incubated for 20 hours at 30.0° (pulse-infection) and 8 days at 40.0°. All petri dishes (50 mm) were inoculated with 200 cells from monolayers treated with interferon and challenged with VSV as follows: (A) 54 units/ml of interferon, no virus; (B) no interferon, $m_{CKP} = 6$; (C) 54 units/ml of interferon, $m_{CKP} = 6$.

and other group *I* mutants with heat-labile virion-associated transcriptase to kill cells at 40° is compatible with this explanation. However, ts0-5(*I*), a group *I* mutant with a heat-stable transcriptase in the virion [9], kills cells at 40°. To reconcile this observation, the UV-survival curve data and the evidence for a single site of initiation for transcription, we postulate that the defect of ts0-5 protein accommodates primary transcription at 40° and an additional critical function required for the production of cell-killing factor. Possibly a replicative event is involved, since several investigators have postulated a multifunctional role for this large viral protein [40]. Experiments are in progress to test this hypothesis.

8. Interferon Action Can Prevent the Expression of Cell Killing by VS Virus. We have examined the effect of interferon action in preventing cell killing by VS virus because of our earlier studies which indicated that interferon-mediated interference may inhibit primary transcription [24], an observation subsequently confirmed [20]. Indeed, interferon action may represent the only biologic system capable of preventing cell killing by viruses. For example, homotypic interference induced by defective-interfering particles [3,26] and intrinsic interference [13,31] both act in an all-or-none manner to block the replication of infectious virus, yet they do not prevent the expression of cell killing by the infecting virions. It is important to note that in both of these examples of viral interference, primary transcription is extant [12,13,34], thus fulfilling our initial postulate for cell killing.

The single-cell survival curves in Figure 10 demonstrate that cells exposed to increasing concentrations of interferon become increasingly refractory to cell killing by VS virus. Thus, GMK-Vero cells treated with 54 PR_{50} (VSV) units/ml of interferon for 24 hours become totally resistant to cell killing by VS virus. The photographs in Figures 11 A–C graphically illustrate this important point, and show that cells treated with this dose of interferon and infected with an otherwise lethal multiplicity of VS virus form colonies with the same plating efficiency as uninfected control cells.

ON THE REQUIREMENTS FOR CELL KILLING BY VESICULAR STOMATITIS VIRUS

Data acquired through use of the single-cell survival procedure have clarified several aspects of cell killing by vesicular stomatitis virus and established that: (1) the virion per se is not intrinsically toxic, (2) cell killing does not depend upon complete viral replication and (3) conventional [–]RNA defective-interfering particles do not kill cells. (Cell killing observed at $m_{DIP} \geq 50$–100 must be interpreted which caution since it is difficult to exclude the action of single contaminating cell-killing particles.)

Studies in progress with DI particles which contain a single-strand of covalently linked [±]RNA with complementary sequences [16], and potentially capable of forming dsRNA in cells, demonstrate that under conditions where a quantum yield of interferon can be induced by a single particle there is no cell killing [29]. However, cell killing by VS virus is expressed in viral interference under two different conditions where primary transcription is extant: (1) DI-particle-induced homotypic interference [12,34] and (2) the heterotypic intrinsic interference [13], but not under conditions where virion-directed transcription is inhibited significantly by the interferon system [20,24,28,30]. These findings led us to hypothesize that virion-associated (primary) transcription was a requisite, though not sufficient, event for cell killing by VS virus. In support of this hypothesis, we observed that cell-killing particle activity and *in vivo* virion-associated transcriptase activity are both lost at essentially equivalent rates upon subjecting the virion to (1) heat at 50°, or (2) UV radiation, and following infection (3) of interferon-treated cells, or (4) by temperature-sensitive mutants with a thermolabile virion transcriptase. Clearly, then, virion-associated transcriptase must be functional to initiate the molecular event(s) required for cell killing by this virus.

Surviving activity curves for CK particles following UV radiation provide an insight into how much of the viral genome need be intact to express cell killing. Since the capacity of VS virus to kill cells is about 5 times more resistant to UV radiation than is infectivity, it follows that about one-fifth of the viral genome must remain undamaged, presumably in order to produce primary transcripts. These transcripts are thought to function ultimately as messenger and/or template RNA to generate a putative cell-killing factor. Although the UV survival data do not distinguish between the need for full or partial transcription of a special sequence representing one-fifth of the genome, or one chosen at random, our studies on cell killing by the ts-mutants provide the insight to resolve this point. First, the apparent requirement for integrity of one-fifth of the genome is incompatible with a need for newly synthesized L protein in the cell-killing reaction. (As noted above, this protein is so large as to require coding by one-half of the genome. Furthermore, the location of the L protein gene at the 5′ terminus of the genome and the establishment of a mandatory initiation point at the 3′ end [1] means that a uracil dimer induced by UV radiation [33] anywhere in the genome would prevent transcription of L protein mRNA). Secondly, the lack of cell killing by some mutants of complementation groups *II* and *IV* at nonpermissive temperature indicates that the polypeptides of these two viral genes must be minimally functional to bring about cell killing. In this context, we note that the two genes coding for proteins NS and N (complementation groups *II* and *IV*, respectively) not only constitute one-fifth of the genome length, but they are the two genes most proximate to the single initiation site at the 3′ end and therefore are the only sequence of two genes that would present a target size proportional to their combined molecular weights (cf. [1]).

According to our knowledge to date, we propose as a working hypothesis that the following viral functions and reactions are required for the lethal action of VS virus on cells: (1) functional viron transcriptase to (2) synthesize transcripts derived from genes *IV* (N protein) and *II* (NS protein) which (3) must be translated into minimally functional polypeptides that (4) interact with minimally functional virion-transcriptase (L protein) and genome RNA from whence (5) formation of a putative cell-killing factor ensues and (6) cell killing follows. It is not clear whether the reaction postulated in step (4) involves or culminates in an early aberrant or abortive replicative event involving the genome RNA as a reactant, or the production of dsRNA—a possible candidate for cell-killing factor because of its profound effects on many biologic systems [4], including cytotoxicity [5]. Apart from its production during normal infection, the putative cell-killing factor of VS virus is produced under a wide range of conditions—upon exposure of cells to (1) defective CK particles, (2) UV-inactivated virions, (3) infectious particles in the presence of DI particle induced interference, (4) infectious particles in cells manifesting intrinsic interference, and (5) certain ts-mutants at nonpermissive temperatures. Since in each of these cases viral replication was aborted, though primary transcription was extant, we have clearly established that VS virus can kill cells in the absence of viral replication, but only when primary transcription ensues. In the single instance where primary transcription is demonstrably inhibited i.e., by interferon action [20,24,30] cell killing can be blocked totally [28].

In this resumé of our most recent studies we have sought to determine which viral components and functions are required for cell killing by vesicular stomatitis virus. Studies are now in progress to determine (1) the nature of the putative cell-killing factor and (2) how cell-killing factor acts to bring about the demise of the cell.

REFERENCES

1. Ball LA, White CN: Order of transcription of genes of vesicular stomatitis virus. Proc Natl Acad Sci USA 73: 442–446, 1976.
2. Baltimore D, Huang AS, Stampfer M: Ribonucleic acid synthesis of vesicular stomatitis virus. II: An RNA polymerase in the virion. Proc Natl Acad Sci USA 66: 572–576, 1970.
3. Bellet AJD, Cooper PD: Some properties of the transmissible interfering component of vesicular stomatitis virion preparations. J Gen Microbiol 21: 498–509, 1959.
4. Carter WA, DeClercq E: Viral infection and host defense. Science 186: 1172–1178, 1974.

5. Cordell-Stewart B, Taylor MW: Effect of double-stranded viral RNA on mammalian cells in culture: cytotoxicity under conditions preventing viral replication and protein synthesis. J Virol 12: 360–366, 1973.
6. Desmyter J, Melnick JL, Rawls WE: Defectiveness of interferon production and of rubella virus interference in a line of African green monkey kidney cells (Vero). J Virol 2: 955–961, 1968.
7. Dulbecco R: Production of plaques in monolayer tissue cultures by single particles of an animal virus. Proc Natl Acad Sci USA 38: 747–752, 1952.
8. Emerson SU, Yu Y: Both NS and L proteins are required for *in vitro* RNA synthesis by vesicular stomatitis virus. J Virol 15: 1348–1356, 1975.
9. Flammand A, Bishop DHL: Primary *in vivo* transcription of vesicular stomatitis virus and temperature sensitive mutants of five vesicular stomatitis complementation groups. J Virol 12: 1238–1252, 1973.
10. Franklin RM: Studies on the growth of vesicular stomatitis virus in tissue culure. Virology. 5: 408–425, 1958.
11. Huang AS: Defective interfering viruses. Ann Rev Microbiol 27: 101–117, 1973.
12. Huang AS, Manders ER: RNA synthesis of vesicular stomatitis virus. IV. Transcription by standard virus in the presence of defective interfering particles. J Virol 9: 909–916, 1972.
13. Hunt JM, Marcus PI: Mechanism of Sindbis virus induced intrinsic interference with vesicular stomatitis virus replication. J Virol 14: 99–109, 1974.
14. Hunt DM, Wagner RR: Location of the transcription defect in Group *I* temperature-sensitive mutants of vesicular stomatitis virus. J. Virol. 13: 28–35, 1974.
15. Lafay F: Envelope proteins of vesicular stomatitis virus: Effect of temperature-sensitive mutations in complementation groups *III* and *V*. J Virol 14: 1220–1228, 1974.
16. Lazzarini RA, Weber GH, Johnson LD, Stamminger GM: Covalently linked message and anti-message (genomic) RNA from a defective vesicular stomatitis virus particle. J Mol Biol 97: 289–307, 1975.
17. Luria SE: Méthodes statistiques appliquées à l'étude du mode d'action des ultravirus. Ann Inst Pasteur 64: 415–436, 1940.
18. Luria SE, Delbrück M: Interference between inactivated bacterial virus and active virus of the same strain and of a different strain. Arch Biochem 1.207–218, 1942.
19. McAllister PE, Wagner RR: Differential inhibition of host protein synthesis in L cells infected with RNA^- temperature-sensitive mutants of vesicular stomatitis virus. J Virol 18: 550–558, 1976.
20. Manders EK, Tilles JG, Huang AS: Interferon-mediated inhibition of virion-directed transcription. Virology 49: 573–581, 1972.
21. Marcus PI: Host-cell interaction of animal viruses. II. Cell-killing particle enumeration: Survival curves at low multiplicities. Virology 9: 546–563, 1959.
22. Marcus PI: Symposium on the biology of cells modified by viruses or antigens. IV. Single-cell techniques in tracing virus-host interactions. Bacteriol Rev 23: 232–249, 1959.
23. Marcus PI, Cieciura SJ, Puck TT: Clonal growth *in vitro* of epithelial cells from normal human tissues. J Exp Med 104: 615–628, 1956.
24. Marcus PI, Engelhardt DL, Hunt JM, Sekellick MJ: Interferon action: Inhibition of vesicular stomatitis virus RNA synthesis induced by virion-bound polymerase. Science 174: 593–598, 1971.
25. Marcus PI, Puck TT: Host-cell interaction of animal viruses. I. Titration of cell-killing by viruses. Virology 6: 405–423, 1958.
26. Marcus PI, Sekellick MJ: Cell killing by viruses. I. Comparison of cell-killing, plaque-forming, and defective-interfering particles of vesicular stomatitis virus. Virology 57: 321–338, 1974.

27. Marcus PI, Sekellick MJ: Cell killing by viruses. II. Cell killing by vesicular stomatitis virus: A requirement for virion-derived transcription. Virology 63: 176–190, 1975.
28. Marcus PI, Sekellick MJ: Defective interfering particles with covalently linked [±]RNA induce interferon. Nature, in press, 1977.
29. Marcus PI, Sekellick MJ: Interferon induction. I. Interferon induction by single stranded [+] RNA defective-interfering particles of vesicular stomatitis virus. Submitted.
30. Marcus PI, Sekellick MJ: Interferon action. III. Interferon-interference affects vesicular stomatitis virus primary transcription rate, but not mRNA stability. Submitted.
31. Marcus PI, Zuckerbraun HL: Newcastle disease virus RNA synthesis: Inhibition by the action of heterologous viral polymerase (intrinsic interference). *In:* Barry RD, Mahy BWJ (Eds): The Biology of Large RNA Viruses. New York, Academic Press, 1970, pp 455–481.
32. Marvaldi JL, Lucas-Lenard J, Sekellick MJ, Marcus PI: Cell killing by viruses. IV. Cell killing and the inhibition of cell protein synthesis require the same gene functions of vesicular stomatitis virus. Virology, in press, 1977.
33. Miller RL, Plagemann PGW: Effect of ultraviolet light on mengovirus. Formation of uracil dimers, instability and degradation of capsid and covalent linkage of protein to viral RNA. J Virol 13: 729–739, 1974.
34. Perrault J, Holland JJ: Absence of transcriptase activity or transcription-inhibiting ability in defective-interfering particles of vesicular stomatitis virus. Virology 50: 159–170, 1972.
35. Puck TT, Marcus PI: A rapid method for viable cell titration and clone production with HeLa cells in tissue culture: The use of X-irradiated cells to supply conditioning factors. Proc Natl Acad Sci USA 41: 432–437, 1955.
36. Puck TT, Marcus PI, Cieciura SJ: Clonal growth of mammalian cells *in vitro.* Growth characteristics of colonies from single HeLa cells with and without a "feeder" layer. J Exp Med 103: 273–284, 1956.
37. Szilágyi JF, Pringle CR: Effect of temperature-sensitive mutations on the virion-associated RNA transcriptase of vesicular stomatitis virus. J Mol Biol 71: 281–291, 1972.
38. Unger JT, Reichman ME: RNA synthesis in temperature-sensitive mutants of vesicular stomatitis virus. J Virol 12: 570–578, 1973.
39. Vogt M, Dulbecco R: Properties of a HeLa cell culture with increased resistance to poliomyelitis virus. Virology 5: 425–434, 1958.
40. Wagner RR: Reproduction of rhabdoviruses. *In:* Fraenkel-Conrat H, Wagner RR (Eds): Comprehensive Virology. New York, Plenum Press, 1975, pp 1–93.
41. Wagner RR, Schnaitman TC, Synder RM: Structural proteins of vesicular stomatitis virus. J Virol 3: 395–403, 1969.
42. Wong PKY, Holloway AF, Cormack DV: Characterization of three complementation groups of vesicular stomatitis virus. Virology 50: 829–840, 1972.

Studies on Infants and Children with Sex Chromosomal Abnormalities

Arthur Robinson, M.D.

With the development of technics for studying the chromosome constitution of man, a new departure in human genetics was instituted which led to the realization that chromosomal nondisjunction in man is a common cause of fetal wastage and human disease. Thirty to fifty per cent of spontaneously aborted fetuses are aneuploid [3]. At least 0.5% of live births have aneuploid karyotypes, most of which are associated with conditions which will interfere with normal growth and development.

In an attempt to determine possible causes of nondisjunction, an epidemiologic study was instituted, utilizing a sex chromatin examination of the amniotic membrane obtained from the placentas of 40,000 consecutive newborn babies at two Denver hospitals. This method of screening, always checked by a complete chromosomal analysis in any baby with a discrepant chromatin finding, proved to be both sensitive and accurate. It therefore became a reliable method of monitoring nondisjunction of the sex chromosomes in a defined population over the period of study (1964–1974). The over-all incidence of sex chromosomal anomalies in this population was determined to be about 0.3%. By following the occurrence of trisomy 21 (Down's syndrome) in the same population of newborns, it became possible to compare the frequency of nondisjunction of the sex chromosomes with that of an autosome [6].

The incidences, classified according to karyotype, are listed in Table 1.

When the births of the aneuploid newborns are listed according to month of birth, it can be noted that, if the year is divided into two six-month periods (May–October and November–April), there is a large disparity in the incidence of aneuploid births (Table 2). This was true for babies with trisomy 21 born to mothers less than 35 years of age, as well as for those

From the National Jewish Hospital and Research Center, Denver, Colorado, and Department of Biophysics and Genetics (contribution no. 688), University of Colorado Medical Center, Denver, Colo.

This study was supported by grants from the U.S. Public Health Service, 5R01-HD00622, and The Genetics Foundation.

The assistance of Joyce Borelli, M.D., Walter Goad, Ph.D., Michael Hudson, M.D., Mary Puck, M.A. and Katherine Tennes, M.A. is gratefully acknowledged.

TABLE 1. Denver Chromosomal Study (1 July 1964–30 June 1974)

Total number of newborns screened 40,371
(Only the last 15,641 screening tests included Y-body detection)
Males = 20,666
Females = 19,705
Ratio (M/F) = 1.05

Sex Chromosomal Abnormalities

In phenotypic males	*Number*	*Number/live birth*
47,XXY	20	.050%
47,XYY	4	.026% (= 4/15,641)
Mosaics	3	.0074%
In phenotypic females		
47,XXX	12	.030%
45,X	11	.027%
Mosaics	16	.040%
Total	66	.180%
Trisomy 21		
To mothers under 35	28	.072%
To mothers 35 and older	15	1.0%*

* Approximately 3.7% of all newborns in the study had mothers 35 years of age or older.

TABLE 2. Seasonal Contrasts

Newborns screened	November–April 19,318	May–October 21,053	$P_{\text{uniform incidence}}$*
Sex chromosomal abnormalities	23	43	6×10^{-3}
Trisomy 21 to mothers under 35	6	22	1×10^{-3}
Sum	29	65	2×10^{-4}
Trisomy 21 to mothers 35 and older	8	7	~.5

* i.e., the probability that were the incidence constant in time, so large a disparity as that observed would occur.

with sex chromosomal anomalies. Statistical analysis of the data indicates that we can be 95% confident that the seasonal increase in abnormalities is at least 1.6-fold, and 99% confident that it is at least 1.3-fold. The data also revealed a disparity in the annual occurrence of aneuploidy during 1969 and 1970 and the remainder of the period (Table 3).

We have concluded that there is a seasonal "epidemic" component of the incidence superimposed on an "endemic" component of approximately constant frequency, and that the epidemic incidence may be present for several consecutive years and be absent for several consecutive years. The cause(s) of this nonrandom occurrence of sex chromosomal anomalies is (are) unknown. Since the data for trisomy 21 suggest some parallelism in occurrence with the sex chromosome anomalies, it may be that all of the chromosomes in the human karyotype may be susceptible to the same factors producing nondisjunction. Since this nonrandom distribution has been found in a genetically stable population, the determination of the environmental factors which may influence disjunction of *all* human chromosomes becomes of great importance to the public health.

Aberrations of the sex chromosomes, which constitute about 50% of all chromosome anomalies, generally have less profound phenotypic effects than do those of the autosomes. However, the frequent association of sex chromosomal anomalies with medical and psychosocial problems in adults has been much discussed in the literature. Most of these patients have been nonrandomly selected by virtue of their having specific physical, intellectual, and emotional problems [2,4,7,9]. Several studies on more normal populations, however, have identified individuals with sex chromosomal aneuploidy who were normal physically, intellectually, and emo-

TABLE 3. Contrast of "Epidemic" Years with 1968–1970

	1968–1970	1964–67 and 1971–74	$P_{\text{uniform incidence}}$
Newborns screened	12,072	28,299	
Sex chromosomal abnormalities	8	54*	4×10^{-4}
Trisomy 21 to mothers under 35	6	22	.15
Trisomy 21 to mothers 35 and older	5	10	~.5

* The four individuals with a 47,XYY karyotype who were identified during the period 1971–1974 are omitted from this calculation, since Y aneuploidy was not screened for during the period 1968–1970.

tionally [1,5,9,12]. Since the incidence of sex chromosomal abnormalities in a variety of institutionalized populations is significantly higher than that of a population of unselected newborn infants, we have concluded that the latter are at a significantly increased but unknown risk for developmental or emotional problems. For this reason we are studying prospectively the developmental course, both physically and psychologically, of these infants. Some, but not all, of these individuals (47,XXX; 45,X; 47,XXY and 47,XYY) can be expected to have intellectual, emotional, or social problems in later life, and it is our hope to develop criteria, as early as possible, for separating those children who are at significant risk from those who are not, so that early preventative therapy may be instituted. In addition, it is important to resolve the current dilemma of the patient and physician faced with these conditions in utero by defining the long-term prognosis of infants with these anomalies.

METHODS

When chromosome analysis of a newborn confirmed that a sex chromosomal anomaly was present, parents were informed by the physician that a chromosomal variation had been found and that this information might be of future importance, although its significance could not be predicted. This candid approach was appreciated by the parents, most of whom were invited and agreed to join a longitudinal study, and signed an informed consent form to that effect.

Our screening program (1964–1974) identified 68 affected infants. Of these, 50 volunteered to enroll in our longitudinal study. It is of interest that 7 additional infants (11%) died in the neonatal period [11]. These latter results emphasize the importance of examining the karyotype of neonatal deaths, and suggest that newborns with sex chromosomal anomalies (especially 47,XXY and 45,X) may not have the favorable early prognosis that has generally been assigned to them.

Good working relationships with the families have been preserved by continuing responsiveness to their needs. Inclusion of all family members in the study helps the proband to feel less selected as "different" and provides siblings as "controls" for both genetic and cultural factors that might also influence growth and development.

As has been reported in detail before [10], the children are seen quarterly during the first year of life, semiannually during the second and third years, and annually thereafter. At these times, physical, developmental, and psychological examinations are performed. A detailed study of physical and emotional aspects of environment is facilitated by home visits and an easy rapport with each family.

RESULTS

The following are our findings at this point in these longitudinal studies. Many of the conclusions, it will be seen, are tentative, and suggest areas for further inquiry.

47,XXY Boys

Of the 20 newborns determined to have a 47,XXY karyotype, 3 died in the neonatal period (15%). Of the sixteen currently being followed, one is 10½ years old, twelve are 3–10 years old, and three are under 3. A summary of some findings in this group follows:

1. Small, soft testes were present in 87% by 2 years of age. Clinodactyly was another minor anomaly (Table 4). There was a notable absence of epicanthic folds and high arched palate.
2. Major anomalies were found in the three who died [11]. One additional boy had an inguinal hernia.
3. Head circumference at 4 years of age was within normal limits but skewed to the left (Fig. 1), with 13 out of 16 boys having a circumference less than the 50th percentile.
4. Height revealed a tendency to get relatively taller with increasing age (Fig. 2). This would fit in with the tall "eunuchoid" build of the adult with a 47,XXY karyotype. However, at 2 years of age these children are not tall.

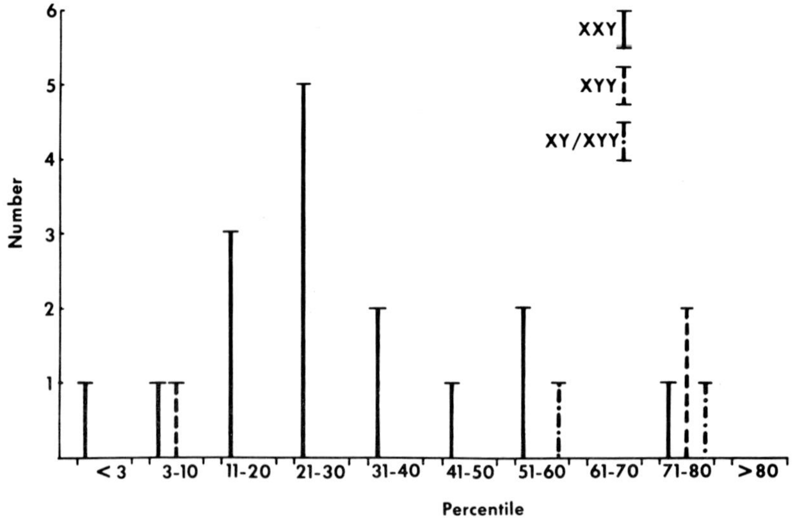

FIG. 1. Head circumference of male probands at 4 years of age by percentile [8].

TABLE 4. Minor Physical Abnormalities

Karyotype	No.	Clino-dactyly (5th finger)	Simian Lines	High Palate	3rd Fonta-nelle	Epi-canthic Folds	Wide-set Eyes	Testes (small, soft)
47,XXY	16	8	1	0	2	0	2	14
47,XYY	4	0	0	0	0	0	0	0
Mosaic Males	2	0	0	0	0	0	0	0
47,XXX	11	8	0	1	1	4	5	
45,X	9	0	4	6	2	2	0	
Mosaic Females	8	1	0	2	1	0	3	

5. Intelligence quotients varied over a wide range, with a mean score of 97. Although the ten sibs old enough to have intelligence tests had a mean score of 109, the difference is not significant (Fig. 3).

6. Language development was delayed an average of six months in 6 out of 15 probands (40%).

7. With one exception, these boys are showing normal emotional development. Half of them are more placid and "docile" than their siblings. The oldest child, coming from a very disturbed environment, is sufficiently immature to cause concern.

47,XXX Girls

There were a total of 12 newborns with this karyotype. Of these, one died in the neonatal period. Of the eleven remaining girls, two are over 10 years, and nine are 3 to 10 years old.

1. These too had a variety of minor defects (Table 4). Striking is the frequency of clinodactyly, epicanthic folds, and wide-set eyes. One had an atrophic kidney following pyelonephritis, and the one who died had marked immaturity and a birth weight of 785 grams.

2. At 4 years of age, nine of the girls had a head circumference which was less than the 50th percentile, and only two were above (Fig. 4).

3. Intelligence quotients ranged between 68 and 112 (Fig. 5). The mean I.Q. of the 11 probands who could be tested was 88, while that of 15 sibs was 108 (a significant difference ($p < .001$)).

4. Six of the girls have had speech problems.

5. Five of the eleven, all with difficult family relationships, have emotional disturbances which can best be described as inability to relate to others. On the other hand, 3 of 23 siblings have emotional problems.

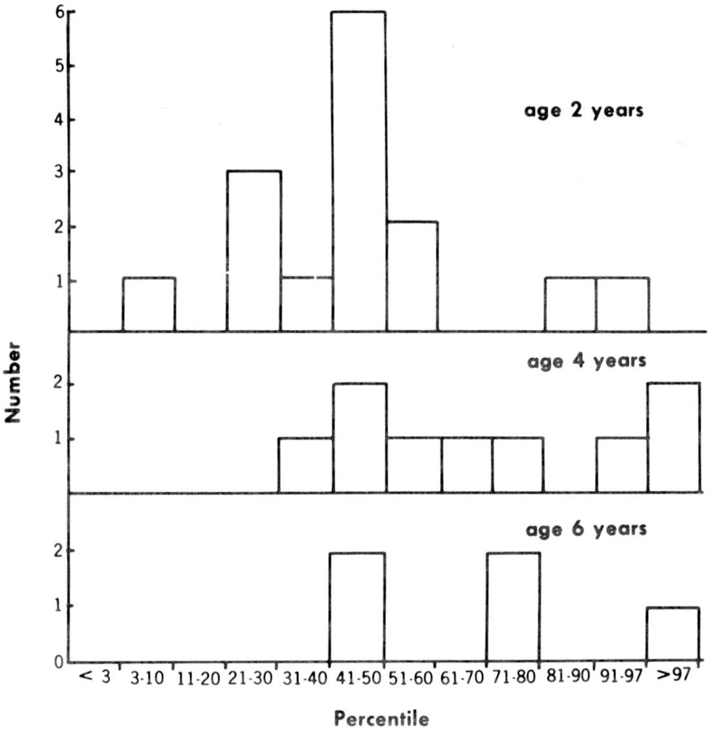

FIG. 2. Height of 47,XXY males at 2, 4, and 6 years of age by percentile [8].

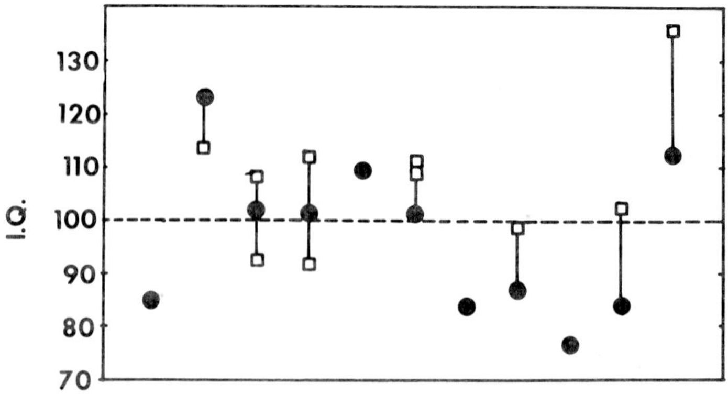

FIG. 3. Intelligence test scores for boys with 47,XXY karyotype (●) compared with siblings (□) to which they are connected by a vertical line. Individuals are arranged chronologically by birthdate left to right. Probands—Mean score 97. Siblings—Mean score 109.

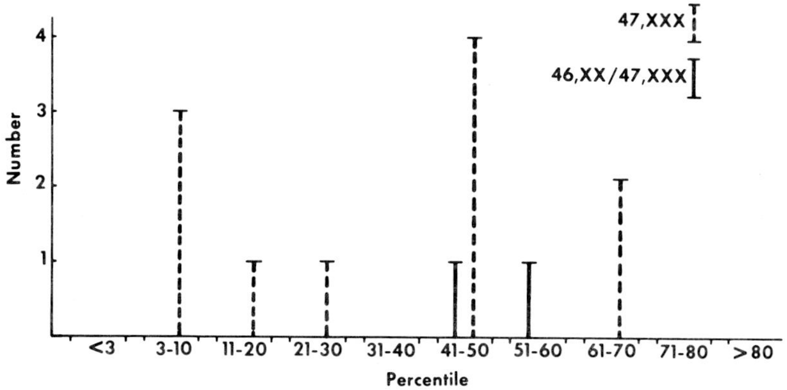

FIG. 4. Head circumference of nine girls with 47,XXX karyotype and two with 46,XX/47,XXX mosaicism by percentile [8] at 4 years.

Sex Chromosomal Mosaic Females

This group of girls may be considered as a control group, since one would suppose that any pathology associated with the completely abnormal karyotype would be modified by the presence of cells with a normal karyotype.

These consist of ten girls with a 45,X/46,XX karyotype, of whom only four remain in the study, one with a 45,X/46,XX/47,XXX karyotype, three

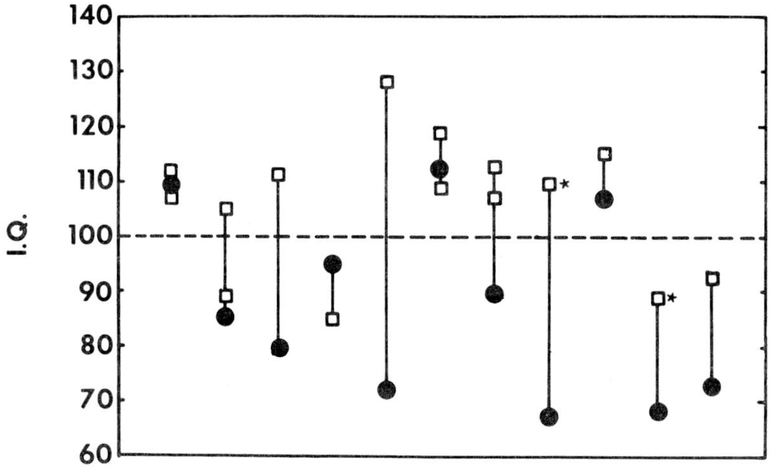

FIG. 5. Intelligence test scores for girls with 47,XXX karyotype (●) compared with siblings (□) to which they are connected by a vertical line. Individuals are arrayed chronologically by birthdate left to right. Probands—Mean score 88. Siblings—Mean score 108.

with a 46,XX/47,XXX karyotype (one not in the study), and one with a 45,X/47,XXX karyotype. The total number of female mosaics being followed, therefore, is eight. Two of these girls are over 10 and six are between 3 and 10 years old.

1. There were no major malformations in this group, and relatively few had minor malformations (Table 4).

2. Intelligence quotients of these girls (Fig. 6) varied between 80 and 122, with the mean score being 104. The mean for 11 siblings was 107.

3. Two of the probands have had emotional problems, as have two of fifteen siblings.

45,X (6) 46,XXq- (2), 46,XXr (1)

These have been lumped together since they all are monosomic for all or part of the long arm of X, a situation which has been associated with Turner's syndrome.

Three additional 45,X females expired in the newborn period. Nine girls are being followed. Two are over 10 years, six are 3 to 10 years, and one is under 3 years.

1. As would be expected, minor malformations are common in this group (Table 4). Additional findings unique to these girls are pedal edema in six and webbing of the neck in two. Striking was the frequency of pedal edema, simian lines and high arched palate. Four of the girls had major malformations (2 coarctations, 1 dextrocardia, 1 inguinal hernia). In addition, all three of those who died had major malformations [11].

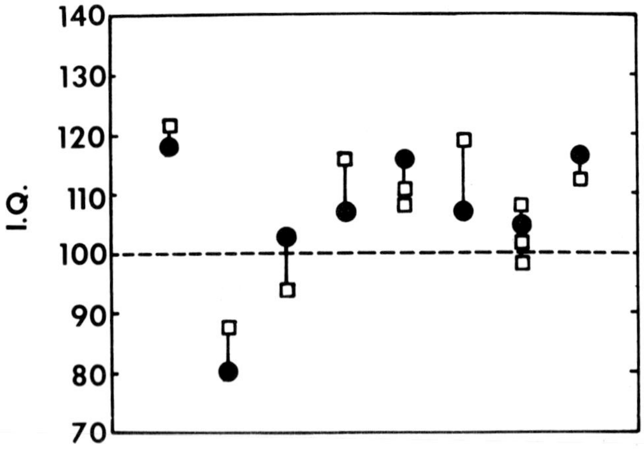

FIG. 6. Intelligence test scores for girls with a mosaic karyotype (●) compared with siblings (□) to which they are connected by a vertical line. Individuals are arrayed chronologically by birthdate left to right. Probands—Mean score 104. Siblings—Mean score 107.

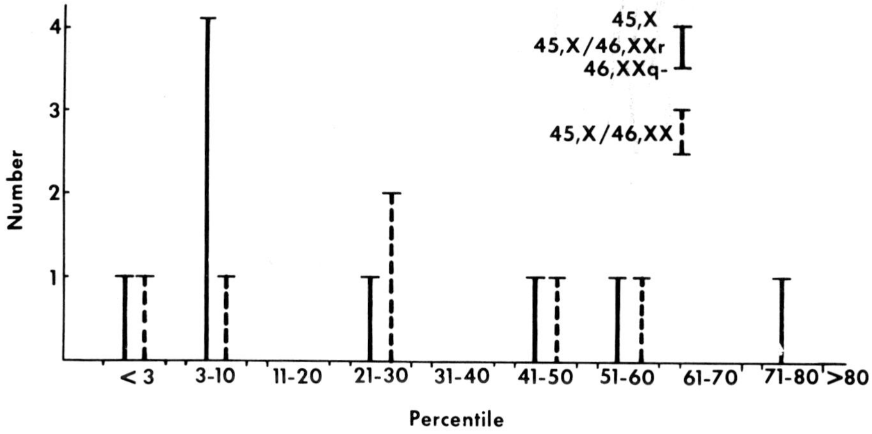

FIG. 7. Head circumference of girls with 45,X or variant and 45,X/46,XX mosaicism at 4 years of age by percentile [8].

2. Head circumference at four years of age was within normal limits, but again was skewed to the left, with six out of nine children being less than the 30th percentile. Four of seven 45,X/46,XX mosaics showed a similar distribution (Fig. 7).

3. The mean I.Q. of seven girls was 89, as compared to eight siblings with a mean score of 109 (Fig. 8).

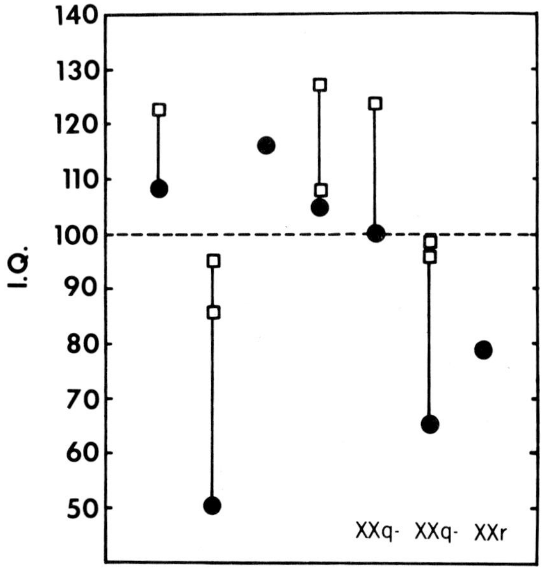

FIG. 8. Intelligence test scores for girls with 45,X, 46,XXq- and 46,XXr karyotypes (●) compared to their siblings (□). Probands—Mean score 89. Siblings—Mean score 109.

4. In general, these girls are well-adjusted and lacking in behavior problems.

XYY Boys

There are only four boys in this group, ages 3 to 6. One of these boys had both a congenitally dislocated hip and inguinal hernia. Otherwise, there were no anomalies. Head circumferences were within normal limits. Height is within normal limits for three boys. However, the fourth is over the 97th percentile. Speech development was delayed in two boys.

DISCUSSION

This is an interim report of a prospective study aimed at defining the incidence and course of the various types of sex chromosomal abnormalities in children. The predisposition for these karyotypes to occur in "epidemics" suggests environmental factors which are currently unknown.

The literature suggests that a significant number, but not all, individuals with these karyotypes will have intellectual, emotional, or social problems in later life. These studies should illuminate the pathological picture, if any, in these individuals.

Most of the children being followed are developing within normal limits. One might speculate that for each karyotype (omitting the 47,XYY group with very limited data) it is possible to divide the probands into two groups: those above a developmental median and those definitely below the median. It may be that the latter are the ones with a poor ultimate prognosis.

The intellectual development of the diagnostic groups, with the exception of the 47,XYY group, whose number is too small, is in general below the mean (Table 5). The mosaics, on the other hand, mirror the siblings in

TABLE 5. Percentage of Children with I.Q.'s below 90 in Each Subgroup Compared with Their Siblings

Probands	%	Siblings %
47,XXX	54	21
47,XXY	45	0
45,X	43	13
Female Mosaics	14	10

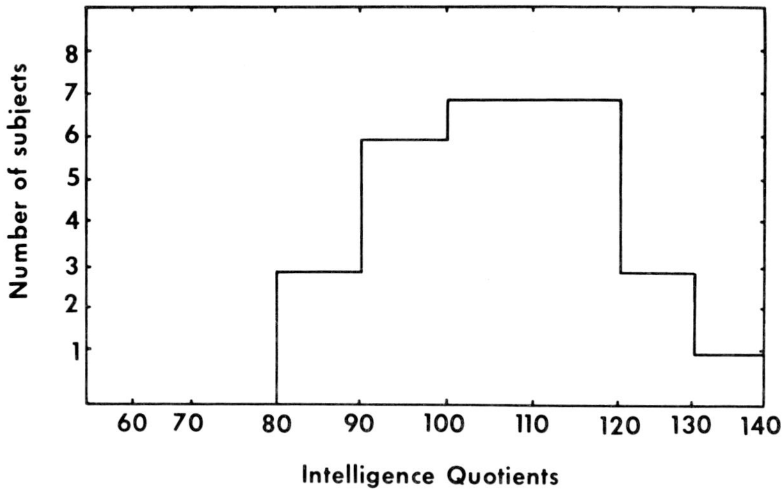

FIG. 9. Distribution of intelligence quotients of siblings of children with sex chromosomal anomalies (no more than one sibling per family).

revealing the normal distribution that one might expect in a random population (Fig. 9). This would argue against the "prophesy-fulfilling" concerns of some, since the mosaics are followed in exactly the same way as the other probands.

In summary, it is possible, as has been suggested by others [12], that sex chromosome imbalance may nonspecifically depress intellectual function, but not to the same degree as imbalance of the autosomes. The location of head circumference to the left of the median for many of these children may be a related phenomenon.

These children are entering adolescence. It remains to be seen how they will adjust to its normal strains. The data on adolescents and adults with these karyotypes have been obtained from biased populations of markedly deviant people. We as yet have no knowledge of the ability of less severely involved individuals living in a variety of different environments to adjust to the stresses of life.

REFERENCES

1. Baikie AG, Garson OM, Weste SM, Ferguson J: Numerical abnormalities of the X chromosome—Frequency among inpatients of a general hospital and in a general population. Lancet 1: 398–400, 1966.
2. Barr ML, Sergovich FR, Carr DH, Shaver EL: The triplo-X female: An appraisal based on a study of 12 cases and a review of the literature. Can Med Assoc J 101: 247–258, 1969.

3. Carr D: Chromosomes and abortion. *In:* Harris H, Hirschhorn K (Eds): Advances in Human Genetics. New York, Plenum Press, 1971.
4. Casey MD, Street DRK, Segall LJ, Blank CE: Patients with sex chromatin abnormality in two state hospitals. Ann Hum Genet 32: 53–63, 1968.
5. Close HG: Two apparently normal triple-X females. Lancet 2: 1358–1359, 1963.
6. Goad WB, Robinson A, Puck TT: Incidence of aneuploidy in a human population. Am J Hum Genet 28: 62–68, 1976.
7. Maclean N, Mitchell JM, Harnden DG et al: A survey of sex chromosome abnormalities among 4514 mental defectives. Lancet 1: 293–296, 1962.
8. McCammon RW: Human Growth and Development. Springfield, Charles C Thomas, 1970.
9. Polani PE: Abnormal sex chromosomes and mental disorder. Nature 223: 680–686, 1969.
10. Puck M, Tennes K, Frankenburg W et al: Early childhood development of four boys with 47,XXY karyotype. Clin Genet 7: 8–20, 1975.
11. Robinson A: Neonatal deaths and sex chromosome anomalies. Lancet 1: 1223, 1974.
12. Witkin HA, Mednick SA, Schulsinger F et al: Criminality in XYY and XXY men. Science 193: 547–555, 1976.

The Growth of HeLa Cells in a Serum-Free Hormone-Supplemented Medium

Gordon H. Sato, Ph.D.

Twenty years ago, in the midst of the development of the single cell plating technic, Ted Puck decided that tissue culture lacked quantitative rigor and that it would be a good idea to recruit phage workers into the field. At the time, I had just finished my graduate work with Max Delbruck and was working with Gunther Stent on the mechanism of replication of phage DNA. I am grateful today for the chain of accidental circumstances which brought me to Colorado to participate in events which would prove to have an important impact on cell biology. It is especially gratifying for me to take part in these proceedings because the present work of my laboratory is a direct extension of the work started in Colorado twenty years ago. At that time, Harold Fisher and I, under the guidance of Ted Puck, attempted to isolate the factors in serum necessary for the growth of HeLa cells. We were also to show that the serum could be replaced by a combination of albumin and alpha 1 globulin (Fetuin) [5].

My laboratory has been led back to this problem by our attempts to culture hormone-dependent cells. We were able to establish an ovarian cell line which we knew required hormonal conditioning of animal hosts for growth *in vivo* [3]. Nevertheless, we were unable to demonstrate hormonal dependency *in vitro* until we devised methods for depleting the serum component of the medium of hormones, a carboxy methyl cellulose column method for removing basic peptides from serum and a charcoal extraction procedure for removing steroids [2,12]. When the serum was appropriately depleted, the ovarian cells required the addition of crude luteinizing hormone to the medium for growth. Since luteinizing hormone was required at very high concentrations, we suspected that the active component was an impurity. The active component was isolated and named ovarian growth factor, or OGF [6]. Armelin, in our laboratory, showed that 3T3 cells

From the Department of Biology, University of California at San Diego, LaJolla, Calif.

This work was supported by a Human Biochemical Genetics Program Project Grant from the N.I.G.M.S. We thank Stanley Cohen for the gift of epidermal growth factor. The luteinizing hormone was provided by the National Institutes of Health hormone distribution program.

The careful performance of the experiments reported here by Ms. Sharon Hutchings is gratefully acknowledged.

required insulin, hydrocortisone and high concentrations of LH [1]. This led to a reexamination of the fractions generated in isolating OGF for growth activity for 3T3, and fibroblast growth factor was found [7]. The lesson was clear. To show hormonal requirements in culture, the serum had to be depleted of hormones, and some of the requirements for cells in culture would be novel peptides. This led us to hypothesize that serum could be replaced by hormones, some of which would be peptides, such as the somatomedins [13]. Experimental support for this notion was provided by Izumi Hayashi in our laboratory when she showed that a rat pituitary cell (GH_3) could grow in F12 medium supplemented with triiodothyronine, parathormone, transferrin, TSH releasing hormone, and a somatomedin preparation obtained from Dr. Knut Uthne [9]. We now know that the somatomedin preparation can be replaced with insulin and highly purified peptides from blood (unpublished observations).

In this report we give evidence that the serum in the medium of HeLa cells can be replaced by a complex of hormones.

MATERIAL AND METHODS

The procedures have been previously described [9]. In our laboratory we have introduced two modifications which are extremely important for experiments with serum-free media. The first is the use of acidic permanganate distillation of water in an all-glass still. We distill our water in a 5-liter boiling flask to which we add 50 grams of potassium permanganate and 5 ml of concentrated sulfuric acid. Our results are highly variable if we make our media with water from commercial stills and highly reproducible if we use the permanganate-distilled water. We also prepare our own F12 medium as described by Ham [8]. Again, we find this to be necessary because commercial powdered media, in our hands, give highly variable results, when we grow cells in serum-free media supplemented with hormones.

RESULTS

In Figure 1 are presented the growth curves of HeLa cells in F12 medium supplemented with 8%, 1% and .1% fetal calf serum, in serum-free F12 media, and in serum-free media supplemented with transferrin, insulin, epidermal growth factor and NIH luteinizing hormone. In serum-free media, HeLa cells do not grow, and, in fact, slowly die, whereas in hormone supplemented media they grow almost as well as they do in serum-supplemented media. We now know that if this media is further supplemented with hydrocortisone, the growth is equivalent to that in serum and the cells can be

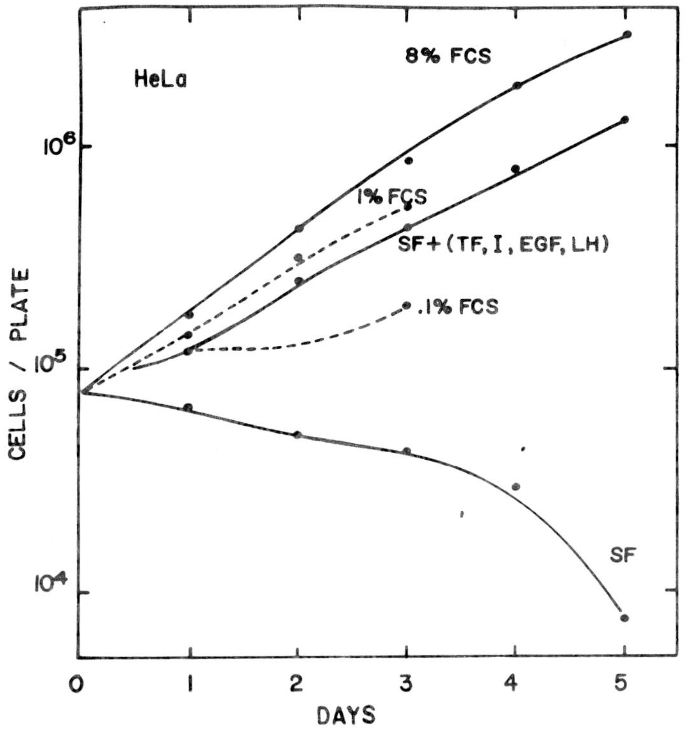

FIG. 1. Comparison of growth of HeLa cells in F12 medium supplemented with hormones versus that with serum. The cells were innoculated at 8×10^4 per 60 mm tissue culture dish into F12 medium supplemented with the indicated concentrations of fetal calf serum (FCS), no addition, or a combination of transferrin (TF, 5μg/ml), insulin (I, 10μg/ml), epidermal growth factor (EGF, 10ng/ml) and NIH luteinizing hormone (LH, 1μg/ml).

propagated indefinitely in serum-free, hormone-supplemented media (unpublished results).

In Figure 2 are presented the number of cells after five days of growth in serum-supplemented medium, hormone-supplemented media and hormone-supplemented media from which each of the individual hormones is deleted. Deletion of either epidermal growth factor, transferrin, luteinizing hormone or insulin results in diminished growth. If no hormones are added to the serum-free medium, the cell number is below the initial plating level.

DISCUSSION

A serious obstacle to progress in cell culture has been the inability to achieve completely defined media in which every constituent is under con-

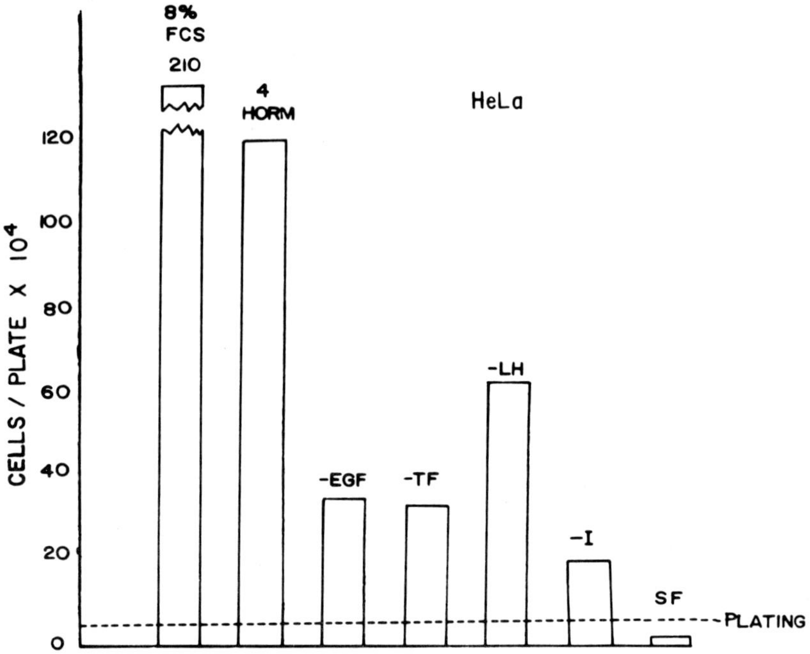

FIG. 2. Cells were plated at a density of 8×10^4 cells per 60 mm tissue culture dish and counted on day 5. Growth is shown in F12 medium supplemented with 8% FCS, 4 hormones (EGF, TF, I, LH) in the concentrations given in Figure 1; only three of the above hormones with the deletion of the indicated hormone; or no addition (SF).

trol. This has hampered the establishment of certain cell types in culture and has left unrealized many of the contributions which tissue culture could make to our understanding of integrated physiology. Due to the work of Eagle, Ham and many others [4,8] our knowledge of the nutritional requirements of cells is nearing completion, and attention should be turned toward understanding the undefined component of the medium, namely, serum. Our hypothesis is that serum provides hormones, and a few accessory factors such as transferrin. This hypothesis leads to a distinctive experimental approach in analyzing the components of serum necessary for cell growth. Some of the hormones are well known, classic hormones which can be purchased from biochemical supply houses. In fact, the bulk of the serum requirement can be met by such hormones. Thus, rat pituitary cells, GH_3, were found to grow in 0.2% serum when the medium was supplemented with transferrin, triiodothyronine, parathormone and TRH. To obtain further reduction in serum concentration, it became necessary to turn to another class of hormones typified by the somatomedins. These cannot be bought from biochemical supply houses, and in the case of somatomedins have no storage organ in the body and so must be isolated from blood. Other than somatomedins and a number of less well defined factors, this class of hormones would include ovarian growth factor (OGF), nerve growth factor (NGF), epidermal growth factor (EGF) and fibroblast growth factor (FGF) [10–17]. Although the physiologic roles of these substances have not yet been firmly established, they are found in blood and, when purified, are active at very low concentrations. We therefore surmise that they operate as regulators and can be provisionally classified as hormones. Furthermore, we would contend that many hormones remain to be discovered and that their discovery will come in large part through culture experiments.

In the light of our hypothesis and experience we can now suggest a simple explanation for the failure of early attempts to replace whole serum with various fractions from serum or tissues. In essence, the problem was that early investigators were trying to isolate one or a few factors from whole serum, when the serum in fact was required to furnish both a number of classic hormones and novel peptides. Many classic (and novel) hormones are present in very low concentrations in serum and often bind to serum proteins. Therefore, not only truly novel but also classic hormones may appear as "growth factors," the true nature of which is difficult to discern in classic serum fractions; other growth-promoting entities such as transferrin and essential fatty acids also would appear as serum "growth factors." Our contention is that before seeking after such "factors" one should first determine the requirements of the cells for classic hormones.

Twenty years ago I was assigned the problem of finding the components of serum necessary for the growth of HeLa cells. I apologize for the extreme tardiness and I thank Ted Puck for being so patient.

REFERENCES

1. Armelin HA: Pituitary extracts and steroid hormones in the control of 3T3 cell growth. Proc Natl Acad Sci USA 70: 2702–2706, 1973.
2. Armelin HA, Sato G: Cell cultures as model systems for the study of growth control. World Symposium on Carcinogenesis, John Hopkins University Press, 1973.
3. Clark JL, Jones KL, Gospodarowicz D, Sato G: Growth response to hormones by a new rat ovary cell line. Nature [New Biol] 236: 180–181, 1972.
4. Eagle H: Nutrition needs of mammalian cells in tissue culture. Science 122: 43–46, 1955.
5. Fisher N, Puck TT, Sato GH: Molecular growth requirements of single mammalian cells. Proc Natl Acad Sci USA 42: 900, 1956.
6. Gospodarowicz D, Jones K, Sato G: Purification of a growth factor for ovarian cells from bovine pituitary glands. Proc Natl Acad Sci USA 71: 2295–2299, 1974.

7. Gospodarowicz D: Purification of a fibroblast growth factor from bovine anterior pituitary gland. J Biol Chem 250: 2515–2520, 1975.
8. Ham RG: Clonal growth of mammalian cells in a chemically defined synthetic medium. Proc Natl Acad Sci USA 53: 288–293, 1965.
9. Hayashi I, Sato G: Replacement of serum by hormones permits the growth of cells in a defined medium. Nature 259: 132–134, 1976.
10. Levi-Montalcini R: Growth control of nerve cells by a protein factor and its antiserum. Science 143: 105–110, 1964.
11. Luft R, Hall K (Eds): Somatomedins and some other growth factors. *In:* Advances in Metabolic Disorders, Vol. 8, 1975.
12. Nishikawa K, Armelin H, Sato G: Control of ovarian cell growth in culture by serum and pituitary factors. Proc Natl Acad Sci USA 72: 483–487, 1975.
13. Sato G: The role of serum in culture. *In:* Litwack G (Ed): Biochemical Actions of Hormones, Vol. III. New York, Academic Press, 1975.
14. Smith GL, Temin HM: Purified multiplication-stimulating activity from rat liver cell conditioned medium: comparison of biological activities with calf serum, insulin and somatomedin. J Cell Physiol 84: 181–192, 1974.
15. Taylor JM, Cohen S, Michell WM: Epidermal growth factor: high and low molecular weight forms. Proc Natl Acad Sci USA 67: 164–171, 1970.
16. Tormey DC, Imrie RC, Mueller GC: Identification of transferrin as a lymphocyte growth promoter in human serum. Exp Cell Res 74: 163–169, 1972.
17. Van Wyck JJ, Underwood LE, Hintz RL et al: The somatomedins: a family of insulin-like hormones under growth hormone control. Rec Prog Hormone Res 30: 259–318, 1974.

Laboratory Workshops

LABORATORY WORKSHOP CONTRIBUTORS

Raja B. Bahu, Ph.D., Northwestern University Medical School, Chicago, Illinois.

Diane Bass, Department of Pathology, University of California, School of Medicine, San Diego, LaJolla, California.

Henry Burr, Ph.D., Department of Biological Sciences, Stanford University, Stanford, California.

Carlo Croce, M.D., Wistar Institute of Anatomy and Biology, Philadelphia, Pennsylvania.

Dean H. Hamer, Ph.D., Department of Biological Chemistry, Harvard Medical School, Boston, Massachusetts.

George H. Keller, Ph.D., Department of Biological Chemistry, The Milton S. Hershey Medical Center, The Pennsylvania State University, Hershey, Pennsylvania.

D. Scott Linthicum, Ph.D., Department of Pathology, University of California, School of Medicine, San Diego, LaJolla, California.

Paul K. Nakane, Ph.D., Department of Pathology, University of Colorado, School of Medicine, Denver, Colorado.

Indu Parikh, Ph.D., Department of Molecular Biology, Burroughs Wellcome Company, Research Triangle Park, North Carolina.

Guido Pontecorvo, Ph.D., Imperial Cancer Research Fund Laboratories, Lincoln's Inn Fields, London, England.

John Scholes, M.D., Department of Pathology, College of Physicians and Surgeons, Columbia University, New York, New York.

Stewart Sell, M.D., Department of Pathology, University of California, School of Medicine, San Diego, LaJolla, California.

David Shapiro, Ph.D., Department of Biochemistry, University of Illinois, Urbana, Illinois.

Enid Sisskin, M.S., Department of Pathology, College of Physicians and Surgeons, Columbia University, New York, New York.

Susan F. Slovin, M.S., Department of Pathology, College of Physicians and Surgeons, Columbia University, New York, New York.

John M. Taylor, Ph.D., Department of Biological Chemistry, The Milton S. Hershey Medical Center, The Pennsylvania State University, Hershey, Pennsylvania.

Charles A. Thomas, Jr., Ph.D., Department of Biological Chemistry, Harvard Medical School, Boston, Massachusetts.

Barbara Wilson, M.S., Department of Pathology, University of Colorado, School of Medicine, Denver, Colorado.

Synthesis and Applications of Complementary DNA

John M. Taylor, George H. Keller, David Shapiro, Henry Burr, John Scholes

The hybridization of complementary nucleic acid strands is an essential experimental tool for the study of many fundamental problems in eukaryotic gene expression. Hybridization methodology has been developed into a quantitative assay system through the efforts of many investigators, including R. J. Britten, J. O. Bishop, J. Paul, S. Spiegelman and their associates. These technics have been applied to a wide variety of eukaryotic and prokaryotic DNAs and RNAs. In the case of the eukaryotic genome, DNA reassociation kinetics have revealed three general sequence complexity classes [1,2]: highly reiterated components present in thousands of copies per genome (including satellite DNA); moderately repetitive sequences varying from tens to hundreds of copies per genome; and nonrepetitive (unique) sequences assumed to be present in only one copy per haploid genome. The functional significance of these DNA classes is generally not understood. Some highly repetitive sequences appear to be concentrated in the heterochromatin surrounding the chromosome centromeres, whereas the moderately repetitive sequences are interspersed throughout the genome [1–3]. The highly repetitive sequences may have a structural role in chromosome organization [4]. Moderately repetitive sequences have been postulated to have a regulatory role in gene expression and appear to consist of families of related sequences [1,2]. However, the genes coding for ribosomal RNA and transfer RNA belong to the middle repetitive DNA class [5,6]. Certain messenger RNA (mRNA) genes such as histones [7,8] and feather keratins [9] have also been clearly identified in the middle repetitive class. Furthermore, as much as a fourth of the cytoplasmic poly(A) containing mRNA may have been transcribed from middle repetitive DNA [10,11]. The great majority of cytoplasmic mRNA, however, appears to arise from the unique sequence DNA. Specific mRNAs, including those coding for globin [12,13], silk fibroin [14] and ovalbumin [15,16], have been shown to be transcribed from the nonrepetitive DNA class. Although early studies on gene transcription products employed radioactively labeled mRNAs in hybridization assays, the more recent studies have depended upon the availability of specific DNA copies of mRNA sequences.

The preparation of radioactively labeled complementary DNA (cDNA) copies of specific mRNAs has provided a sensitive tool to investigate the question of gene reiteration, as well as a wide variety of other problems concerning eukaryotic gene expression. The initial preparations of cDNA copies of a eukaryotic mRNA in the laboratories of Baltimore [17] and Spiegelman [18] stimulated the rapid exploitation of these hybridization probes. Specific cDNAs have been employed to quantitate mRNA levels in cell development and differentiation [19,20], and in hormone-stimulated tissue differentiation [21,22]. The use of human globin cDNA to investigate the molecular nature of thalassemia [23–27] indicates the potential use of nucleic acid hybridization in the study of disease. The use of cDNAs as screening agents for the detection of tumor virus sequences has recently been reviewed [28]. Specific cDNAs have also been employed as substrates for the nucleotide sequence determination of the corresponding mRNAs [29,30]. More general studies with cDNAs are directed towards understanding the sequence complexity of total cell mRNA populations and their relationships with nuclear RNA and mRNAs from different tissues [31–36].

The purpose of this review is to outline technics in the preparation and use of complementary DNAs. Radioactively labeled cDNA is made from a messenger RNA template via a viral RNA-dependent DNA polymerase, which requires a primer for its activity. In practice, an oligo(dT) primer is annealed to the 3′-terminal poly(A) segment of an mRNA preparation, and the primer-template is incubated with the enzyme in the presence of deoxynucleotide substrates and required reaction components. The cDNA product is readily isolated and employed in hybridization studies. The various methods discussed below were presented recently at a laboratory workshop in the Given Institute of Pathobiology, Aspen, Colorado.

Reverse Transcriptase

The most commonly used enzyme for cDNA synthesis is the RNA-dependent DNA polymerase (reverse transcriptase) from avian myeloblastosis virus (AMV). Properties of the enzyme and details of its assay conditions have been reviewed [37–39]. A partially purified enzyme preparation is available commercially (Schwarz-Mann, Boehringer), and it also can be obtained from Dr. J. W. Beard (Life Sciences Inc., St. Petersburg, FL) through a special NIH program. Purification of the enzyme from AMV virions is generally performed according to procedures modified from Kacian and Spiegelman [37], which involve detergent lysis of the virions followed by DEAE-cellulose and phosphocellulose column chromatography. Storage of the enzyme at −20°C in a buffer containing 0.2 M potassium phosphate at pH 7.2, 2 mM dithiothreitol, 0.2% Triton X-100, and 50% glycerol (to prevent freezing), will maintain activity for at least a year. Repeated freezing and thawing should be avoided, since this treatment readily denatures the enzyme. It may be desirable to check the enzyme activity with a poly(rA)$_n$ template and an oligo(dT) primer, where one unit of activity is equivalent to one nmole of dTMP incorporated into an acid precipitable product in 10 minutes at 37° under standard assay conditions [37].

Synthesis of cDNA

Sterile technic is recommended during the early phase of cDNA synthesis to prevent the degradation of mRNA. Glassware and reagents can be autoclaved and sterile H_2O should be used. Nucleases can be inactivated by treatment of materials with 0.5% diethylpyrocarbonate. A convenient reaction vessel to use is a 1 ml glass vial with a tapered conical bottom and a teflon-lined cap (e.g., Reacti-vials from Kontes). The vessel should be siliconized (Siliclad, from Clay-Adams) to prevent adsorption of protein and nucleic acid to the vessel wall.

Radioactively labeled cDNA with a specific activity of greater than 10^4 cpm per ng can be obtained by employing [^{3}H]dCTP (about 25 Ci per mmole) in the substrate mixture. A cDNA with this label is adequate for most hybridization purposes. However, more sensitive probes (higher specific radioactivity) can be prepared by also employing [^{3}H]dGTP and [^{3}H]dATP, or by using ^{32}P-labeled substrates. The use of radioactive dTTP is not recommended because nonspecific poly(A) sequences would be scored in hybridization assays. Shortly before the reaction, the labeled substrates should be added to the reaction vessel at 0° and evaporated to dryness under a stream of nitrogen (to evaporate the ethanol in which the radioactive materials are stored). Subsequent additions to the reaction vessel should be performed at 0–4° until the final incubation temperature is to be employed.

Deoxyribonucleotide substrate concentrations of approximately 50 μM or greater have been found to be required for the synthesis of long cDNAs (close to a full-length copy). In the case of dTTP, higher concentrations (100–200 μM) may be required, possibly due to some copying of the poly(A) regions of mRNA templates. If only dCTP is to be radioactively labeled, it is convenient to adjust the reaction mixture to a final concentration of 200 μM in dTTP, dGTP, and dATP. Higher substrate levels have not been found to be required in most cases.

The dried radioactive substrates are then dissolved in the reaction buffer which contains (final concentration) 50 mM Tris at pH 8.3, 10 mM dithiothreitol, 6 mM $MgCl_2$, and 100 mM NaCl. There appear to be allowable variations in these reaction conditions that do not significantly affect the length of the final product. Among the more complete studies on cDNA reaction conditions is the work of Monahan et al. [40]. Generally, 50 mM Tris over a range of pH 7.5–8.5 is used. Dithiothreitol has been employed over a range of 1–20 mM and is preferable to 2-mercaptoethanol. The magnesium ion concentration is more critical with an optimum range of 6–8 mM. Both NaCl and KCl have been employed over a range of 50–150 mM. Low salt concentrations lead to a decreased yield of acid-precipitable cDNA. However, low salt concentrations appear to facilitate the synthesis of longer cDNA, possibly due to a destabilization of some inherent secondary structure of the mRNA template.

A particular property of reverse transcriptase is its requirement for a DNA primer in the synthesis of a cDNA to an mRNA template [37–39]. The location of a poly(A) segment at the 3′-terminus of most eukaryotic mRNAs allows oligo(dT) to be conveniently employed as a primer [17,18]. Ribosomal RNA, a frequent contaminant of mRNA preparations, is not copied by the enzyme under these conditions. Oligo(dT) primer over a range of 1–50 μg per ml has been employed, and 10–30 μg per ml is an effective concentration for the reaction conditions described here. An excess of oligo(dT) is recommended in order to minimize copying of the poly(A) region of the template. Commercially available primer of 12–18 nucleotides in length is satisfactory for cDNA synthesis. It may be possible to prepare cDNAs against internal portions of the template by employing other primers such as oligo(A) or oligo(U), but the location of these internal initiation points would be difficult to determine.

Since the reverse transcriptase possesses a significant DNA-dependent DNA polymerase activity, actinomycin D should be added to prevent the formation of double stranded material. A concentration of greater than 50 μg per ml should be used to block this activity, although as much as 200 μg per ml can be included without interfering with single stranded DNA synthesis [41]. Since actinomycin D is light-sensitive, it should be protected from direct light.

The most frequently employed templates for cDNA synthesis have been poly(A)-containing eukaryotic mRNAs, including viral RNAs isolated from eukaryotic host cells. A wide range of concentrations have been used, depending largely on the availability of the template. A level of 20–30 μg per ml, in water or dilute buffer, is convenient and effective for cDNA production with about 50–100 units per ml of enzyme. The mRNA should be dissolved in water or very dilute buffer, and it should be added to the reaction vessel after all other components, with the exception of the enzyme.

Reverse transcriptase is then added to the mixture and the reaction vessel incubated at 37° for 20 min. Longer incubation times tend to result in the

TABLE 1. Reaction Mixture for cDNA Synthesis

Component	Final Concentration
Tris, pH 8.3	50 mM
Dithiothreitol	10 mM
$MgCl_2$	6 mM
NaCl	100 mM
[^{3}H]dCTP, 25 Ci/mmole	50 μM
dTTP	200 μM
dATP	200 μM
dGTP	200 μM
Oligo$(dT)_{12-18}$	30 μg/ml
Actinomycin D	50 μg/ml
mRNA	20 μg/ml
AMV reverse transcriptase	100 units/ml

production of double stranded material. Under the above conditions, most cDNA synthesis is completed by about 10 minutes. Higher incubation temperatures, such as 46°, have been suggested to destabilize secondary structure of the mRNA and thereby yield longer cDNAs [40]. Synthesis of cDNA is completed earlier in this case.

The concentrations of the reaction components which have been found to be essentially optimal for the production of long cDNAs to rat albumin mRNA and chicken ovalbumin mRNA are summarized in Table 1.

Following the incubation period, the reaction mixture is adjusted to about 1% in sodium dodecyl sulfate and 10 mM EDTA at pH 7.0 to stop the reaction. The final reaction mixture is then chromatographed on Sephadex G50 in a 0.7 × 40 cm column which has been equilibrated with 100 mM NaCl, 10 mM EDTA, and 20 mM Tris at pH 7.1. The excluded fraction (containing ^{3}H-labeled cDNA free of unreacted substrates and reaction salts) is collected (Fig. 1). Complete elution of the column can be accomplished with about 4 bed volumes of buffer before reuse. Sephadex G75 or G100 is just as effective and can also be employed. Narrow columns should not be siliconized, because the resultant hydrophobic surface will interfere with the buffer flow properties. However, if glass wool is placed beneath the gel, it should be siliconized to prevent cDNA adsorption.

The cDNA fraction from the Sephadex column is adjusted to 0.2 M NaOH and heated at 75° for 15 minutes to destroy the mRNA template, then quickly cooled to 0°. If RNase is employed to destroy the template, particular care is essential to remove or inactivate any DNase that usually contaminates RNase preparations. A 5–10 μl aliquot of the cDNA solution

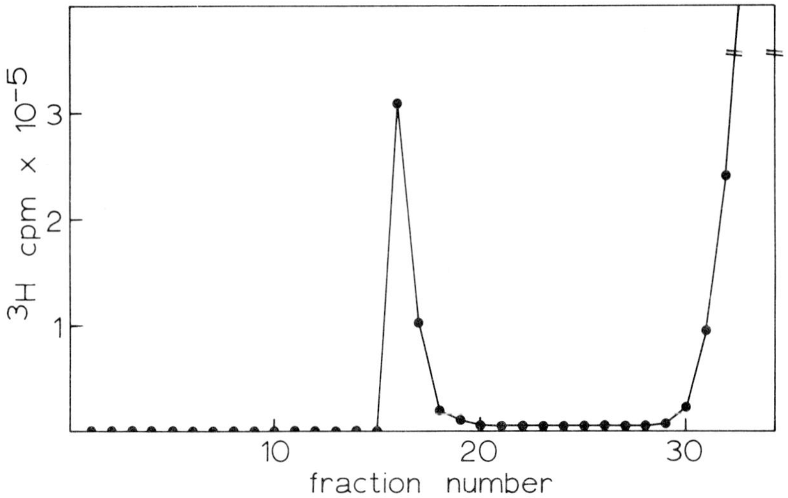

FIG. 1. Gel filtration of cDNA reaction mixture. The completed final reaction mixture (adjusted to 1% in sodium dodecyl sulfate and 10 mM EDTA) was applied to a 0.7 × 40 cm bed of Sephadex G-50 which has been equilibrated in 1X column elution buffer. The sample was eluted with 1X column buffer under a 50 cm pressure head and approximately 0.5 ml fractions were collected. Aliquots of 10 μl from each fraction were added to a scintillation fluid and the radioactivity was determined.

is removed to determine the radioactivity; and the approximate yield of cDNA product can be estimated from the specific radioactivity of the initial labeled substrate, assuming a base composition of 25% for each base and an average deoxynucleoside monophosphate molecular weight of 325. An additional aliquot containing 10,000–20,000 cpm of radioactivity is usually employed to examine the size distribution of the cDNA product by sedimentation on isokinetic alkaline sucrose gradients (Fig. 2). The use of these gradients for the determination of sedimentation coefficients has been extensively investigated by McCarty et al. [42], and the relationship between the sedimentation of DNA and its molecular weight have been examined by Studier [43]. The size of each cDNA preparation should be examined, since there is a significant effect of length on hybridization reaction kinetics [40,44]. Linear sucrose gradients (Fig. 2) can also be used to examine the size distribution of a cDNA preparation, but they do not provide direct information regarding molecular weights. The remainder of the cDNA preparation should then be neutralized with 1 M HEPES at pH 7 and 100–200 μg of carrier RNA can be added. After adjusting the solution to 0.2 M in NaCl, the nucleic acids are precipitated with 2.5 volumes of ethanol at −20° overnight.

Additional methods of size estimation can be employed when more detailed information on the length of a cDNA preparation is required. Electrophoresis of labeled cDNA with unlabeled DNA markers of known length in polyacrylamide gels under the denaturing conditions of 98% formamide is a very useful technic [41]. When longer mRNAs have been copied, the cDNA contour length on electron photomicrographs allows a direct measurement of size distribution [40]. An alternative approach to estimate cDNA lengths involves a titration of radioactively-labeled mRNA with the cDNA, followed by nuclease digestion to determine the extent of cDNA protection of the mRNA [40].

Synthesis of cDNA

1. Reaction vials are siliconized and autoclaved before use. Place vials in an aluminum block and cool the block on ice. Add 2.8 μg of [^{3}H]dCTP (about 25 Ci/mmole) and evaporate to dryness under a stream of nitrogen in a hood.

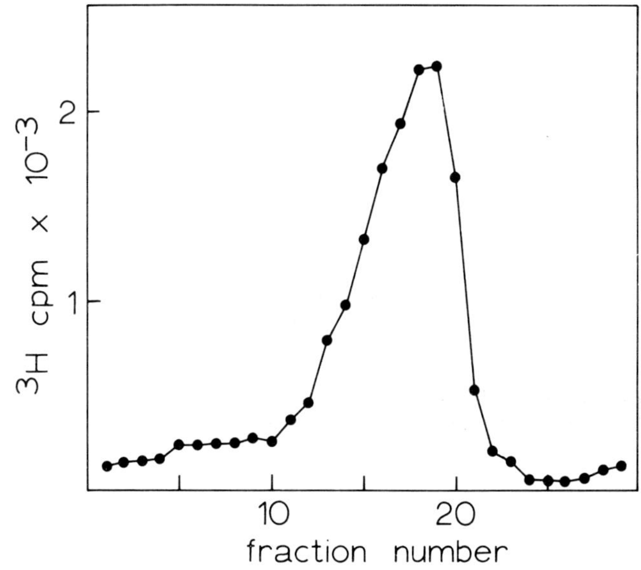

FIG. 2. Sedimentation of a cDNA preparation on alkaline sucrose gradients. An albumin cDNA preparation was alkali denatured, and an aliquot containing about 30,000 cpm was overlaid on a 5–20% linear sucrose gradient in 0.2 M NaOH, 0.1 M NaCl, and 0.01 M EDTA. The gradients were sedimented at 280,000 g_{max} for 20 hr at 4°. Fractions of about 0.4 ml were collected and the radioactivity was determined. The direction of sedimentation shown is from left to right.

2. Add the following components at 0°C with sterile technic in the indicated order to a final reaction volume of 100 μl:

10 X Reaction buffer	10 μl
Stock nucleotides	12 μl
Oligo(dT) solution	15 μl
Actinomycin D solution	13 μl
mRNA solution and H_2O to give	50 μl

3. Mix thoroughly with a Vortex mixer and let stand on ice for 5 min.
4. Add 2 μl of AMV reverse transcriptase (containing about 10 units of enzyme activity). Mix thoroughly and incubate at 37° for 20 min.
5. Add 15 μl of 10X SDS-EDTA buffer and mix. The reaction mixture may be stored at −20° until further use.
6. Add the final reaction mixture to a sterile column of Sephadex G50 (0.7 × 40 cm) that has been equilibrated in 1X column elution buffer. Elute the column with this buffer and collect fractions of about 0.5 ml in sterile glass disposable tubes.
7. Measure the radioactivity of 10 μl aliquots from each fraction. Combine the 3–4 fractions (1.5–2.0 ml) at the void volume containing the peak of cDNA radioactivity. Before application of another reaction mixture, elute the entire column with about 4 column volumes of buffer until all remaining radioactivity has been removed.
8. Adjust the pooled cDNA fraction to 0.2 M NaOH with 10 M NaOH. Incubate at 75° for 15 min. Cool the solution quickly to 0°.
9. Measure the radioactivity of a 10 μl aliquot. Adjust another aliquot containing about 20,000 cpm to 1X in NaCl-EDTA buffer and 0.2 M NaOH. Neutralize the remainder of the cDNA preparation with 1 M HEPES at pH 7, add 150 μg of carrier RNA, adjust to 0.2 M NaCl, add 2.5 volumes of ethanol, and precipitate at −20° overnight.
10. Examine the 20,000 cpm aliquot on isokinetic sucrose gradients in 0.9 M NaCl, 2 mM EDTA, and 0.1 M NaOH, or on linear sucrose gradients in 0.1 M NaCl, 2 mM EDTA, and 0.1 M NaOH.

Reagents for cDNA Synthesis

10X Reaction Buffer

500 mM	Tris, pH 8.5 at 25°
1000 mM	NaCl
60 mM	Mg acetate

Autoclave. Then adjust to 100 mM in dithiothreitol.

Stock Nucleotides Substrate
3.0 mg each of dTTP, dATP, and dGTP dissolved in 3.0 ml H_2O. Store at −70°.

Oligo(dT) Solution, 0.20 mg per ml
Oligo$(dT)_{12\text{-}18}$ (from Collaborative Research or P-L Biochemicals) dissolved in H_2O. Store at −70°.

Actinomycin D Solution, 0.40 mg per ml
Actinomycin D (from P-L Biochemicals or Calbiochem), 200 μg per bottle. Add 0.5 ml sterile H_2O via syringe in dark. Store at $-70°$.

10X SDS-EDTA Buffer, pH 7.0

100 mM	EDTA·Na_2
10%	SDS

Adjust to pH 7.0. Autoclave.

10X Column Elution Buffer, pH 7.5

0.1 M	EDTA·Na_2
0.2 M	HEPES
1.0 M	NaCl

Adjust to pH 7.5. Autoclave.

5X NaCl-EDTA Buffer, pH 8.0

10 mM	EDTA·Na_2
4.5 M	NaCl

Adjust to pH 7.0. Autoclave. Use for isokinetic alkaline sucrose DNA gradients.

Carrier RNA Solution
1.5 mg per ml RNA in 1X SDS-EDTA buffer

Nucleic Acid Hybridization

The rate of hybridization or reassociation of complementary nucleic acid sequences depends upon several reaction parameters: the cation concentration, incubation temperature, initial nucleic acid concentration, and nucleic acid fragment length. These parameters will be discussed briefly as they apply to the specific procedures discussed here. A more detailed treatment of the principles and kinetics of nucleic acid hybridization has been presented in several excellent reviews [1,2,45–49]. This discussion will review only certain practical considerations involved in performing the hybridization assays. The kinetic considerations are presented elsewhere [45–49].

The salt concentration and the temperature of incubation determine the stringency (also referred to as criterion [1,2,45]) of complementary sequence matching in a hybridization reaction. For example, increasing the cation (e.g., Na^+) concentration and decreasing the temperature increases the reaction rate by facilitating the association of complementary nucleic acid strands. However, these less stringent conditions also facilitate sequence mismatching. Although the relationships between divergent sequences can be assessed by altering the criterion of association, the precise detection of specific sequences requires more stringent conditions

[2]. In the case of mRNA:DNA hybridization, the potential degradation of the RNA generally leads to a compromise in the reaction parameters. Higher salt concentrations, which minimize the repulsion of the negatively charged strands, are often chosen to increase reaction rates, and a modest increase in temperature minimizes sequence mismatching. Since higher temperatures facilitate RNA strand scission, many investigators include formamide (up to 50%) in the hybridization buffer which lowers the optimum incubation temperature [50]. The optimum temperature for DNA reassociation has been shown to be about 20–30° below the Tm (the temperature at which 50% of the reacting strands are in duplex structures) [45]. The effect of temperature, as well as cation concentration, is generally assumed to be the same for RNA:DNA hybridizations as well.

The rate of hybridization is usually measured in terms of the parameter Cot (for DNA reassociations) or Rot (for RNA hybridizations). This term is defined as the product of time in seconds and the initial concentration of the nucleic acis in moles of nucleotides per liter [1,2,45]. For convenience, one can assume that one mole of deoxynucleoside monophosphate is 325 g, so that Cot = (g of nucleotide per liter/325 g per mole) (elapsed reaction time in sec). The $Cot_{1/2}$ (the value of Cot at which the reaction of a particular species has proceeded to half completion) is a useful parameter for comparing different species in hybridization reactions. The $Cot_{1/2}$ is proportional to the sequence complexity of the reactants (the complexity of a nucleic acid preparation is the length of the longest composite nonrepeating sequence measured in nucleotide base pairs). A measurement of $Cot_{1/2}$ is made directly from an analysis of the experimentally determined Cot curves, which also provide relative information on the abundance of the different species in the reaction population. Hybridization data are frequently expressed as the equivalent Cot, in which the experimental reaction rate is normalized to the rate that would occur if the reaction had been performed in 0.18 M monovalent cation at 25° below the Tm. A table of salt correction factors is available [45]. However, correction of experimentally determined values to an equivalent Cot (or Rot) in situations where several parameters have been changed may not be valid.

The most common technic currently in use for the detection of hybrids involve either single strand specific nuclease (S1 nuclease) digestion or hydroxyapatite (HAP) chromatography. S1 nuclease from *Aspergillus oryzae,* which is available commercially, degrades single stranded DNA with both exo- and endonuclease activity. The double stranded hybrids remaining after S1 nuclease digestion contain no single stranded material and reflect the number of base pairs participating in the hybrid. In the case of hydroxyapatite, hybrid structures containing more than about 50 base pairs adsorb to HAP in about 0.12 M sodium phosphate buffer at pH 6.8, whereas single stranded material does not adsorb. Hybrids can then be eluted with 0.4 M phosphate buffer. The duplexes isolated by HAP chromatography can contain significant amounts of single stranded material, and reflect the number of molecules participating in hybrids. Thus, the S1 nuclease and HAP technics provide somewhat different experimental information due to the nature of the hybrids detected.

Assay of RNA : DNA Hybrids with S1 Nuclease

The hybridization of mRNA and cDNA is performed in a sterile buffer containing 0.5 M NaCl, 2 mM EDTA, 0.1% sodium dodecyl sulfate, and 25 mM HEPES at pH 7.4. A range of 0.3–0.6 M NaCl is commonly employed to increase reaction rates by facilitating hybrid formation. The EDTA chelates divalent metal ions which catalyze the hydrolysis of RNA and activate nucleases. Hybridization buffers may also be treated with an ion-exchange resin such as Chelex 100 (BioRad Laboratories) to remove metal ions. Sodium dodecyl sulfate is included to inactivate low levels of nucleases that might be present. Additional nonreacting carrier RNA is frequently included in hybridization buffers to further protect the mRNA under investigation from degradation or loss. Buffers which are commonly employed include HEPES, Tris, and PIPES; but the higher temperature coefficient of Tris makes it the least desirable buffer. Phosphate buffers should be avoided since they inhibit S1 nuclease.

At the high salt concentration employed, a hybridization incubation temperature of 68° is a recommended optimum for specific hybrid formation. However, these high temperatures facilitate RNA degradation at the longer reaction times (i.e., 48 hours or more) which might be required to achieve high Rot values. To avoid very long reaction times, it may be desirable to use stock RNA solutions of higher concentration. Many investigators avoid the higher temperatures by including 50% formamide in the hybridization buffer, which allows the reaction to be performed at 43°. The Tm of a hybrid is reduced about 0.7° for each 1% of formamide [50].

The incubation should be performed in small polypropylene tubes or siliconized glass tubes to prevent adsorption losses. If test tubes are used, solutions should be covered with mineral oil to prevent evaporation. Alternatively, the hybridization may be performed in siliconized glass capillary tubes which are sealed by brief flaming at each end after filling. It is convenient to begin the hybridizations for each time point at the same starting time. When the predetermined incubation times are reached, the reactions are immediately terminated by immersing the tubes in an alcohol-dry ice bath. The tubes are stored frozen until all samples have been collected, and the hybrids can then be assayed under the same conditions.

Conditions for digesting single stranded DNA by S1 nuclease without introducing breaks in double stranded DNA have been described in detail [51]. The enzyme shows greatest activity in 10–200 mM NaCl and 0.01–1.0 mM Zn^{++} at a pH of 3.5–5.0. A half inhibition of activity is caused by 2 mM phosphate, 20 μM pyrophosphate, 85 μM dAMP, and 1 μM dATP. Double stranded digestion activity is enhanced by the lower pH range because of acid depurination. This problem can be essentially overcome by increasing the pH of the digestion to pH 5.0 and by using 100–200 mM NaCl. Under appropriate conditions, S1 nuclease from *Aspergillus* can cut duplex strands at the site of mismatched base pairs, the strand opposite a nick in duplex DNA, superhelical duplex DNA, and heteroduplex loops. The enzyme requires a minimum concentration of single stranded nucleotides, and denatured DNA is usually added to cDNA solutions to assure effective digestion.

The *Aspergillus* S1 nuclease is available commercially (Miles, Sigma) but each preparation may vary somewhat in its reaction properties, and the appropriate amount of enzyme required to effectively digest single stranded DNA under the chosen reaction conditions must be examined. Various amounts of enzyme are incubated for 60 minutes with a constant amount (500–1,000 cpm) of cDNA and 10–50 μg per ml of denatured DNA. (This DNA can be from any source, e.g., calf thymus, *E. coli,* etc.) Acid precipitable radioactivity is determined and compared to an enzyme-free control sample. Single stranded specificity of an S1 nuclease preparation can be investigated with native and denatured (by boiling and quick cooling) radioactively labeled DNA. Following incubation with the enzyme, native DNA should be greater than 95% double stranded, whereas denatured DNA shoud be less than about 5% nuclease resistant.

Following sample collection in the mRNA:cDNA hybridization reaction, to each sample add S1 nuclease buffer containing 100 mM NaCl, 0.5 mM Zn acetate, and 100 mM Na acetate at pH 5.0. When test tubes have been used, the hybridization reaction sample must be mixed thoroughly with the added solutions, then centrifuged briefly to separate oil and aqueous phases. In the case of capillaries, they must be rinsed thoroughly with the S1 nuclease buffer. Adjust the sample to 20–50 μg in denatured calf thymus DNA, add an appropriate amount of enzyme, and incubate for 60 minutes at 37°. The remaining hybrid material is then acid-precipitated in the presence of an additional 200 μg of calf thymus DNA carrier, and the precipitates are collected on nitrocellulose (Millipore) or fine glass fiber filters (Whatman GF/C). A typical hybridization kinetic curve is shown in Figure 3.

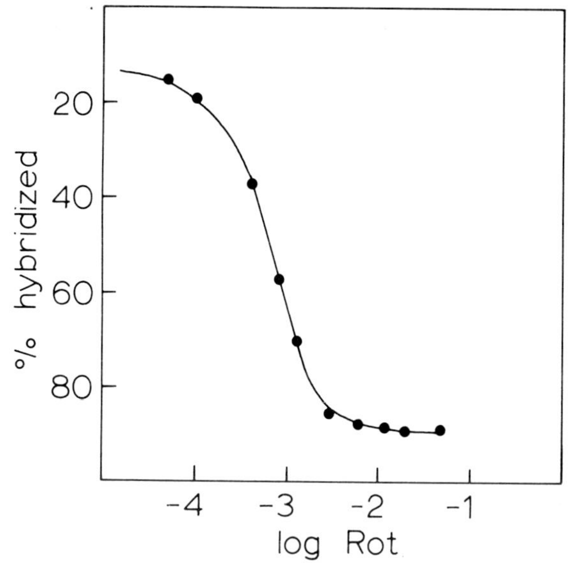

FIG. 3. Hybridization kinetics of a specific cDNA and its mRNA template. Each hybridization reaction mixture contained 5 ng albumin mRNA and 0.04 ng ^{3}H-labeled cDNA. Incubation times varied from 30 sec to 8 hr. The hybrids were assayed with S1 nuclease.

Procedure for mRNA:cDNA Hybridization and S1 Nuclease Assay

1. Combine the following components in the indicated order in a small test tube:

50 μl	2X Hybridization buffer
10 μl	Carrier RNA
10 μl	[^{3}H]cDNA; 5,000–10,000 cpm

RNA sample and H_2O to a final volume of 100 μl.

2. Mix thoroughly and immediately dispense 10 μl of the hybridization mixture into each of 9 polypropylene tubes.
3. Overlay the mixture with 8 drops of mineral oil, cap the tubes, centrifuge to 2,000 xg_{max}, and remove the tubes from the centrifuge as quickly as possible.
4. Immerse the tubes in the 68° water bath and incubate for the desired length of time. Terminate the reactions by immersing the tubes in an ethanol dry ice bath, and store the tubes at −20° until they are to be assayed.
5. Add 400 μl of S1 nuclease buffer containing 20 μg of denatured calf thymus DNA to each tube and mix. (Appropriate amounts of buffer and stock DNA solution can be mixed together just prior to dispensing, e.g., 4.0 ml buffer and 200 μl of DNA solution.)
6. Add predetermined amount of S1 nuclease solution (usually 10–50 μl) to each tube and mix thoroughly. Centrifuge to 2,000 xg_{max} briefly.
7. Incubate the samples at 37° for 60 min.

8. Add 200 μl of calf thymus DNA solution to each tube and mix.
9. Add 2 ml of cold 12% TCA to each tube, mix thoroughly, and let stand at 0° for 30 min.
10. Collect the precipitates by filtration through glass fiber filters (Whatman GF/C) previously moistened with 10% TCA. Wash the incubation tube and the sides of the filtration apparatus three times with cold 5% TCA.
11. Dry the filters in glass vials, add 0.75 ml of solubilizer and dissolve the precipitates, then add 10 ml of scintillation fluid and measure radioactivity.

Reagents

2X Hybridization Buffer

1.0 M	NaCl
50 mM	HEPES
4 mM	EDTA
0.2%	SDS

Adjust to pH 7.4. Autoclave.

S1 Nuclease Buffer

100 mM	Na acetate
0.5 mM	Zn acetate
100 mM	NaCl

Adjust to pH 5.0. Autoclave.

Denatured Calf Thymus DNA
1 mg per ml in 100 mM NaCl. Place in boiling water bath for 10 min, then cool quickly to 0°. Store at 4°.

Carrier RNA
5 mg per ml of tRNA or any suitable rRNA, dissolved in 10 mM HEPES, pH 7.4 containing 1% SDS. Store at −20°.

Carrier DNA
1 mg per ml of calf thymus DNA in 50 mM NaCl and 5 mM EDTA, pH 8.

S1 Nuclease Solution
1 mg per ml in S1 nuclease buffer containing 20% glycerol. Store at −20° (do not subject to repeated freezing and thawing).

12% TCA, 5% TCA, mineral oil (Squibb)

Introduction to DNA:cDNA Hybridization

Much of our knowledge of the principles and technics of eukaryotic DNA reassociation comes from the pioneering work of Britten and his co-workers. Several detailed and excellent reviews on this methodology have been published [1,2,46,48,49], and they should be consulted before the beginner embarks on his journey into nucleic acid hybridization analysis.

The following discussion reviews only the practical aspects of DNA:cDNA hybridization in certain basic experiments as presented in the recent workshop.

Among the fundamental information that is important in understanding eukaryotic gene expression is a knowledge of the reiteration frequency of the structural gene under investigation. Although radioactively labeled mRNAs have been employed, a cDNA to the specific mRNA is better suited to this study. Gene frequency experiments which utilized specific cDNAs were among the first applications of these hybridization probes [15,16]. The reliability of these studies, however, depended on the fidelity of the cDNA sequence matching to their specific complementary genomic DNA sequences. Examination of base pair fidelity in hybrids was accomplished by thermal melting studies. Protocols for gene frequency analysis and thermal melting analysis will be reviewed, and the use of hydroxyapatite (HAP) in hybridization analysis will be introduced. Hybridizations are performed in solutions with relatively high DNA concentrations, because reactions with immobilized DNA do not readily achieve the high values of Cot required for eukaryotic gene frequency experiments.

DNA Fragmentation

The fragment size (length) of eukaryotic DNA must be controlled in order to obtain interpretable data in reassociation analysis. Recent studies show that gene sequences with a wide range of reiteration frequencies are interspersed in all eukaryotic genomes examined [2,3,52–54]. These observations are interpreted to mean that unique (single-copy) sequences are alternated with various families of repeated (i.e., middle repetitive, highly repetitive, and pallidromic) sequences [2]. The implication of this model is that long DNA fragments are likely to contain repetitive sequence elements capable of a relatively rapid formation of networks upon reassociation. Therefore, the length of the DNA fragments employed is important to the interpretation of DNA:cDNA hybridization studies. A variety of fragment lengths are useful for sequence interspersion studies. For gene frequency studies in particular, short fragments of 300–500 nucleotides in length have been generally employed because they are conveniently obtained [2]. There are three methods for shearing DNA into short fragments that are currently in use.

A convenient method for shearing DNA into short fragments is by passage of DNA solutions through a small orifice under high pressure. Special high pressure cells (Aminco) and presses (Aminco, Carver) are available commercially. Fragment size is primarily a function of pressure and orifice configuration, which determines the flow rate. Solutions of 0.1–2.0

mg per ml of DNA in 0.1–0.2 M buffer are passed through the cell twice. Pressures of 12,000–16,000 psi generally yield fragments of 400–500 nucleotides mean length; however, each cell should be calibrated. Lower pressures result in longer fragments.

High speed blending of DNA solutions is another useful shearing technic [45]. The fragment size is a function of the blending speed, solution viscosity, and configuration of blades and vessel. To produce short fragments, viscosities are increased by adjusting solutions to about 70% glycerol and blending at about 50,000 rpm for 30–40 minutes, with the vessel cooled in a dry ice-ethanol bath.

Sonication with a microprobe tip is also widely used for DNA fragmentation. Intermittent bursts of the sonifier and cooling solutions in an ice bath minimize excessive damage of DNA molecules [55]. The DNA fragments produced with this technic are often more heterogeneous in length than with the above technics.

All methods of shearing native DNA into short fragments yield material that is heterogeneous in length over a 4- to 5-fold range or greater. The kinetics of hybridization do not appear to be significantly affected by this size distribution problem. However, fragment lengths should be determined for each preparation. DNA sizes are conveniently estimated by sedimentation on alkaline sucrose gradients, agarose gel filtration, or agarose gel electrophoresis, and comparison with standard markers of known length.

Since divalent cations can activate nucleases (possible trace contaminants) or catalyze nucleic acid degradation (especially at higher temperatures), the sheared DNA solutions (as well as all buffers) can be passed over Chelex-100 (BioRad) to remove the contaminating ions. DNA fragments can be concentrated by adjusting solutions to 0.2 M sodium acetate and precipitating the DNA with 2 volumes of ethanol at $-20°$ for a few hours. Phosphate buffers (PB) should be removed by dialysis in this case, since phosphate is precipitated by ethanol.

Conditions for DNA : cDNA Hybridization

The reactions are conveniently carried out in small polypropylene tubes (Falcon—6 ml, or Beckman—1.5 ml), and the solutions are overlaid with a few drops of mineral oil to prevent evaporation. Alternatively, large glass capillaries (50–100 μl) may be used.

DNA samples (including the cDNA) should be denatured by heating for about 3 minutes in a boiling H_2O bath, or at least 10° above the Tm for the particular buffer and salt concentration in use. Samples are immediately cooled to 60° for hybridization to begin.

The standard conditions of salt and incubation temperature (the standard criterion) for DNA reassociation are 0.12 M sodium phosphate, pH 6.8 (PB) at 60° (45). The optimum reassociation rate lies between 15° and 30° below the Tm. If higher salt concentrations are employed to increase the reaction rate, then the incubation temperature should be increased accordingly to maintain the criterion. This consideration is especially important if an analysis of repeated sequences is to be undertaken, since changes in conditions significantly alter the amount of repetitious DNA detected [2,45]. However, changes in the criterion can provide information on the possible relationships between families of divergent sequences. Sodium dodecyl sulfate (0.1%) may be added to the buffer to inactivate possible contaminating nucleases. Chelating agents such as 1–2 mM EDTA are also sometimes included in the buffer to minimize degradation. However, their effects on subsequent hybrid assay conditions, particularly in the case of hydroxyapatitie, should be checked.

For measuring reassociation with hydroxyapatite, the reaction can be stopped by diluting the sample with 0.12 M PB and quick frozen with a dry ice-ethanol bath and stored at −20° until further analysis. Slow freezing can result in the formation of aggregates which are not effectively assayed on HAP.

Hydroxyapatite

Hydroxyapatite has the apparent structure $Ca_{10}(PO_4)_6(OH)_2$ and can be made by mixing $CaCl_2$ and Na_2HPO_4, then heating the intermediate product (brushite, $CaHPO_4 \cdot 2\ H_2O$) under carefully controlled conditions [56,57]. However, good HAP is also available commercially (BioRad, Clarkson). DNA adsorbs to HAP through an apparent interaction of DNA phosphate and HAP Ca^{++} groups, although the binding may be more complex. Elution of DNA is accomplished with phosphate ions, which apparently compete for the Ca^{++} groups. An empiric relationship between phosphate ion concentration and DNA secondary structure allows HAP to be employed for reassociation (and hybridization) assays [45,56]. In low phosphate concentrations (10–30 mM PB) at 60°, both single stranded and double stranded DNA adsorb to HAP. Intermediate buffer concentrations (0.12–0.14 M PB) prevent single stranded DNA from binding, or elute bound single stranded DNA. High buffer concentrations (0.4–0.5 M PB) elute double stranded DNA. Elution of double stranded DNA can also be accomplished by raising the temperature to above 97° and washing the HAP with the intermediate PB.

Each batch of HAP is variable in its binding properties with DNA, and its capacity and elution behavior must be checked. Binding properties can be checked by adsorbing ^3H-labeled, sheared, denatured (single stranded) bacterial DNA to the column at 10–30 mM PB, eluting with increasing concentrations of PB (in the range of 0.12–0.18 M) until no more DNA is eluted, then washing with 0.4–0.5 M PB to check for quantitative recovery of DNA. Nonspecific binding (single stranded DNA that binds to HAP at intermediate PB) can be reduced by boiling HAP in intermediate PB for 5–15 min, or by including 0.1–0.4% sodium dodecyl sulfate in the binding buffer. A small amount of DNA may be irreversibly bound as well (not eluted by 0.4–0.5 M PB or high temperature). The total amount of non-specifically and irreversibly bound DNA should be less than 2–3% of the single stranded material. The useful capacity of HAP for DNA is usually about 200 μg of DNA per ml of HAP. At these levels, the DNA in a sample is quantitatively adsorbed. If this level is exceeded, only a portion of the additional DNA will bind. HAP that does not meet these general specifications should be discarded.

Columns to be used should be completely jacketed or mounted in a water bath assembly which allows easy sample loading and elution. Glass wool or glass fiber filters are recommended for the bottom of the column. A 1 cm diameter column is adequate for 1 ml of HAP (0.5 g HAP is equivalent to about 1 ml gel). Larger volumes of HAP require wider columns. Syringes (3 ml–20 ml) serve as convenient columns when they are mounted in a water bath assembly connected to a circulating water bath apparatus (Haake, Lauda, etc.). Convenient assay columns are prepared just before use by making a slurry of 1 g of HAP in 5 ml of 0.12 M PB, heating the slurry in a 100° water bath for 5 minutes, and adding to a 20 ml syringe containing a glass fiber filter. Allow the column to equilibrate at 60°, then wash with 5 ml of 0.12 M PB. Let the gel stand overlaid with 1 ml of 0.12 M PB.

For most experimental purposes, it is recommended to perform the hybridization under the standard conditions of 0.12 M PB at 60°, and to load the diluted sample onto the HAP column under these same conditions where the single stranded DNA does not bind. The column is then washed with 5–10 volumes of 0.12 M PB, and double stranded DNA is eluted with 0.4 M PB. The assays are reproducible and relatively free of problems only if a given set of procedures is strictly followed without change during the project under study. Difficulties arise usually when a "minor" change is made for the sake of expediency. A fresh HAP column should be prepared for each assay, since DNA-binding properties may change slightly with each run. It is permissible to stir the HAP during a run to prevent channeling or if the column has run dry. Rapid flow rates are acceptable and gentle air pressure may be applied to the column. If small amounts of radioactive DNA are to be assayed, the sample should be supplemented with additional nonhomologous DNA to 50 μg per ml of HAP bed volume to minimize the problem of nonspecific binding.

There are experimental advantages to using HAP in addition to the practical aspects discussed above. In particular, the hybrid can be recovered for further studies. Reassociations can be extended to various Cot values and broad sequence frequency classes can be separated on HAP. Thermal stabilities of hybrids can be examined conveniently on HAP. For many purposes, relatively small amounts of radioactive DNA can be employed, minimizing the need for optical technics.

Typical Reassociation and Hybrid Assay Protocol

1. Adjust a mixture of [^{3}H]cDNA and sheared driver DNA to a concentration of 10 mg per ml in 0.12 M PB.
2. Distribute 50 μl portions of the DNA into small polypropylene tubes, add 0.12 M PB, and mix, as in the following hypothetical example:

Desired Cot	*Co*	*Incubation Time*	*0.12 M PB*	*Final Reaction Volume*
zero time blank	—	0 min	350 μl	400 μl
3	3.8×10^{-3}	13 min	350 μl	400 μl
10	7.7×10^{-3}	22 min	150 μl	200 μl
30	7.7×10^{-3}	64 min	150 μl	200 μl
100	3.1×10^{-2}	54 min	—	50
300	3.1×10^{-2}	2.7 hr	—	50
1000	3.1×10^{-2}	9.0 hr	—	50
3000	3.1×10^{-2}	27 hr	—	50

3. Overlay the solution with a few drops of mineral oil and cover the tubes.
4. Denature the samples by immersing the bottom portion of the tubes in a 100° water bath for 3–5 min, then immediately place tubes in 60° water bath.
5. At the end of each incubation period, dilute sample to a final volume of 1 ml with 0.12 M PB at 60°. Quickly freeze the sample in a dry ice-ethanol bath and store at −20°. If the samples are to be assayed immediately (do not freeze), add the DNA solution to a prepared 1 g HAP column containing an additional 1 ml of 0.12 M PB at 60°. Rinse tubes with 1 ml of 0.12 M PB and add to the columns. Mix the diluted sample into the HAP thoroughly with a glass rod.
6. Allow the column to drain and collect the effluent. Wash the column with 5–10 ml of prewarmed 0.12 M PB into the same collection tubes. This fraction contains the single stranded DNA of the reaction.
7. Elute the hybrid DNA with 5–10 ml of prewarmed 0.4 M PB. Cool these samples on ice, adjust to 5% in trichloroacetic acid, and let stand for 30 min. Collect the precipitates on nitrocellulose or glass filters and measure radioactivity.
8. After the low salt fractions have cooled to room temperature, measure the absorbance at 260 and 320 nm. Subtract the A_{320} values from the A_{260} value to correct for light scattering due to any HAP fines in the solution. Correct for the hyperchromicity of the denatured DNA and calculate the relative amount of DNA in the reaction that was single stranded at the particular Cot value assayed.
9. Plot the data as percent HAP-bound vs. log Cot. Comparison of the $Cot_{1/2}$ for cDNA hybridization (radioactivity) with the $Cot_{1/2}$ for the unique sequence portion

(from absorbance readings) of the driver DNA reassociation will give an estimate of the reiteration frequency of the particular gene under study.

10. Reiteration frequencies (F) of the experimental sample under study (exp.) can also be estimated from a comparison with *E. coli* DNA reassociation kinetics according to the relationship [47]:

$$F = \frac{Cot_{1/2}\,(E.\ coli)}{Cot_{1/2}\,(exp.)} \times \frac{\text{Genome Complexity (exp.)}}{\text{Genome Complexity } (E.\ coli)}$$

Thermal Chromatography on Hydroxyapatite

A comparison of the thermal stability of DNA hybrids to that of native DNA duplexes from the homologous source indicates the base-pairing precision (sequence-matching fidelity) between the strands of the hybrid. For example, experimental evidence suggests that a 1° difference in Tm is an indication of 1.5% mismatched bases in the hybrid [58]. Thermal stability of a duplex is also reduced by shorter molecular lengths and decreased G-C base composition. A determination of the thermal stability of a cDNA:DNA hybrid can therefore indicate the fidelity of mRNA copies by the reverse transcriptase. In the more general case, an examination of the melting profiles of reassociated DNA can provide information on the sequence divergence (or relatedness) of gene families in both homologous and heterologous preparations [2].

The stability of hybrids can be conveniently investigated by thermal chromatography on hydroxyapatite. DNA hybrids are bound to HAP columns under the standard conditions of 0.12 M PB at 60°. As the temperature of the column is increased, the duplex DNA melts and the single strands can be eluted by 0.12 M PB. When radioactively labeled samples are employed, small amounts of material can be analyzed. Since this technic is nondestructive, hybrid components can be separated on the basis of their sequence relatedness for further study. Comparison to a known standard DNA can be accurately made by including a reference DNA (e.g., ^{14}C-labeled homologous native DNA) in the experimental sample (e.g., ^{3}H-labeled cDNA: unlabeled homologous DNA in a hybrid). In order to obtain comparable and reproducible thermal melting profiles, the elution and temperature change procedures must be carefully standardized (i.e., a constant volume of elution buffer per constant temperature increment). An examination of widely different DNAs must take into account possible differences in base composition as well as the fraction of nucleotides involved in base pairs.

Hydroxyapatite may have an apparent stabilization effect on duplex DNA, particularly for longer molecules. The HAP Tm of long DNA can be

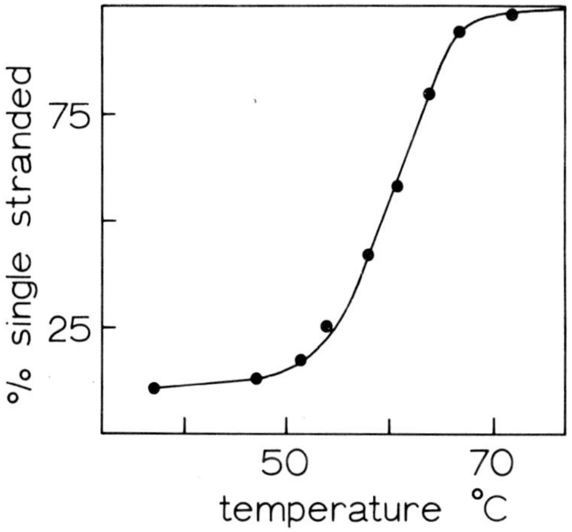

FIG. 4. Thermal denaturation of a DNA:cDNA hybrid. Heat-denatured DNA was incubated with cDNA to Cot = 10,000 M sec/liter under standard conditions. The final hybridization mixture was adjusted to 0.18 M equimolar phosphate buffer in 50% formamide, and adsorbed to a 1 gm HAP column at 35° which had been equilibrated in this solution. The temperature was raised in increments, and following temperature equilibration, the column was eluted with 5 ml of 0.18 M phosphate buffer in 50% formamide. The radioactivity in each fraction was determined and plotted as the cumulative percentage eluted at each temperature.

about 5–8° higher than its optically determined Tm [59]. However, short DNA fragments show almost the same Tm as determined by both methods.

Formamide is frequently included in the elution buffer when RNA:DNA hybrid structures are examined to minimize a possible HAP catalyzed degradation of RNA at higher temperatures. For the range of 45–90°, every 1% formamide in a reassociation reaction reduces the Tm of duplex DNA by 0.72° [50,60]. Many investigators have used this empirical relationship on the assumption that duplex dissociation is affected in the same manner. However, for RNA:DNA hybrids, the relationship between formamide concentration and Tm may not be linear [60]. Thus, the usefulness of formamide in thermal chromatography is probably limited to studies in which known internal standards are included in the experimental samples. A typical thermal melting profile performed with 50% formamide is shown in Figure 4.

Procedure for Thermal Chromatography of cDNA:DNA Hybrids

1. Incubate ^{3}H-labeled cDNA with sheared homologous driver DNA to Cot = 10,000 M·sec/liter under standard criterion conditions. Then add ^{14}C-labeled homologous DNA to the sample. Adjust the solution to 0.18 M sodium phosphate buffer at pH 6.8 and 50% formamide (v/v).

2. Load the sample on a HAP column at 35° which has been equilibrated in 0.18 M PB and 50% formamide elution buffer.

3. Raise the column temperature in 3–5° increments and elute column with 10 ml of elution buffer when the column temperature has equilibrated.

4. Continue the elution procedure until the temperature has reached 80°.

5. Cool the elution fractions on ice and add carrier calf thymus DNA to a concentration of about 100 μg per ml.

6. Adjust the fractions to 5% in trichloroacetic acid and let stand on ice for 30 min. Collect the precipitates on nitrocellulose filters or glass fiber filters and measure radioactivity.

7. Plot the data as cumulative percentage of total radioactivity eluted per fraction vs. temperature of elution of the fraction.

REFERENCES

1. Britten RJ, Kohne DE: Science 161: 529–540, 1968.
2. Davidson EH, Britten RJ: Quant Rev Biol 48: 565–613, 1973.
3. Lewin B: Cell 1: 107–111, 1974.
4. Lewin B: Gene Expression, vol. 2. New York, Wiley, 1974.
5. Birnsteil ML, Sells BH, Purdom IF: J Mol Biol 63: 21–39, 1972.
6. Brown DD, Sugimoto K: Cold Spring Harbor Symp Quant Biol 38: 507–513, 1974.
7. Kedes LH, Birnsteil ML: Nature [New Biol] 230: 165–169, 1971.
8. Kedes LH: Cell 8: 321–331, 1976.
9. Kemp DJ: Nature 254: 573–577, 1975.
10. Campo MS, Bishop JO: J Mol Biol 90: 649–663, 1974.
11. Ryffel GU, McCarthy BJ: Biochemistry 14: 1385–1389, 1975.
12. Bishop JO, Rosbash M: Nature [New Biol] 241: 204–207, 1973.
13. Harrison PR, Hell A, Birnie GD, Paul J: Nature 239: 219–239, 1972.
14. Suzuki Y, Gage LP, Brown DD: J Mol Biol 70: 637–649, 1972.
15. Sullivan D, Palacios R, Stavnezer J et al: J Biol Chem 248: 7530–7539, 1973.
16. Harris SE, Means AR, Mitchell WM, O'Malley BW: Proc Natl Acad Sci USA 70: 3776–3780, 1973.
17. Verma IM, Temple GF, Fan H, Baltimore D: Nature [New Biol 235: 163–166, 1972.
18. Kacian DL, Spiegelman S, Bank A et al: Nature [New Biol] 235: 167–170, 1972.
19. Chan LN, Wiedman M, Ingram VM: Devel Biol 40: 174–185, 1974.
20. Orkin SH, Swan D, Leder P: J Biol Chem 250: 8753–8760, 1975.
21. McKnight GS, Pennequin P, Schimke RT: J Biol Chem 250: 8105–8110, 1975.
22. Harris SE, Rosen JM, Means AR, O'Malley BW: Biochemistry 14: 2072–2081, 1975.
23. Kan YW, Dozy AM, Varmus HE et al: Nature 255: 155–156, 1975.
24. Houseman D, Forget BH, Skoultchi A, Benz EJ Jr: Proc Natl Acad Sci USA 70: 1809–1813, 1973.
25. Kacian DL, Gambino R, Dow LW et al: Proc Natl Acad Sci USA 70: 1886–1890, 1973.
26. Forget BG, Benz EJ, Skoultchi et al: Nature 247: 379–381, 1975.
27. Kan YW, Holland JP, Dozy AM, Varmus HE: Proc Natl Acad Sci USA 72: 5140–5144, 1975.
28. Gallo RC, Reitz MS Jr: Int Rev Exp Path 16: 1–58, 1976.
29. Salser W: Ann Rev Biochem 43: 923–939, 1974.
30. Proudfoot NJ, Brownlee GG: Nature 263: 211–214, 1976.
31. Monahan JJ, Harris SE, O'Malley BW: J Biol Chem 251: 3738–3748, 1976.

32. Galau G, Klein WH, Davis MM et al: Cell 7: 487–505, 1976.
33. Axel R, Feigelson P, Schutz G: Cell 7: 247–254, 1976.
34. Young BD, Birnie GD, Paul J: Biochemistry 15: 2823–2829, 1976.
35. Ryffel GU, McCarthy BJ: Biochemistry 14: 1379–1385, 1975.
36. Rosbash M, Campo MS, Gummerson KS: Nature 258: 682–686, 1975.
37. Kacian DL, Spiegelman S: Meth Enzymol 29: 150–173, 1974.
38. Verma IM, Baltimore D: Meth Enzymol 29: 125–120, 1974.
39. Leis J, Hurwitz J: Meth Enzymol 29: 143–150, 1974.
40. Monahan JJ, Harris SE, Woo SLC et al: Biochemistry 15: 223–233, 1976.
41. Keller GH, Taylor JM: Manuscript in preparation.
42. McCarty K Jr, Vollmer RT, McCarty K Jr: Anal Biochem 61: 165–183, 1974.
43. Studier F: J Mol Biol 11: 373–380, 1965.
44. Weiss GB, Wilson GN, Steggles AW, Anderson FW: J Biol. Chem. 251: 3425–3431, 1976.
45. Britten R, Graham P, Neufeld B: Meth Enzymol 29: 363–418, 1974.
46. Bishop JO: Karolinska Symposium on Research Methods in Reproductive Endocrinology, vol. 5, (ed. by E. Diczfaluzy and A. Diczfaluzy). Stockholm, Karolinska Institutet, 1972, p 247–276.
47. Purdom I, Williamson R, Birnsteil M: Nucleic Acid Hybridization in the Study of Cell Differentiation (ed. by H. Ursprung). New York, Springer-Verlag, 1972, p 25–36.
48. Smith MJ, Britten RJ, Davidson EH: Proc Natl Acad Sci USA 72: 4805–4809, 1975.
49. Britten RJ, Davidson EH: Proc Natl Acad Sci USA 73: 415–419, 1976.
50. McConaughy BL, Laird CD, McCarthy BJ: Biochemistry 8: 3289–3295, 1969.
51. Wiegand RC, Godson GN, Radding CM: J Biol Chem 250: 8848–8855, 1975.
52. Goldberg RB, Crain WR, Ruderman JV et al: Chromosoma 51: 225–251, 1975.
53. Walbot V, Dure LS III: J Mol Biol 101: 503–536, 1976.
54. Firtel RA, Kindle K, Huxley MP: Fed Proc 35: 13–22, 1976.
55. Saunders GF, Shirakawa S, Saunders P et al: J Mol Biol 63: 323–334, 1972.
56. Bernardi G: Proc Nucl Acid Res 2: 455–499, 1971.
57. Miyazawa Y, Thomas C: J Mol Biol 11: 223–237, 1965.
58. Laird CL, McConaughy BL, McCarthy BJ: Nature 224: 149–154, 1969.
59. Kohne D, Britten R: Proc Nucl Acid Res 2: 500–514, 1971.
60. Schmeckpepper BJ, Smith KD: Biochemistry 11: 2344–2358, 1972.

Cell Fusion

Guido Pontecorvo, Carlo Croce, Enid Sisskin

Cell fusion has been found to be a useful tool in investigating various aspects of cell biology. Hybrids can be studied to answer questions about dominance of cellular control processes, nuclear-cytoplasmic interactions, determination of control of differentiation and malignancy, virus replication and rescue, and chromosome mapping. In fact, its most fruitful use to date has been in the mapping of human genes.

In 1960, Barski et al [1] were the first modern authors to note spontaneous fusion of two divergent cell lines grown together. Fusion and isolation of hybrids was extensively developed by Ephrussi [2] and used for cell biological work. It was mostly a matter of luck and tedious effort until 1964, when Littlefield [6] introduced a biochemical selection system using Szybalski's et al. [9] HAT (hypoxanthine-aminopterin-thymidine) medium, and then in 1965 when Harris and Watkins [3] showed that one can increase the rate of fusion by treating the cells to be fused with UV-inactivated Sendai virus, a modification of a procedure introduced by Okada in 1958 [7].

Since that time, there have been improvements in the technics for selection and characterization of hybrids, and refinement in fusion per se, but for the most part the only consistently reliable method has been use of inactivated Sendai virus for fusion. Recently, however, in 1975 a new technic for fusion of animal cells using polyethylene glycol (PEG) was introduced by Pontecorvo [8], adapting a procedure used for fusion of plant protoplasts since 1974 [4]. Since then, in 1976, Hales [5] further refined the technic for fusion of cells in suspension by means of PEG in DMSO-containing solutions.

Because of widespread interest in the fusion technics, a program was organized at the Given Institute of Pathobiology in Aspen, Colorado this past summer for lectures and laboratory workshops utilizing both procedures. At this workshop, a group of students successfully fused cells using both PEG and Sendai virus, and compared results.

The procedures used are as follows:

I. Sendai virus

A. Cells in a Monolayer—Amounts of media calculated for fusion to be done in a 25 cm² T-flask.

1. Cells should be seeded densely; proportions vary according to cell lines, but cell mixture should be at a density high enough to insure intimate cell contact. This is done the day before fusion.
2. The next day, drain medium.
3. Add 1 ml of medium, without serum, containing 1000 HAU inactivated Sendai virus to each flask.
4. Refrigerate flasks at 4°C for 15 minutes, and
5. Gently rock every 5 minutes.
6. Incubate flasks for 30 min at 37°C.
7. Add 4 ml medium containing serum.
8. Incubate, change medium next day.

B. Cells in Suspension

1. Prepare a suspension of cell lines to be used (we used TK^- mouse cells × SV-40 transformed $HGPRT^-$ human cells) in serum-free medium buffered at pH 8.0–8.2 as follows:
   a. Trypsinize.
   b. Add medium and serum.
   c. Centrifuge at 1000–1200 rpm for 5 min.
   d. Resuspend in medium without serum, buffered at pH 8.0 to 8.2.
2. Mix the two cell types (in glass or plastic tubes) 2 ml final volume.
3. Add 0.2 ml inactivated Sendai virus (2000 HAU per 10^7 cells). (Note: within a few seconds, clumps are noticeable.)
4. Keep the fusion mixture on ice for 20 minutes.
5. Transfer the tubes to a 37°C shaking water bath for 30 minutes to break up large clumps.
6. Seed the cells either in flasks or Petri dishes. (We used 10^6 cells/60 mm Petri dishes.)
7. Add HAT or other selection medium to flasks or dishes.
8. Change medium often.

II. Polyethylene Glycol (PEG)

A. Monolayer—PEG 6000 m.wt.

1. Inoculate a 25 cm^2 Falcon flask with a mixture of two cell types—need a sparser culture than for Sendai fusion—suggest 10–25% confluence. Incubate overnight in medium + serum in 10% CO_2 incubator.
2. The next day drain medium *very thoroughly*.
3. Add away from the cells, 3–5 ml PEG 1 + 1 (w + v) (preparation below). Cover the monolayer uniformly by rocking for 30–40 seconds and then start draining. Draining must be completed within one minute of adding PEG.
4. Add 5 ml PEG 1 + 3, rock and mix well with the thin layer of 1 + 1 still coating the monolayer. Drain.
5. Add 5 ml PEG 1 + 7, rock and mix well. Drain.
6. (Optional) Repeat with PEG 1 + 15. Drain.
7. Wash two times with Dulbecco's with or without serum.
8. Incubate in Dulbecco's with serum and change medium 1–2 hours later.

   Preparation of PEG solutions:

   A weighed amount of PEG 6000 m.wt. is sterilized and molten by autoclaving. Dulbecco's medium without serum, preheated to 60°–70°C is added quickly with swirling to the still hot molten PEG.

Solutions: 1 + 1 = 10g PEG + 10 ml Dulbecco's.
1 + 3 = 10g PEG + 30 ml Dulbecco's, etc.

(Note: All solutions are used at 37°C.)

B. Monolayer—PEG 1000 m.wt. + 15% DMSO.

1. Monolayer is prepared as for PEG 6000 and incubated overnight.
2. Drain medium thoroughly.
3. Overlay for one minute with 3 ml PEG 1000 1 + 1.4 (41.6%)/15% DMSO (Preparation below) as in step 3 of PEG 6000.
4. Replace with 5 ml PEG 1000 1 + 3 (25%) without DMSO. Drain.
5. Replace with 5 ml PEG 1000 1 + 7 without DMSO. Drain.
6. Wash twice with 5 ml each Dulbecco's.
7. Incubate in Dulbecco's with serum at 37°C.

Preparation of PEG solutions.

A weighed amount of PEG 1000 is sterilized by autoclaving, medium is heated to 60°–70°C and added to the molten PEG as described for PEG 6000.

Solutions: 1 + 1.4/15% DMSO = 10g PEG + 14 ml Dulbecco's prepared with 15% DMSO in it.
1 + 3 = 10g PEG + 30 ml Dulbecco's without DMSO, etc.

(Note: All solutions are used at 37°C.)

C. Suspension—PEG 1000 m.wt. + 15% DMSO (technic by Anne Hales).

1. Centrifuge mixed suspension of the two kinds of cells to be fused: total no. of cells 1–5 × 10^6. Preferably use plastic Universal Container with conical bottom. Remove supernatant and drain quite thoroughly.
2. Add 0.5 ml PEG 1000 1 + 1.4/15% DMSO, resuspend pellet gently with a pipet, rock to agglutinate for no more than one minute.
3. Add 0.5 ml PEG 1000 1 + 3 and rock to agglutinate for 2–3 minutes.
4. Add 4.0 ml of Dulbecco's and rock.
5. Take the 5 ml of suspension up very gently with a large bore pipet to avoid disruption of the agglutinates and seed drop by drop into Petri dishes containing Dulbecco's medium plus serum which have been pre-incubated to 37°C and gassed with CO_2. Shake gently to distribute the cells in a total volume of medium in the Petri dishes of at least 20 ml.
6. After two hours, change medium.

(Note: Use all solutions at 37°C.)

One difference between the two technics is the density of cells used for fusion in a monolayer. When using Sendai, the best results are obtained from cells just about at confluence, while with PEG, sparser cultures are preferable. This relates to the mechanisms of their fusion (Fig. 1). Sendai is thought to form bridges between adjacent cells, and the more cell contacts, the more fusion. With PEG, the fusion of cells in monolayer is mainly not along the smooth edge of the cells, but at the tips of filamentous processes. Here the two cells join, fuse almost immediately after addition of PEG, and

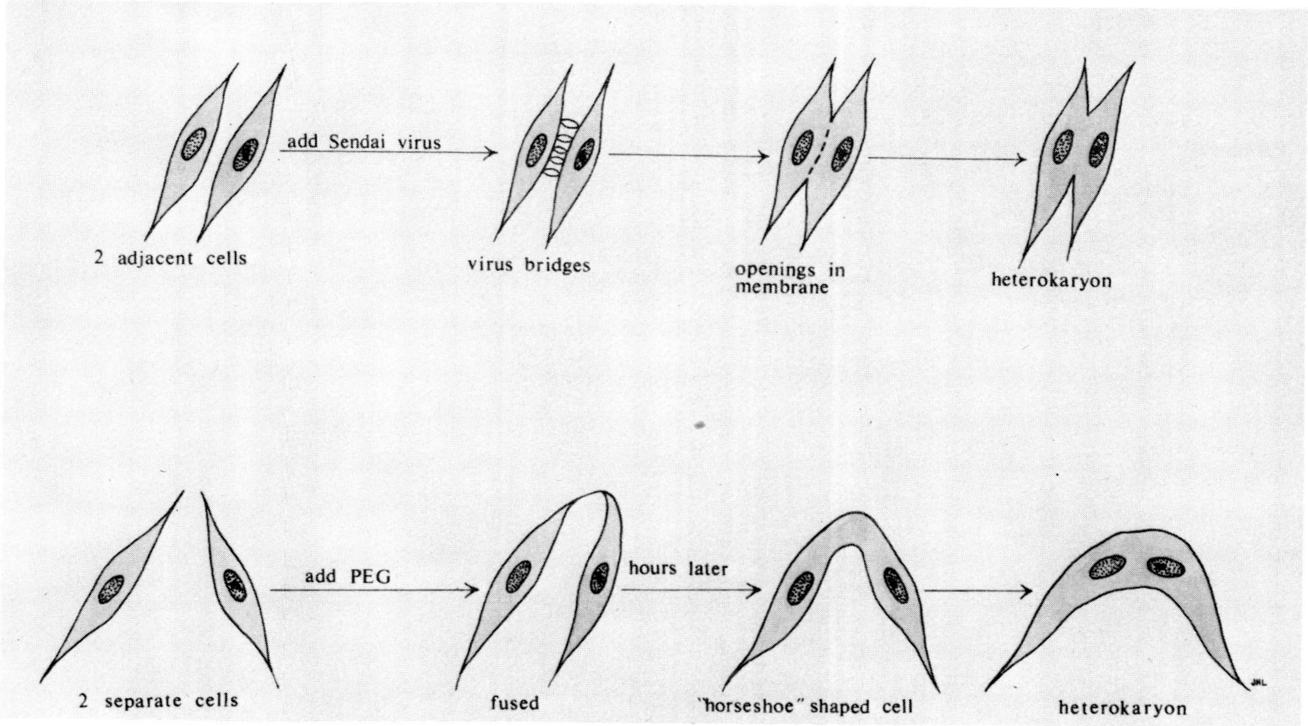

FIG. 1. Mechanisms of cell fusion using Sendai virus and PEG. Sendai virus requires intimate cell contact because the virus forms intercellular bridges, while PEG causes fusion at the tips of the cellular processes.

the fusion becomes visible after several hours when their cytoplasms have had time to run together and their nuclei have become juxtaposed.

The results we obtained support the two mechanisms of cell fusion as well as demonstrate their effectiveness. Because the workshop lasted only five days, we could not determine what proportion of fusion events yielded hybrid cells. We could, however, determine the number of bi- and multinucleate cells formed using the two technics, and the number of nuclei each cell contained. This is important because cells containing more than two nuclei rarely go through mitosis, so to get a good yield of hybrids a lot of preferably binucleate heterokaryons are needed. The ratio of heterokaryons to uninucleate hybrids, whatever the method of fusion used, is seldom higher than 100–1000:1. The real bottleneck in cell hydridization is not the rate of cell fusion but the proportion of heterokaryons which fuse their nuclei and form uninucleate hybrids.

With Sendai fusion in monolayer, there were many more bi- and multinucleate cells than with PEG. However, there was a high proportion of cells containing more than three nuclei. With PEG fusion in monolayer, there were fewer fused cells immediately visible. Several hours after the fusion, and even the next day, however, many "horseshoe" shaped cells were apparent. Most, although not all, of the PEG-fused cells contained two or three nuclei, giving them a good chance of developing into hybrids.

REFERENCES

1. Barski G, Sorieul S, Cornefert F: Production dans des cultures *in vitro* de deux souches cellulaires en association, de cellules de caractere "hybride." C R Acad Sci (Paris) 251: 1825–1827, 1960.
2. Ephrussi B: Hybridization of Somatic Cells. Princeton, N.J., Princeton University Press, 1972.
3. Harris H, Watkins JF: Hybrid cells derived from mouse and man: artificial heterokaryons of mammalian cells from different species. Nature 205: 640–646, 1965.
4. Kao, KN and Michayluk, MR: A method for high frequency intergeneric fusion of plant protoplasts. Planta (Berlin) 115: 355–367, 1974.
5. Hales A: Unpublished personal communication, 1976.
6. Littlefield JW: Selection of hybrids from mating of fibroblasts *in vitro* and their presumed recombinants. Science 145: 709–710, 1964.
7. Okada Y: The fusion of Ehrlich's tumor cells caused by HVJ virus *in vitro*. Biken's J 1: 103–110, 1958.
8. Pontecorvo G: Production of mammalian somatic cell hybrids by means of polyethylene glycol treatment. Somat Cell Genet 1: 397–400, 1975.
9. Szybalski W, Szybalska EH, Regnie G: Genetic Studies with Human Cell Lines. National Cancer Inst. Monogr. No. 7:75–89, 1962.

An Introduction to Affinity Chromatography

Indu Parikh, Susan F. Slovin

Affinity chromatography is a type of reversible adsorption chromatography which derives its adsorptive properties from the covalent binding of a ligand to an insoluble matrix, usually agarose. The usefulness of this procedure as a means of purification was recognized as early as 1910, when insoluble starch was used as adsorbent to purify amylase [1]. However, the major advance in the procedure came with the purification and isolation of enzymes such as staphylococcal nuclease and carboxypeptidase A, first illustrated by Cuatrecasas [2]. As a result, affinity chromatography has gained a foothold in the isolation and purification of viruses, estrogen and insulin receptors, plant auxin acceptor proteins, lectins, the galactose repressor of *E. coli,* as well as specific antigens and antibodies [3,4]. It facilitates the study of the mechanism of action of certain enzymes by employing immobilized polynucleotides such as single-stranded DNA and RNA [5], as well as double-stranded DNA coupled to cyanogen bromide activated agarose (unpublished observations). Its role in the specific fractionation of immune cell populations is only now being recognized [6].

Beaded agarose (usually Sepharose 4B) has been the most commonly employed insoluble matrix material in affinity chromatography. It is very stable, inert, porous with resulting greater surface area, permits greater ligand substitution and hence a greater capacity. Polystyrene and glass beads, which tend to be hydrophobic, nonspecifically adsorb most proteins. The usefulness of hybrid polymers (polystyrene beads coated with hydrophilic material) is limited.

The design and experimental strategy of any affinity chromatographic system would require individual attention in the selection of a matrix, a ligand, as well as selection of the buffer conditions for the adsorption and desorption processes. In most instances of affinity chromatography the physical parameter of greatest importance is the magnitude of affinity between the protein and its specific ligand (dissociation constant). A knowledge of this parameter would help to determine whether the system will have strong enough association to be useful, and it can dictate how the adsorbed protein will need to be treated to recover it (elution) from the affinity column.

The consideration of spacer arms, small alkyl groups interposed between ligand and matrix to prevent steric interference, often becomes important in the actual chromatographic procedure. The spacer arms tend to greatly increase the effectiveness of the adsorbent. This is especially true for low affinity systems. Diaminodipropylamine is the most frequently recommended small spacer since it has relatively little hydrophobic properties. Such spacer molecules are commonly attached to the matrix before coupling the ligand. It is often considered advantageous by some workers to attach a spacer to the ligand and subsequently couple the spacer-ligand conjugate to the matrix. In this way, adsorbents are less likely to exhibit undesirable properties (such as ionic interactions) which often tend to interfere with affinity chromatography.

It has been recognized that ligands attached to a variety of matrices are slowly released from the matrix. The quantity of ligand released (leakage or bleeding) depends on the temperature, type of buffer, and above all, on the nature of chemical bond involved in the coupling of a ligand to a matrix. During affinity chromatography of extremely small quantities of biologically important protein (eg., hormone receptor proteins), minute quantities of released ligand become a major problem [7,8].

A recently introduced polymeric macromolecular spacer [7,8] serves more purposes than being a simple extension arm between the ligand and the matrix. The concept of macromolecular spacer arises from an intensive search for anchoring reagents which would be able to covalently attach a ligand to a matrix by multipoint attachments, with resultant increased chemical stability of the ligand. For example, if a ligand is released from the matrix with a probability of one part per thousand (0.1%), a ligand attached (by the chemical means) to the matrix at two points might be expected to be released with a probability of one part per million (0.0001%). A variety of water-soluble polyfunctional polymers were found to possess properties suitable for being such anchoring reagents. In addition to increasing the chemical stability of the ligand-matrix conjugate, these anchoring reagents offer other advantages in affinity chromatography. The macromolecular spacers provide appreciably greater separations of ligand from the matrix backbone than can be achieved with conventional spacers. Furthermore, these polyfunctional spacers exhibit minimal hydrophobicity, and the microenvironment created in the vicinity of the ligand substituted on such polyfunctional spacers appears to be especially favorable in many types of interactions encountered in affinity chromatography [7].

Methods of Matrix Activation

Although the cyanogen bromide method of activation for agarose has been universally associated with affinity chromatography, there exists a multitude of choices for proper selection of chemical routes to matrix derivation. Proper choice of an activation method very often depends on the availability of functional groups on the ligand and the matrix selected. An in-depth comparative study of various modes of matrix activation is still lacking. Cyanogen bromide activates agarose by reacting with its vicinal hydroxyl groups to form cyclic (perhaps also acyclic) imido carbonates. Because of high reactivity to nucleophiles, the imidocarbonates readily react with primary aliphatic as well as aromatic amines to form covalent bonds. The method is broadly used when aliphatic diamines as extension arms are desired or when a ligand contains a reactive amino group. The chemical stability of the ligand-agarose bonds formed by this method is recently being questioned.

Sodium periodate is an alternative reagent recently introduced [8] for activation of polysaccharide matrices, including agarose, Sephadex and cellulose. The method is simple, safe, fast, and offers chemically stable ligand-matrix bonds. The oxidation of cis-vicinal hydroxyl groups of a polysaccharide matrix by sodium periodate generates aldehyde groups, which react with primary amines to form so-called Schiff's bases. The latter may be reduced with sodium borohydride ($NaBH_4$) to form stable secondary amines. Although sodium borohydride is efficient, reductive amination by sodium cyanoborohydride ($NaBH_3CN$) is in some cases advantageous [8]. Since the latter reducing agent, at pH 6.0, drives the reaction toward completion, relatively more efficient use of the amino ligand will result. The periodate method in general yields a lower degree of ligand substitution than CNBr method but offers greater stability of ligand-matrix bond. Due to its stability to oxidative conditions, cellulose is more easily and better substituted by the periodate method than sephadex gels.

An interesting and useful derivative is prepared when periodate-activated cellulose is reacted with a symmetrical dihydrazide such as succinic dihydrazide [8]. The endstanding hydrazide group on cellulose may now be reacted not only to proteins, amino- or carboxyl-group containing ligand, but also to glycoproteins which have been subjected to limited periodate treatment. Among other methods for matrix activation are the use of *s*-triazine trichloride [9], benzoquinone [10] and bifunctional oxiranes [11].

Stable Activated Derivatives of Agarose

Agarose derivatives containing stable activated functional groups like N-hydroxysuccinimide esters [12] are convenient for routine applications. The N-hydroxysuccinimide ester and hydrazide derivatives (discussed above) of agarose can be used, by simple and mild procedures, to immobilize proteins and complex group-containing ligands to agarose, without interfering or complicating side reactions. The N-hydroxysuccinimide ester of agarose is easily prepared and is stable for months when stored in anhydrous dioxane or as a dry powder. When coupling is performed in buffers at a pH in the range of 5 to 8.5, the unprotonated form of the amino-ligand reacts very rapidly (5–30 minutes) under very mild conditions (4°) to form stable amide bonds. Above pH 8.5, however, there is rapid hydrolysis of the active ester. Beside an amino-group containing ligand, a ligand containing SH-group may also react with the N-hydroxysuccinimide ester of agarose.

CNBr Activation of Agarose

Beaded agarose is the most frequently used matrix because of its relative stability, biologic inertness, superior chromatographic properties, and facile activation by cyanogen bromide [13]. Activation involves the addition of CNBr to a stirred aqueous slurry of agarose beads. Activation is greatest at alkaline pH, and the reaction generates protons. Previously published procedures require manual titration of the reaction, with the aid of a pH meter, to maintain a basic pH. This type of activation procedure lacks a well defined end point and is difficult to reproduce. In addition, wide fluctuations in pH are difficult to avoid during titration. Addition of cyanogen bromide as a solid requires that this volatile lacrimator be weighed in a closed vessel in a fume hood. Differences in the rate of dissolution of solid cyanogen bromide also add to the variability of activation. If cyanogen bromide is added in a miscible, inert solvent and a buffer is used to maintain an alkaline pH, these difficulties are eliminated. This approach has been included in the so called "Buffer Method" for agarose activation [14].

Titration Method of Activation

Agarose beads, washed thoroughly with large volumes of 0.1 M NaCl and water, are suspended in water to achieve a total volume equal to twice the packed volume of the gel. A volume of the 1:1 slurry is added to a beaker containing a magnetic stirrer and a few chunks of ice. The CNBr is weighed into a glass bottle in a fume hood and capped until used. The base used for titration is poured into a 50 ml buret. For amounts in the range of 50 mg of CNBr per milliliter of packed beads, 2 M NaOH is adequate. When higher degrees of activation are desired, 300 mg of CNBr per milliliter and 8 M NaOH should be used. All the CNBr is added at once, and the rate of flow from the buret is adjusted to maintain the pH near 11. Ice is occasionally added to the slurry to maintain the temperature near 30°. In 10–12 minutes the CNBr dissolves and the rate of base consumption decreases. At this time the slurry is poured into a chilled coarse disc sintered-glass funnel (containing chunks of ice), and the gel is rapidly (2–3 minutes) filtered with suction and washed with about 5–10 volumes of ice-cold distilled water. The gel is filtered to a moist cake, the outlet of the funnel is covered with Parafilm, and the buffer containing the material to be coupled is added to the funnel and mixed immediately with a glass rod. The slurry is poured into a plastic bottle, and the suspension is gently shaken on a mechanical shaker (preferred) or stirred for at least 12 hours at 4°. Agitation of the slurry should be by gentle means, since shattering of the beads may occur upon vigorous stirring. After completion of the coupling reaction, the gel is washed briefly and suspended in a 4-fold volume of 1 M ethanolamine pH 9.5 in order to promote masking of activated groups that may have still remained on the agarose. The suspension is shaken for 4 hours at room temperature.

Buffer Method of Activation

Despite the relative simplicity of the titration method, a faster and safer alternative method has been developed [14]. In this modification, the reaction is performed in carbonate buffer. The coupling efficiencies obtained are comparable to those observed after activation by the titration method. A typical reaction involves mixing one volume each of agarose, water and 2 M sodium carbonate. The slurry is cooled in an ice bath with slow stirring. After cooling to 5°, the rate of stirring is increased and the desired amount of CNBr, dissolved in acetonitrile (2 g of CNBr per milliliter of CH_3CN) is added. After 2 minutes the beaker is decanted into a coarse disk sintered-glass funnel, and the gel is washed and coupled to a ligand as described above. The results are reproducible; the mean deviation in the quantity of ligand or protein coupled in a group of individually activated gels has been as low as 4%. Cooling the reaction before and during activation is important. For example, coupling may be reduced by as much as 35% if the reaction is performed at room temperature.

Sodium Periodate Method of Activation

This rapid, simple and safe alternative method offers chemically stable ligand-agarose bonds. The method is also applicable to other polysaccaride matrices, such as Sephadex and cellulose. The method depends on the oxidation of cis-vicinal hydroxyl groups of polysaccarides by $NaIO_4$ to generate aldehyde groups. These aldehyde functions react with primary amines [8]. To a suspension of 100 ml of agarose in 60 ml water is added 40 ml of 0.5 M $NaIO_4$. The suspension is gently shaken for 2 hours at room temperature. The oxidized agarose, after washing, is added to 1 volume of 0.2 M phosphate buffer, pH 6.0 containing 5 mg/ml protein or 50 mM diaminodipropylamine and 0.5 mM sodium cyanoborohydride. The suspension is gently shaken for 3 days in a closed capped polyethylene bottle at room temperature with a mechanical shaker. The gel is extensively washed and the unreacted aldehyde functions on the matrix are reduced with 1 M $NaBH_4$. The adsorbent in then extensively washed.

Benzoquinone Activation of Agarose

A new and simple method for the covalent attachment of proteins and amino group containing ligands to agarose and other polysaccharide carriers has been devised which uses p-benzoquinone as matrix activating agent [10]. The technic seems to offer a high degree of ligand substitution and good chemical stability of the ligand-matrix bond. Once a matrix has been activated by this reagent, coupling of a ligand may be achieved over a broad pH range.

Equal volumes of prewashed agarose and 0.1 M phosphate buffer, pH 8.0 are treated with a solution of p-benzoquinone in ethanol to give a final concentration of 20% ethanol and 50 mM benzoquinone. The suspension is gently shaken for 1 hour at room temperature. After the activation, the gel is washed on coarse disc sintered-glass funnel with 2 volumes each of 20% ethanol, water, 1 M NaCl and water. An appropriate quantity of the activated gel may then be coupled to an amino group containing ligand in buffers at pH between 8 and 9. The coupling reaction is usually complete within 1 hr at pH 8.

Cyanuryl Chloride Method

Cyanuryl chloride (also known as *s*-triazine trichloride) is perhaps one of the earliest bifunctional reagents known. It is known that replacement of one of the chlorine atom deactivates the remaining two so that a stepwise reaction can be obtained by control of pH and/or temperature [9,15]. The immobilization of various enzymes to cellulose by the use of this reagent is described in the literature. The reagent is equally efficient for agarose and Sephadex (unpublished results).

One volume of agarose (as wet suctioned cake) is added to one volume of an acetone solution (100 mM) of cyanuryl chloride. One volume of water is added and the suspension is gently shaken in a screw capped plastic bottle for 5 minutes at room temperature. The activation reaction is quenched with 2 volumes of 20% acetic acid and the gel is filtered and washed with 3 volumes each of 50% acetone, 20% acetic acid, 50% acetone and water. The activated gel may then be coupled to a ligand in buffers, pH 8.0. The coupling reaction is usually completed within 5 to 10 minutes. The adsorbent is extensively washed as usual before its use for affinity chromatography.

Purification of Staphylococcal Nuclease

A potent competitive inhibitor, 3′-(4-aminophenylphosphoryl)deoxythymidine 5′-phosphate is an ideal ligand for the purification of this enzyme [2]. Furthermore, it is stable over a broad pH range and its 3′-phosphodiester bond is not cleaved by the enzyme. This inhibitor couples to CNBr activated agarose and the resulting adsorbent has a high capacity for the enzyme. Columns containing this specific adsorbent completely and strongly adsorbs samples of nuclease (pure or crude) in buffers containing 10 mM $CaCl_2$, pH 8.0. The elution of the enzyme, after washing the contaminating proteins, is achieved with 0.1 M acetic acid with almost quantitative recovery. In the case of staphylococcal nuclease, there seems to be no necessity of introducing any long extension arm between the inhibitor and the matrix.

Purification of β-Galactosidase

The enzyme can be purified to homogeneity by affinity chromatography [16] when its substrate analogue inhibitor, p-aminophenyl-β-D-thiogalactopyranoside is used as an affinity ligand. The commercially available p-nitrophenyl-β-D-thiogalactopyranoside may be reduced to the amino form by 50 mM sodium dithionite in water at neutral pH. The resulting p-aminophenyl-β-D-thiogalactoside may then be coupled, without isolation, to carboxyl group containing agarose with the use of water-soluble carbodiimide. Small columns (Pasteur pipets) packed with the specific affinity adsorbent can adsorb large quantities (up to 1 mg) of partially purified enzyme (Sigma) or that present in the crude homogenate of *E. coli*. After washing the column with appropriate buffers to remove the contaminating proteins, the enzyme may be eluted in homogeneous form by 0.1 M borate buffer pH 10. It may be noted that buffers (pH 10) other than borate are not effective in elution of the enzyme (unpublished results).

Hydrophobic Interactions and Purification of D-Lactic Dehydrogenase

The observation that hydrocarbon chains of optimum lengths, when covalently attached to a solid matrix, may preferentially adsorb one protein over another has been exploited for purification of proteins and various enzymes [17]. Adsorption and elution of proteins may be accomplished by selecting buffers of proper ionic strength and optimum temperature. One excellent example of such hydrophobic affinity chromatography is provided in the purification of D-lactate dehydrogenase from *E. coli* [18]. This membrane-bound enzyme is solubilized in 0.75 M guanidine-hydrochloride-phosphate buffer, pH 6.6. Adsorbents prepared by coupling n-hexylamine to CNBr-activated agarose seem to give best results during purification of this enzyme. Agarose adsorbents containing alkylamines smaller than six carbon chains are less efficient in the adsorbtion process, while those containing longer chains adsorb the enzyme too tightly to be eluted by reasonable buffers. The elution of this enzyme is most efficiently achieved by 0.1 M phosphate buffer, pH 7.8 containing 1% Triton X-100. The enzyme activity is followed in the presence of 1% Triton.

REFERENCES

1. Starkenstein E: Ueber Fermentwirkung und deren Beeinflusung durch Neutralsalze. Biochem Z 24: 210, 1910.
2. Cuatrecasas P, Wilchek M, Anfinsen CB: Selective enzyme purification by affinity chromatography. Proc Natl Acad Sci USA 61: 636, 1968.
3. Parikh I, Cuatrecasas P: Affinity chromatography, principles and applications. *In:* Catsimpodas N (Ed:) Methods in Protein Separation, vol. 1. New York, Plenum Press, 1975, p 255.
4. Parikh I, Cuatrecasas P: Affinity chromatography in immunology, *In:* Atassi AZ (Ed): Immunochemistry of Proteins, vol. 2. New York, Plenum Press, 1976 (in press).
5. Poonian MS, Schlabach AJ, Weissbach A: Covalent attachment of nucleic acids to agarose for affinity chromatography. Biochemistry 10: 424, 1971.
6. Rutishauser U, Millette C, Edelman G: Specific fractionation of immune cell populations. Proc Natl Acad Sci USA 69: 1596, 1972.
7. Sica V, Parikh I, Nola E et al: Affinity chromatography and the purification of estrogen receptors. J Biol Chem 248: 6543, 1973.
8. Parikh I, March SC, Cuatrecasas P: Topics in the methodology of substitution reactions with agarose. Meth Enzymol 34: 77, 1974.
9. Smith NL III, Lenhof HM: Covalent binding of proteins and glucose-6-phosphate dehydrogenase to cellulosic carriers activated with *s*-triazine trichloride. Anal Biochem 61: 392, 1974.
10. Brandt J, Andersson L-O, Porath J: Covalent attachment of proteins to polysaccharide carriers by benzoquinone. Biochim Biophys Acta 386: 196, 1975.
11. Sundberg L, Porath J: Preparation of adsorbents for biospecific affinity chromatography. J Chromatgr 90: 87, 1974.
12. Cuatrecasas P, Parikh I: Adsorbents for affinity chromatography use of N-hydroxysuccinimide esters of agarose. Biochemistry 11: 2291, 1972.
13. Cuatrecasas P: Protein purification by affinity chromatography. J Biol Chem 215: 3059, 1970.

14. March S, Parikh I, Cuatrecasas P: A simplified method for cyanogen bromide activation of agarose for affinity chromatography. Anal Biochem 60: 149, 1974.
15. Thurston JT, Dudley JR, Kaiser DW et al: Cyanuric chloride derivatives. I. Aminochloro-*s*-triazine. J Am Chem Soc 73: 2981, 1951.
16. Steers E Jr, Cuatrecasas P, Pollard HB: The purification of β-galactosidase from *E. coli* by affinity chromatography. J Biol Chem 246: 196, 1971.
17. Hofstee BHJ: Hydrophobic affinity chromatography of proteins. Anal Biochem 52: 430, 1973.
18. Short SA: Unpublished results. Manuscript in preparation for Methods in Enzymology.

Immunohistologic Technics

Stewart Sell, D. Scott Linthicum, Dianne Bass,
Raja Bahu, Barbara Wilson, Paul Nakane

INTRODUCTION

The precise and exquisitely specific nature of the reaction of antibodies with antigens, when appropriately applied to labeling systems, provides an invaluable tool for identifying and localizing various biologically important molecules in cells and tissues. The purpose of this workshop is to define the conditions and technology which permit specific antibody-antigen reactions to be used for more or less precise localization of antigens in tissue, i.e., immunohistology. Many immunochemical, animal handling and fractionation procedures that may be useful in preparing and testing immunohistologic reagents have recently been reviewed in other texts and will not be presented herein [6,13,14].*

It is not within the scope of this paper to define in detail the structure of antibody or to present an analysis of the specificity of antibody-antigen reactions. Individuals not familiar with these subjects should consult basic texts, such as references [2,4,11]. However, two important points should be brought out. Firstly, in regard to the structure and properties of the antibody used for labeling, almost all labeling reactions utilize the IgG class of antibody. The important structural parts of the molecule are the Fab pieces, which contain the antigenic binding sites and thus provide the specificity of the reaction with the antigen, and the Fc piece, which may cause nonimmune specific binding. As will be discussed in detail below, it is sometimes necessary to label only with Fab pieces, although most labeling is done with the whole IgG antibody. Secondly, the specificity of antigen binding can be extremely precise, as even small parts of molecules can be specifically recognized (see references [5] and [7]).

The authors thank Dr. Donald West King and the Given Institute of Pathobiology, Aspen, Colorado for sponsoring this course and for the publication of the material presented. We also are grateful to Drs. Ricardo Mesa-Tejada, Fred Silva, and Bruce Woda, Department of Pathology, College of Physicians and Surgeons, New York, N.Y. for their editorial assistance in preparing the manuscript.

* Bibliographic references for this Workshop are listed consecutively at the end of each major section: pp. 279, 286, 294, 305.

The ultimate value of any investigation using labeled antibody depends upon the specificity of the reaction with antigen. This specificity will be determined by the care in purification of the immunogen used to make the antibody, the immunization protocol used, and the preparation and testing of the antibody obtained. The accuracy, sensitivity and significance of the actual labeling experiment depend in turn upon the method used for labeling, the controls employed, the processing of the tissues to be labeled and the design of the protocol used for experimental and/or clinical studies. The last factor is, of course, the responsibility of each investigator. The other factors are the subject of this workshop. Before presenting the detailed technics used for immunohistologic labeling, a brief general discussion of these factors will be presented.

1. *Specificity.* Antigenic binding sites are formed by folding of the hypervariable regions of the light and heavy chains of an antibody molecule (Fab piece) which can juxtapose with an antigenic determinant so that close opposition of oppositely charged ionic groups provides noncovalent binding of different degrees of avidity. Antibody molecules can recognize structures as small as tetrapeptides or different configurations found by positional changes of groups on a phenol ring [5,7]. The specificity of a given antiserum depends upon the care used in preparation of the immunogen and absorption (if necessary) of the antiserum obtained.

2. *Preparation of the Immunogen.* As pure a preparation as can be made for a given molecular population should be obtained. In order to follow the results of any fractionation procedure, it is necessary to have a way of identifying small amounts of the material wanted as well as those molecular species that are not wanted. This is usually possible by using immunochemical or physiochemical technics. Fractionation of proteins may be accomplished by physiochemical methods and fractionation of cells by a variety of procedures. Preparation of artificial molecular structures which contain antigenic configurations may also be employed. Each fraction obtained should be tested for the presence of the antigen wanted as well as for contaminants. For molecules too small to be immunogenic, coupling to an appropriate carrier may be required. Highly immunogenic carriers are larger protein molecules, such as albumin, gamma globulins, and, in particular, keyhole limpet hemocyanin. The immunogenicity of a poor immunogen may be increased by coupling of haptens to the molecule [12]. Once the purest immunogenic material possible for a given laboratory to obtain has been made, the selection of an immunization protocol is the next step.

3. *Immunization Protocols.* (a) Species Selection: The selection of the species to be used for immunization depends upon a number of factors and may be critical in obtaining a usable antibody. If possible, larger experimental animals, such as rabbits or goats, should be used so that a

substantial volume of antiserum can be obtained. It has been stated that the strongest immune response will be induced when the immunogen contains none or very few determinants in common with molecules present in the animal to be immunized [1]. Therefore, for poor immunogens or for molecules such as hormones or muscle proteins which may contain many determinants which have been preserved during evolution, the species of choice may be one that is distant phylogenetically from the species from which the immunogen was obtained. For instance, in order to produce antibodies to rat muscle proteins, the species of choice may be the chicken. On the other hand, in order to obtain antisera to determinants that are present in different individuals of the same species (allotypes), immunization of an individual of that same species is usually required. For instance, in order to produce antibodies to rabbit immunoglobulin allotypes (see below), a rabbit which lacks a given allotype is immunized with Ig from a rabbit that has the allotype [3], and in order to produce an anti-H_2 or Ia antiserum in mice, cells from a mouse strain that has a given H_2 or Ia specificity are injected into strains of mice that do not [8]. When small animals such as mice are used, it is often possible to induce ascitic fluid that contains antibody activity, thus enabling harvest of much larger volumes of antibody-containing fluid than is possible with repeated bleeding.

(b) Immunization Procedures: There are seemingly an endless number of ways to immunize an animal. For details of various methods, see reference [2]. In general, many investigators use too much immunogen and give too many injections. The following is the method that we have used for immunization of rabbits, sheep, goats or burros with purified protein antigens:

Thirty to fifty μg of antigen are emulsified in an equal volume of *complete* Freund's adjuvant (Difco) to give a volume of 3–5 ml. Primary injections are given into the foot pads of rabbits (3 ml) or in multiple intramuscular sites in sheep, goats or burros (5 ml). The animals are bled 2–6 weeks after injection and the serum obtained is tested for antibody activity (see below). If an appropriate reaction has been produced, large volumes of the antiserum can now be obtained; if not, a booster immunization should be carried out. This is done by injection of 3–5 ml of an emulsion of antigen in *incomplete* Freund's adjuvant in multiple intramuscular sites. Sera may now be obtained weekly. This procedure almost always results in production of a good specific antiserum to xenogeneic proteins. Further immunizations may be required for a given antigen, but these should always be at least four to six weeks apart; more frequent injections may actually decrease the response. Injection with larger amounts of antigen is not necessary, and the use of more immunogen increases the possibility of inducing an immune response to trace contaminants in the immunogen. For instance, if 5 μg of 1% contaminant is immunogenic, a response to this contaminant will be induced if 500 μg of total antigen is used, but not if 30–50 μg is used. Many other immunization protocols have been successful. It is recommended that individuals seeking to produce antisera but who are not familiar with immunology should consult an immunologist before selecting a given protocol.

4. *Testing of Antisera.* Determination of the specificity of the antibody to be used for labeling is probably the most critical step in validating the results of a given immune-labeling system. Many immunochemical methods may be used, such as agar diffusion, passive hemagglutination or hemagglutination inhibition, binding of radiolabeled antibody and/or antigen quantitation precipitation analysis, complement fixation and cytotoxicity assays [6,13,14]. The method(s) of choice depend upon the system to be studied, but more than one method should be used if feasible. Although thorough testing must be done, specificity testing must eventually be carried out within the labeling system being used (see below).

5. *Absorption of Antisera.* Although it is preferable to use an antiserum that does not require absorption, frequently absorption is used to remove unwanted antibody specificities. Absorption can be accomplished by simply adding soluble antigens bearing the unwanted determinants, or more effective absorption can be obtained by the use of insoluble antigens such as whole cells or antigen coated to insoluble matrices, such as cellulose, sepharose or agarose beads. Each antiserum should be carefully tested by immunochemical methods before and after absorption to insure that the unwanted specificities have been removed and the desired specificity has been retained.

6. *Preparation of Antisera for Labeling.* An antiserum contains at least 25–30 different serum proteins. In order to restrict conjugation of a given marker (see below) so that not all proteins are labeled, fractionation of the antiserum should be carried out prior to labeling. There are three general ways of doing this: (a) salting out, which only partially purifies the active serum proteins; (b) DEAE chromatography, which isolates IgG but does not isolate specific antibody—i.e., such preparations contain mostly non-antibody IgG molecules; and (c) immunoabsorbent columns which isolate the specific antibody. The first two methods are described in detail below, whereas immunoabsorption is the subject of the preceding workshop herein (affinity chromatography). Fractionation or purification of the antibody used eliminates non-antibody proteins which may cause nonimmune specific binding if not removed.

Fab or Fab_2 fragments of IgG or antibody may be required to reduce the size of the specific binding molecule or to remove nonimmune specific binding caused by the presence of the Fc piece (the rationale and methods for these procedures are discussed below). For instance, Fab conjugated with peroxidase has a molecular weight of 90,000. This allows staining of 6–10 μ aldehyde-fixed frozen sections for electron microscopy (EM) prior to epon embedding, avoids the risk of tissue distortion in staining ultrathin sections and permits light microscopic selection of tissue sites to be embedded for EM.

7. *Selection of Labels Used.* The most frequently used labels for light and electron microscopy are listed in Table 1. A further discussion of the properties, usefulness and disadvantages of these labels is presented with the

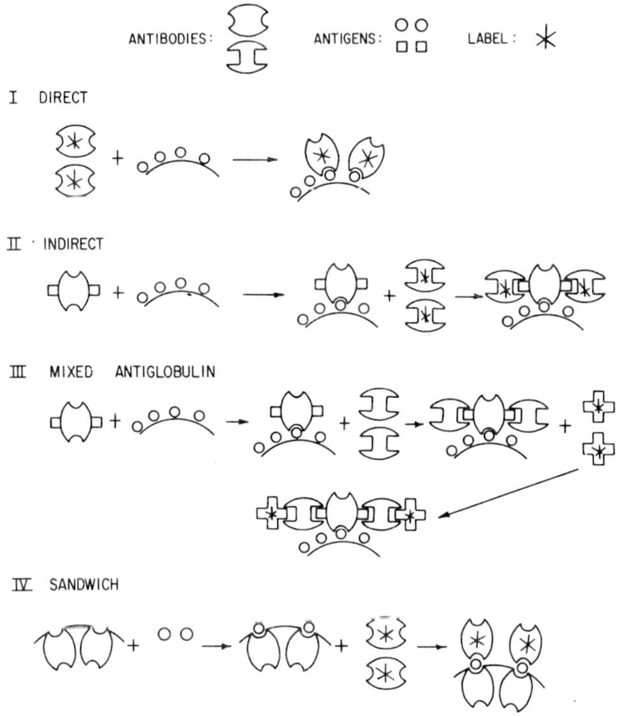

FIG. 1. Immunohistologic Labeling Technics:

I. *Direct Technic.* Specific antibody molecules are labeled (*) and added to tissue sections or cell suspensions containing the antigen.

II. *Indirect Technic.* Unlabeled antibody is reacted with tissue antigen. Labeled antibody to first antibody is then added. The labeled second antibody reacts with the first antibody, which has reacted with tissue antigens. The first antibody provides more binding sites for the second antibody than was provided by the tissue antigen, thus increasing the sensitivity.

III. *Mixed Antiglobulin Technic.* Unlabeled antibody is bound to tissue antigen, which in turn is bound by second unlabeled antibody to the first antibody. Labeled immunoglobulin with antigen determinants in common with the first antibody is then added. The second antibody serves as a bridge binding the labeled Ig to the first antibody. This technic may be used to identify immunoglobulin on tissue by using anti-Ig and labeled Ig.

IV. *Sandwich Technic.* This technic is used to label antibody in tissue rather than antigen. Unlabeled antigen is added to tissue and bound to antibody present in the tissue. Labeled antibody to the antigen is then added, which binds to the antigen fixed by the antibody in the tissue.

TABLE 1. Properties of Some Commonly Used Markers for Immunohistology

Marker	Basis for Use	Molecular Weight, Size	Special Notes
Red blood cells	Can be differentiated from other cell types	5–6 mμ	Visible with light microscope
Fluorescein rhodamine	Glow in light of appropriate wave length	389 FITC 415 TRITC	Used in survey of cell suspensions or tissue sections; can be used for double labeling and vital staining
Horseradish peroxidase	Enzymatic reaction with diaminobenzidine and H_2O_2; produces osmiophilic product electron opaque	40,000	Used for thin sections; can be used for both light and electron microscopy. Sensitivity can be increased by lengthening time of substrate incubation
Ferritin	Electron-opaque core of iron (23% of entire molecule by weight)	750,000 Electron dense diameter 55 Å; Apoferritin diameter 120 Å	Prepared from horse spleen by $CdSO_4$ precipitation and extensive dialysis. Used in thin sections and replicas.
Hemocyanin	Electron-opaque, copper-containing molecule, distinctive shape (cylindrical: square if viewed *en face*)	8,000,000 Diameter 350 Å, length 420 Å	Prepared from hemolymph of keyhole limpet or marine whelk. Used in thin sections, replicas, and SEM.
Latex spheres	Distinctive size and shape	Diameter variable, 200 to 3,400 Å	Used in SEM
Viruses (e.g., tobacco mosaic)	Distinctive size and shape (rod-like—tobacco mosaic)	Variable (180 × 300 Å, tobacco mosaic)	Used in thin sections and SEM
^{125}I	Radioactive material (K-capture emitting auger e^-)	—	Used in EM radioautographic methods

TABLE 2. Some Coupling Procedures Used in Immunomicroscopy

Coupling Reagent	Conditions for Coupling Procedure	Markers Used
Glutaraldehyde	One-step reaction: antibody + marker ± glutaraldehyde at neutral pH	Horseradish peroxidase (HRPO), ferritin, hemocyanin
	Two-step reaction: marker + glutaraldehyde, followed by antibody	Viruses, latex spheres, HRPO
Xylelene diisocyanate	One-step reaction at pH 9.5	Ferritin, fluorescein*, rhodamine*
Toluene diisocyanate	Two-step reaction, since two isocyanate groups are reactive at different pH, 7.5 and 9.5	Ferritin, fluorescein*, rhodamine*, HRPO
FNPS (p,p′ difluoro,m,m′ dinitro diphenyl sulfone)	One-step reaction at pH 10 to 10.5	Ferritin, HRPO
BIS diazotoized dianisidine	One-step reaction at pH 5	Ferritin
Periodate	Carbohydrate groups of markers oxidized to aldehydes with periodate, which then form Schiff bases with amino groups of antibody proteins	Horseradish peroxidase
Antibody binding	Soluble immune complexes in antigen excess	Ferritin, hemocyanin, viruses

* Isothiocyanate derivatives of fluorescein and rhodamine are available commercially.

description of the use of these markers below. This workshop will not include technics using red blood cell markers, e.g., rosette technics.

8. *Methods Used for Conjugating Markers (Labels) to Immune Reagents.* A marker may be combined either to specific antibody or to specific antigen. Various methods used for conjugating markers to immune reactants are listed in Table 2. The technic and application of the most widely used conjugation methods will be described in detail below.

9. *Immunohistologic Labeling Methods.* The various methods of localizing tissue structures with labeled immune reagents is presented in Figure 1. The method of choice depends upon the system being studied and the level of resolution required. For instance, immunofluorescence is useful for identifying either cytoplasmic or cell surface markers at the light microscopic level. The mixed antiglobulin ferritin technic will localize cell surface markers at the resolution of the electron microscope, but is not suitable for cytoplasmic labeling. However, smaller reagents such as HRP-labeled Fab fragments may permit good cytoplasmic localization at the EM level.

10. *Processing of Tissue.* Preparation of tissue for immune labeling has two essential goals: antigenic sites in tissue must be preserved, and structural features of the tissue must be maintained. The nature of the antigenic determinant must be considered in the selection of a fixative. Acetone and ethanol have generally been used for light microscopic studies of proteinaceous antigens. At EM level, fixatives containing paraformaldehyde or glutaraldehyde often successfully preserve proteinaceous antigenicities in addition to ultrastructure. Scanning EM requires special processes (critical point freeze drying). Detailed discussions of various technics used for tissue processing are presented below.

1. Cinader B: Antibodies to Biologically Active Molecules. New York, Pergamon Press, 1967.
2. Eisen HM: Immunology, ed 2. Hagerstown, Md, Harper and Row, 1974.
3. Gill PGH, Kelus AJ: Anti-antibodies. Adv Immunol 6: 461, 1967.
4. Humphrey JH, White RG: Immunology for Students of Medicine, ed 3. Oxford, Blackwell Scientific Publications, 1970.
5. Kabat EA: The nature of an antigenic determinant. J Immunol 97: 1, 1966.
6. Kabat E, Mayer MM: Experimental Immunochemistry, ed 2. Springfield, Ill, CC Thomas, 1971.
7. Kitagawa M, Yagi Y, Pressman D: The heterogeneity of combining sites of antibodies as determined by specific immunoabsorbents. J Immunol 95: 446 and 991, 1965.
8. Klein J: Biology of the Mouse Histocompatibility-2 Complex. New York, Springer-Verlag, 1975.
9. Landsteiner K: The Specificity of Serological Reactions (Rev. Ed.). Cambridge, Harvard University Press, 1945.
10. Nakane PK: Recent progress in the peroxidase-labeled antibody method. Ann NY Acad Sci 254: 203, 1975.

11. Sell S: Immunology, Immunopathology and Immunity, ed 2. Hagerstown, Md, Harper and Row, 1975.
12. Weigle WO: Immunological unresponsiveness. Adv Immunol 16: 61, 1973.
13. Weir DM (Ed): Handbook of Experimental Immunology, ed 2. London, Blackwell Scientific Publication, 1973.
14. Williams CA, Chase MS (Eds): Methods in Immunology and Immunochemistry, vol I. New York, Academic Press, 1967.

FLUORESCENT ANTIBODY METHODS FOR THE LABELING OF INTRACELLULAR AND CELL SURFACE SUBSTANCES

Compounds that are illuminated by short wave length light and that after absorption of energy emit a second light of long wave length are called fluorescent compounds. Coons and associates in 1942 were first to use fluorescent-labeled antibody for studying localization of antigen in tissue [18]. The fluorescent compound was used as a chemically linked marker on the antibody without changing its immunologic reactivity. The most widely used fluorochromes are fluorescein and rhodamine [20,22]. Fluorescein absorbs at 490 nm and has a yellowish-green fluorescence with a maximum at about 520 nm. Rhodamine B absorbs at 550 nm and has a reddish-orange fluorescence with a maximum at about 620 nm.

The fluorescent antibody technic is helpful in the localization of antigen and antibody reactions at the cellular level. Fluorescent methods have been used to study numerous immunologic problems, such as: (a) the detection of specific antibody in patient's sera by use of appropriate antisera and antigen substrate; (b) the localization of hormones and enzymes in tissues; (c) the localization of complement, antibody, antigen, and immune complexes in the host tissues [25]; (d) the determination of the site of localization of injected antigens; (e) the localization and identification of the site of replication of infectious agents in bacteriology, virology and parasitology [17,23]; (f) the identification of substances stored or synthesized in cells; and (g) the detection of antigens on cell surfaces.

PREPARATION, ISOLATION AND CONJUGATION OF ANTIBODIES

A. Preparation of Antibodies

1. Immunization of animals with purified protein antigens. The isolation and characterization of a given antigen and methods used for immunization have been discussed in a general way in the Introduction.

2. Determination of antibody concentration. This is essential to determine if sufficient antibody was produced in the animal or whether further immunization is needed. The antibody concentration may be roughly determined by the units of precipitating antibody. By double diffusion gel method, the reciprocal of the highest dilution of the antiserum giving a precipitin line with 1 mg of the antigen is considered to be the number of units of precipitating antibody. A good antiserum should have greater than 4 units/ml.

B. Isolation of the gamma globulin fraction of antiserum by ammonium sulfate precipitation (salt fractionation) [21].

1. Preparation of Saturated $(NH_4)_2SO_4$ Solution. 375 g analytical grade crystalline $(NH_4)_2SO_4$ added to 500 ml of distilled H_2O while heating over a burner. Cool to 4°C and crystals will form. Take the saturated solution, leaving the crystals out. Adjust to pH 7.0 with diluted NaOH.

2. Materials and Reagents

a. Cold saturated $(NH_4)_2SO_4$ (pH 7.0)
b. Buffered saline solution, 0.01 M PO_4 buffered saline (pH 7.0)
c. 0.85% isotonic saline solution
d. Saturated $BaCl_2$—1–2 drops
e. 0.1 N HCL—1–2 mls
f. Dialysis bags

3. Procedure: $(NH_4)_2SO_4$ Fractionation of serum (gamma globulin)

(1) Measure the volume of serum after defrosting (volume x). Place in ice bath.
(2) Add an equal volume (x) of nonbuffered saline (pH 7.0) in the cold.
(3) Add dropwise a volume equal to that above (2x) of saturated ammonium sulfate $(NH_4)_2SO_4$ (pH 7.0). This is done in the cold with constant stirring.
(4) Keep at 0°C with constant stirring for 30 minutes. The white precipitate is the globulin fraction. The albumin fraction remains in the reddish-brown supernatant.
(5) Centrifuge for one hour at 2200 rpm at 4°C.
(6) Discard supernatant. Dissolve the precipitate with nonbuffered saline, bringing the volume up to 100 ml.
(7) Repeat steps 2, 3, 4, 5 and 6 until the supernatant is essentially colorless.
(8) Dissolve the final globulin precipitate in nonbuffered saline, final volume one-third of the original serum volume (1/3 x).
(9) Dialyze in cold against buffered saline for one day, changing the buffered saline at least once.
(10) Dialyze in cold against nonbuffered saline for one or two more days to remove the buffer.
(11) Continue dialysis until no precipitate is demonstrable in the dialysate upon addition of a few drops of saturated $BaCl_2$, which tests for the presence of $(NH_4)_2SO_4$.
(12) When dialysis is complete, determine the amount of protein by the micro-Kjeldahl technic, or spectrometry, and freeze.

C. Isolation of IgG by DEAE Cellulose Chromatography [24]

The IgG fraction is a much better reagent for fluorescent labeling as it produces less nonspecific fluorescence than $(NH_4)_2SO_4$ fractionation. The pure IgG fraction is obtained by dialyzing the $(NH_4)_2SO_4$ fraction (gamma globulin) against 0.01 M phosphate buffer (pH 7.5) and then passing the protein solution over DEAE (N,N-diethylamino-ethyl cellulose). The IgG fraction is eluted in the first peak.

Preparation of DEAE Cellulose Column

The weight of DEAE cellulose used should be at least 20 times the weight of protein applied to the column.

Reagents:

1 N NaOH—40 gm/liter
1 N HCl—82.6 ml HCl/liter
0.1 M Phosphate Buffer pH 8.0
8 liters 0.1 M Na_2HPO_4 (14.2 gms/liter)
+ 600 ml 0.1 M NaH_2PO_4 (13.8 gms/liter)
0.01 M phosphate buffer pH 8.0 (dilute above 1:10)
0.2 M $NaH_2PO_4 \cdot H_2O$ (27.6 gms/liter)

DEAE Cellulose

1. Preparation of DEAE
   a. Add to 100 gm DEAE 1 liter of 1 N NaOH. Filter through fritted glass funnel under negative pressure. Wash with additional NaOH until no more color can be removed.
   b. Suspend the DEAE in sufficient HCl to make strongly acid (~1 liter of 1.0 N NCl). Filter immediately. Wash with distilled H_2O until free of acid (pH of filtrate around 4.0).
   c. Suspend (in funnel) in 1 liter of 1.0 N NaOH and filter. Wash with distilled H_2O until free of NaOH. Approx. 20 liters H_2O (pH of filtrate around 7.0).
   d. Suspend the DEAE several times in 1–2 volumes of 0.1 M PO_4 buffer, pH 8.0 (approx. 6 liters of buffer), filter. Wash again with approx. 4 liters of distilled H_2O.
   e. Place DEAE in large beaker, suspend in 2 or 3 volumes of 0.01 M phosphate buffer. Adjust pH to 8.0, if necessary, with 0.2 M NaH_2PO_4. Add enough distilled H_2O to return suspension to 0.01 M. Recheck pH. Filter.
   f. Wash three times with 1 or 2 volumes of 0.01 phosphate buffer and bring to volume of 1.5 liters with 0.01 M phosphate buffer.
   g. Let slurry stand for about 1 hour. Aspirate supernatant and resuspend in buffer. Repeat 2–3 times to remove slowly sedimenting particles. If leaving overnight (or longer) add a few drops of sodium azide as a preservative.
2. Packing Column (4.5 × 60 cm column)

   By gravity—Pour slurry into column, letting buffer drain through adapter at bottom of column. If several additions of slurry are needed, stir it in column after each addition to release trapped air bubbles. Fill column 3/4 full and let about 5 ml of buffer stand above column. Place a filter paper over the column.
3. Application of sample and buffer to column: (1 gm protein/10 gm dry DEAE) or (10 ml serum/68 ml wet DEAE)

   If sample is crystalline or dry, reconstitute in 0.01 M phosphate buffer pH 8.0; otherwise, dialyze sample against the starting buffer for 48 to 72 hours with several changes before applying to column.
   a. Using a pipet, place tip close to filter paper and add sample very slowly. Do not disturb surface of column.
   b. Let sample run into column.
   c. Just before the last of the sample has moved in the column, begin adding 0.01 M phosphate buffer in the same manner the sample was added in *a*. Fill column with buffer.

4. Collection of Sample
   a. Allow 100–200 ml to run through column; then begin collecting 7–9 ml fractions and determine protein concentration by spectrophotometry.
   b. Combine collected volumes which show strong protein concentration. Since considerable dilution takes place the gamma globulin may be concentrated by negative pressure dialysis if desired. Dialyze sample against distilled H_2O before lyophilization or storage.
5. Recycling DEAE

   Wash with several volumes 0.5 N HCl (~2 liters) then wash with distilled H_2O (4–6 liters). Wash with 1–2 volumes 95% alcohol (1–1.5 liters) then wash with distilled H_2O (4–6 liters). Wash with 2–3 volumes 0.5 N NaOH (~2 liters) then wash with distilled H_2O (10 liters).

D. Preparation of Fluorescein Conjugate

Isothiocyanate derivates of fluorescein (FITC) and tetra-methylrhodamine (TRITC) are the commonest and most stable labeling compounds [16,24].

Reagents and Materials:

1. Carbonate buffer stock solution pH 9.5, 0.5 M Na_2Co_3 13.25 gm dissolved in 250 ml distilled H_2O. $NaHCO_3$ 10.50 gm dissolved in 250 ml distilled H_2O. Before conjugation, add $NaHCO_3$ to Na_2CO_3 to about 1:9 ratio.

Conjugation Reaction Mixture:

1. 5 ml. of 50 mg/ml (250 mg) of purified immunoglobulin dialyzed overnight in nonbuffered saline.
2. Add 10% (0.5 ml) by volume of carbonate buffer, pH 9.5.
3. Add, dropwise, 3.5 mg of FITC dissolved in 5 ml of 2% $NaHCO_3$ (0.014 mg FITC/mg protein).
4. Stir the mixture at 4°C overnight.
5. Dilute with phosphate buffer, pH 7.5 (10 times dilution).
6. Remove the free fluorescein by sephadex G-25 or G-50 column.

E. Separation of Labeled Protein from Unlabeled Fluorescein by Sephadex G-25 or G-50 column:

Suspend in 250 ml physiologic saline solution, 25 grams of G-25. Mix the suspension and allow to swell and settle, repeating this procedure three times. Pour the suspension into 2.5 cm. diameter glass column with a sintered-glass disc and fill the column to at least a 25 cm height. This column will handle 10–50 ml labeled protein. The Sephadex column G-25 or G-50 is set up so that the ratio of the volume of the column to the volumn of conjugate is at least 5:1, and the conjugate is layered carefully on the top. The labeled protein can be concentrated and stored in aliquots in a freezer.

In the fluorescein-conjugated antibody solution there are substances which can cause nonspecific tissue staining. Such substances can be removed by absorption with tissue powder or by the use of DEAE cellulose chromatography which will isolate the specific fluorescent antibodies.

F. Isolation of Specific Fluorescein-Labeled Antibodies with DEAE Cellulose

Conjugation of fluorescein with globulin changes the electrical charge of the protein molecule. When few fluorescein groups are coupled, the conjugate does not emit enough light for visualization; whereas, if excess numbers of fluoresceinated molecules are coupled to the protein the stain is bright and there will be nonspecific fluorescence. DEAE cellulose chromatography may be used to isolate the optimally coupled labeled protein.

Procedure:

1. The conjugated antibody is diluted 10-fold with 0.01 M sodium phosphate buffer and applied to the DEAE cellulose column, equilibrated with the same buffer.
2. The loosely coupled protein does not bind to the DEAE at this molar concentration and is eluted with the 0.01 M buffer.
3. Stepwise elution is done with 0.03 M, 0.05 M, and 0.1 M sodium phosphate buffer.
4. The fractions are collected, concentrated, and divided into small portions for use or for storage in the freezer (−20°C). The fraction that has the optimally conjugated globulin must be determined spectrophotometrically (see below). Usually 0.03 M fraction is the best.

Calculations

1. Protein (immunoglobulin) concentrations in the fluorescein conjugates can be determined spectrophotometrically by reading the absorbancies (optical densities) of the solution at 280 nm (UV) and 495 nm (vis) with a pathlength of 1 cm and using $E^{1\,mg/ml\,Ig}_{280nm;\,1cm} = 1.4$.
   The following formula has been devised by Wood et al. [27].

$$\text{Protein (mg/ml)} = \frac{A_{280} - (0.35 \times A_{495})}{1.4}$$

2. To determine the suitability of a given conjugate, the dye:protein labeling ratio must be determined spectrophotometrically. Satisfactory conjugates have ratios which approach unity. Values below 0.5 indicate low labeling and conjugates giving values over 1.5 would be expected to give problems with nonspecific staining For fluorescein conjugates (IgG) the absorbancy of solutions are determined at 495 nm (FITC peak) and 280 nm (tyrosine peak). The folowing formula has been used to determine molecular F:P ratios.

$$\text{Molecular F:P} = \frac{2.87 \times A_{495}}{A_{280} - (0.35 \times A_{495})}$$

G. Procedure for Conjugation of Antibody Globulin with Tetramethylrhodamine (TRITC)

Tetramethylrhodamine can be conjugated to antibody globulin in a similar procedure outlined for fluorescein isothiocyanate. The amount of rhodamine which can be used in the direct conjugation method is 20–40 micrograms/mg protein [16].

The unreacted rhodamine may be separated from the coupled protein by passage through a Sephadex G-50 column. This step does not need the dilution required for the fluorescein isothiocyanate.

Following the separation of the unlabeled and labeled conjugated globulin, the material is fractionated on DEAE cellulose column similar to the fluorescein isothiocyanate.

H. Evaluation of Conjugates

All conjugates should be evaluated in order to determine their suitability for use in the specific system of the study by:

1. Physicochemical characterization, which determines dye:protein ratio, and evaluation of the specific antibody content and other proteins present. The dye:protein molar ratio is simply calculated by measuring the optical densities of a solution at 495 nm (FITC), 550 nm (TRITC) and at 280 nm (Tyrosine). Satisfactory conjugates have molar ratio of one (see above). The specific antibody content can be determined by immunoelectrophoresis and simple titration in agar.
2. Performance testing which gives information on the range of optimal working dilutions of the conjugate. This tests the presence of antibody specificity and also the optimal working dilution which tests the sensitivity. The specificity of anti-immunoglobulin conjugates are determined by testing them on known bone marrow cells obtained from characterized myeloma patients. The sensitivity is measured by block titration and, hence, the extent to which the conjugates can be diluted and still retain their specific, while losing their nonspecific, staining capability.

PREPARATION AND IMMUNOHISTOLOGIC PROCESSING OF TISSUES AND CELLS

For demonstration of antigen in cells, two conditions must be fulfilled. One, preservation of antigenic activity and, secondly, maintenance of a good morphological condition. The preparation of the tissue to be studied may be carried out in several ways [15,19]. Snap-frozen tissue has frequently been used since the tissue morphology as well as the immunologic reactivity of the desired material is preserved. Fresh tissue should be cut 1–2 mm thick, and placed in a test tube. Quick freezing is accomplished by immersing the tube in a mixture of dry ice and butyl alcohol. The tissue is cut in a cryostat, 4–7 microns thick, and placed on a microscope slide without thawing. The slides can be placed in a mild fixative, e.g., an equal volume mixture of absolute ethyl alcohol and dry ethyl ether at room temperature for 30 minutes followed by 95% ethyl alcohol at 37°C for 15 minutes. Other mild agents such as acetone, methyl alcohol and ethyl alcohol can be used.

Immediate freezing of tissue is not always possible and fixation becomes necessary in some situations. The method that we found obtained good results for intracellular antigenic determinants is that of Sainte Marie [26].

Acetone Fixation

1. Acetone (99 Mol % Pure)
   a) 4 hours at 4°C
   b) 2–4 hours at 4°C
   c) 4 hours or overnight at room temperature (RT)
2. Benzene (99 Mol % Pure)
   Three changes in benzene at RT, total time 1 to 1-1/2 hours.
3. 1:1 Benzene/Parafin for 20–30 min.
4. Embed in 54–56°C melting point parafin (change 3 times).

A. Indirect Immunofluorescence of Tissue Sections

1. The first antibody is added on the slide in drop-like fashion to cover the whole section and incubated at room temperature for 30 minutes in a humidified chamber.
2. The section is washed with PBS, 3 changes.
3. The second fluoresceinated antibody is added for 30 minutes and kept at room temperature in a humidified chamber.
4. The slides are washed 5 times in cold PBS.
5. Glycerol mounting media 2–3 drops is put on the slide and a thin coverslip is used.

The slides are ready for further examination by the fluorescent microscope.

B. Vital Fluorescent Staining of Cells

Lymphocytes are isolated as described in the immunoelectronmicroscopy section. The cells are suspended in PBS buffer, pH 7.5 and washed 3 times. The best concentration of cells in our experience is 1×10^6 cells/ml. The experiment is run O–4°C and/or 0.1% Na azide added to the cells or buffer to discourage endocytosis and capping. After washing, the cells are incubated with the 1st antibody for 30 minutes. The cells are washed 3 times with cold buffer containing azide and incubated with the fluoresceinated 2nd antibody for 30 minutes. Washing with the same buffer is again done for 3–5 times and the cells are suspended and put on a glass slide and examined with the fluorescent microscope.

C. Fluorescent Microscopy

Microscopy is carried out by using a standard microscope with achromatic or apochromatic objectives. Dark field or bright field condensers can be used. A high intensity ultraviolet light source is essential at least for fluorescein. HBO 200 (Osram) or a mercury vapor lamp is satisfactory. Two types of filters must be used. The first, called the excitor filter (BG-36 + S546 for rhodamine and BG-12 for fluorescein), is placed between the strong ultraviolet light source and the microscope. It blocks all wave lengths except the ultraviolet and dark blue. A second filter, called the barrier filter (S-525 for fluorescein and K 620 for rhodamine), is placed between the microscope slide and the viewer's eyes. It allows the color of the fluorescent materials to pass through but blocks any ultraviolet light from passing to the eye.

Photography may be carried out with any standard equipment. Good results are obtained with Ektachrome 125 ASA and exposure time of 1–2 minutes, depending on the strength of the fluorescence.

15. Arnold W, Kalden JR, Mayersback H: Influence of different histologic preparation methods on preservation of tissue antigens in the immunofluorescent antibody technique. Ann NY Acad Sci 254: 27, 1975.
16. Cebra JJ, Goldstein G: Chromatographic purification of tetramethylrhodamine-immune globulin conjugates and their use in the cellular localization of rabbit gamma globulin polypeptide chains. J Immunol 95: 230–245, 1965.
17. Cherry WB, Moody MD: Fluorescent-antibody techniques in diagnotic bacteriology. Bacteriol Rev 29: 222, 1965.

18. Coons AH, Creich HJ, Jones RN, Berliner E: The demonstration of pneumococcal antigen in tissues by the use of fluorescent antibody. J Immunol 45: 159, 1942.
19. Feltkamp-Vroom TM: Preparation of tissues and cells for immunohistochemical processing. Ann NY Acad Sci 254: 21, 1975.
20. Goldman M: Fluorescent Antibody Methods. New York, Academic Press, 1968.
21. Heide K, Schwirck HG: Salt fractionation of immunoglobulins. *In:* Weir DM (Ed): Handbook of Experimental Immunology, ed 2. Oxford, Blackwell Scientific Publications, 1973, p. 1, chap. six.
22. Johnson GD, Holborow EJ: Immunofluorescence. *In:* Weir DM (Ed): Handbook of Experimental Immunology, ed 2. Oxford, Blackwell Scientific Publications, 1973, chap. 18.
23. Kawamura A.: Fluorescent Antibody Techniques and Their Application. Baltimore, University Park Press, 1969.
24. Levy HB, Sober HA: A simple chromatographic method for preparation of gamma globulin. Proc Soc Exp Biol Med 103: 250, 1960.
25. Nakamura, R.M.: Fluorescent antibody methods. *In:* Immunopathology—Clinical Laboratory Concepts and Methods. Boston, Little, Brown and Company, 1974, p. 620.
26. Sainte-Marie G: A paraffin embedding technique for studies employing immunofluorescence. J Histochem Cytochem 10: 250, 1962.
27. Wood BT, Thompson SH, Goldstein G: Fluorescent antibody staining. III. Preparation of fluorescein-isothiocyanate-labeled antibodies. J Immunol 95: 225, 1965.

CURRENT METHODS FOR IMMUNOLOGIC STAINING WITH PEROXIDASE-LABELED ANTIBODIES

Numerous methods for the histochemical demonstration of endogenous enzymes, including peroxidase, have been known for a long time [36]. In hematology laboratories, for example, leukocyte granules are routinely visualized through the oxidation of benzidine (catalyzed by peroxidase). Further application of this reaction was discouraged by the carcinogenic properties of benzidine until Graham and Karnovsky [28] in 1966 reported the ultrastructural localization of exogenous horseradish peroxidase (HRP), using 3-3′-diaminobenzidine (DAB), a benzidine derivative, as substrate. The immunohistochemical application of the HRP-DAB system was soon found, using both HRP-labeled [35] and unlabeled [38] antibodies. The many advantages of the immunoperoxidase technic in light microscopy (increased sensitivity, permanent staining, bright-field visualization), as well as its adaptability to electron microscopy, have rapidly established it as the method of choice in much immunohistochemical work.

This workshop outlines methods for the labeling of antibodies with HRP and for the use of labeled antibodies for the localization of antigens with light and electron microscopy. Specific references are cited for most procedures, to which the reader is directed for further details. Special attention has been given to tissue preparation and the problem of background staining (staining due to nonspecific affinities of the tissue or reagents or to endogenous peroxidase activity). Such nonspecific staining is frequently a bothersome problem, which may interfere significantly with the correct interpretation of results. The magnitude of the problem varies with the tissue or reagents in a given system, and suggestions for its solution are given below.

Finally, the importance of appropriate specificity controls for the correct interpretation of results cannot be overstressed. Controls should be designed to determine the specificity of the positive staining as well as the specificity of the reagents. Even after the staining procedure has been standardized for a given system, controls for staining specificity should be included in each experiment.

GENERAL APPROACHES TO TISSUE PREPARATION FOR STAINING WITH PEROXIDASE-LABELED ANTIBODIES

Ideally, a fixative should preserve both antigenicity and morphology. The best fixative for a particular system is chosen empirically, considering the properties of the antigen and tissue as well as the resolution desired.

At the light microscopic (LM) level, the mildest fixatives are acetone and ethanol, which are excellent for preserving antigenicity, but tissue morphology is little more than adequately preserved. Zamboni's [37] and Bouin's [30] solutions are superior in preserving morphology and can even be used for electron microscopy (EM) preparations for certain antigens.

For immunoelectron microscopy the conventional fixative, glutaraldehyde, is often not suitable because it interacts strongly with proteins and frequently denatures protein antigens. Paraformaldehyde reacts to a lesser degree. Periodate-lysine paraformaldehyde (PLP) [31], on the other hand, primarily stabilizes carbohydrate moieties with adequate preservation of most antigenicity and ultrastructure.

The following are typical methods of tissue preparation.

For Light Microscopy

1. Fresh frozen sections, fixed for ten minutes in cold 100% acetone or ethanol.
2. Fresh frozen sections, fixed for one hour in cold p-formaldehyde, PLP or Zamboni's solution.
3. Frozen sections, of fixed tissues* (e.g., p-formaldehyde, PLP, Zamboni's).
4. Paraffin sections of fixed tissues (e.g., Zamboni's, PLP, Bouin's), rehydrated through xylene and ethanols to PBS.
5. Single cells in monolayer culture or suspension:
   a. Grow cells on slide, fix and stain in situ.
   b. Fix cells in suspension, wash, centrifuge, embed pellet in O.C.T.™ (Ames), cut frozen sections and stain.
   c. Deposit cells onto microscope slides using a cytocentrifuge. Fixation may be done during or after the centrifugation.

* Frozen sections of fixed tissues must be put on albumin-coated slides (see further).

For Electron Microscopy

1. Frozen sections of fixed tissues (e.g., PLP, Zamboni's).
2. Single cells in monolayer culture or suspension: same as for LM. An alternate method to 5.b. is to keep cells in suspension for all reactions.

PLP FIXATION AND SUBSEQUENT WASHING PROCEDURES

For Light Microscopy

1. Fix small pieces of tissue in PLP for four hours at 4° C on a shaker.
2. Wash in *0.05 M sodium phosphate buffer pH 7.2 (PB)* containing 10% sucrose for about 36 hours at 4° on a shaker (three changes of about 12 hours each).
3. Embed in O.C.T. embedding medium (Ames) and quick-freeze in dry ice-ethanol. We shape a heavy aluminum foil boat over the end of a thick pencil, about 1 cm in diameter, place the tissue inside, fill with O.C.T., and freeze, being very careful not to let the ethanol seep into the boat.
4. Cut 6 μ frozen sections onto albumin-coated slides.
5. Air dry 30 minutes.
6. Wash in PBS: 3 $\times$ 5 minutes and stain.

For Electron Microscopy

1. Fix small pieces of tissue in PLP for four hours at 4°C on a shaker.
2. Wash in PB containing 10% sucrose and 4 $\times$ 10^{-5} M digitonin to improve membrane fixation (Sigma). Heat slightly to dissolve digitonin overnight at 4°C on a shaker.
3. Wash in PB containing 15% sucrose for four hours to overnight.
4. Wash in PB containing 20% sucrose for four hours.
5. Wash in PB containing 20% sucrose and 10% glycerol for one hour.
6. Embed in O.C.T. and quick freeze, as described for LM.
7. Cut 6-10 μ frozen sections, onto albumin-coated slides. Sectioning will be difficult, but the high sucrose-glycerol content is required for good ultrastructural preservation.
8. Air dry 30 minutes.
9. Wash in PB containing 10% sucrose: 3 $\times$ 10 minutes and stain.

Sample Staining Protocols:

LM: Mouse kidney basement membrane (Indirect Technic)

Cut from fresh frozen sections (6 μ) of mouse kidney. Air-dry (30 minute) fix in cold 100% acetone (10 minutes), and wash in PBS (0.01 M phosphate-buffered saline, pH 7.2).

1. Make a moist chamber.
2. Dry off excess PBS from around the tissue section and apply about 20 μl antiserum or PBS as follows:
   slides #1 & 2: Rabbit antibasement membrane serum
   slide #3: Normal rabbit serum
   slide #4: PBS

Mix the antisera thoroughly with residual PBS over tissue.
Place in moist chamber for 15 minutes.

3. Rinse antisera off the slides with PBS.
   Wash in PBS: three changes of five minutes each.
4. Dry off excess PBS from around the tissue section and apply about 20 μl HRP-sheep antirabbit IgG immunoglobulin to each slide. Place in moist chamber for 15 minutes.
5. Wash in PBS: 3 $\times$ 5 minutes.
6. Immerse in DAB-H_2O_2 solution for ten minutes.
7. Wash in PBS: 3 $\times$ 5 minutes.
8. Dehydrate in graded ethanols and xylene. Mount.

Results: Brown reaction product of DAB substrate is seen on tubular and glomerular basement membrane (slides 1 and 2). Compare with slides 3 (nonspecific background staining) and 4 (endogenous peroxidase-activity of tissue).

LM: Rat anterior pituitary ACTH and GH (Indirect Double Staining Technic) [32]

Cut four frozen sections (6–10 μ) of PLP-fixed rat anterior pituitary onto albumin-treated slides. Air-dry for 30 minutes and wash in PBS. Sections #1 and #2 will be stained for ACTH first and then for GH. Sections #3 and #4 will be stained for GH first and then for ACTH. The first substrate, DAB, will yield a brown reaction product. The second substrate, 4-Cl-1-naphthol, will give a blue reaction product.

1. Make a moist chamber.
2. Dry off excess PBS and apply about 20 μl antiserum as follows:
   slides #1 & 2: rabbit anti-ACTH
   slides #3 & 4: rabbit anti-GH
   Mix the antisera thoroughly with residual PBS. Place in moist chamber for 15 minutes.
3. Rinse antisera off slides with PBS.
   Wash in PBS: three changes in five minutes each.
4. Dry off excess PBS and apply about 20 μl HRP-sheep anti-rabbit IgG immunoglobulin. Place in moist chamber 15 minutes.
5. Wash in PBS: 3 $\times$ 5 minutes.
6. Immerse in DAB-H_2O_2 for 10 minutes.
7. Wash in PBS: 3 $\times$ 5 minutes.
8. Remove the first antibody and conjugate by washing in 0.1 M glycine-HCl pH 2.2 for ½–1 hour. Stir gently during this time. The DAB reaction product will remain on the tissue.
9. Wash in PBS: 3 $\times$ 5 minutes.
10. Dry off excess PBS and apply about 20 μl antiserum as follows:
    slides #1 & 2: rabbit anti-GH
    slides #3 & 4: rabbit anti-ACTH
    Place in moist chamber for 15 minutes.
11. Rinse antisera off slides with PBS. Wash in PBS: 3 $\times$ 5 minutes.
12. Repeat steps 4 and 5.
13. Incubate in 4-Cl-1-naphthol—H_2O_2 for ten minutes.
14. Wash in PBS: 3 $\times$ 5 minutes.
15. Mount in glycerol: PBS (9:1). Do not dehydrate since the 4-Cl-1-naphthol reaction product is soluble in ethanol.

Results: Cells labeled with anti-ACTH are stained brown on slides 1 and 2 and blue on slides 3 and 4, while cells labeled with anti-GH are stained blue on slides 1 and 2 and brown on slides 3 and 4.

EM: Mouse Kidney Basement Membrane

Cut four frozen sections of PLP-fixed mouse kidney onto albumin-coated slides. Air-dry for 30 minutes and wash in 0.05 M sodium phosphate buffer pH 7.2 containing 10% sucrose (PB-sucrose). Use PB-sucrose for all subsequent wash steps and add 10% sucrose to all conjugates and to the DAB solutions.

1. Make a moist chamber.
2. Dry off excess PB-sucrose from around section with gauze. With a capillary pipet apply about 20 μl HRP-labeled antibody as follows:
   slides #1 & 2: HRP—rabbit antibasement membrane Fab′
   slides #3 & 4: HRP—rabbit Fab′ (control)
   Mix conjugate thoroughly with residual PB-sucrose over tissue, being careful not to scratch the section. Place in moist chamber for 3–4 hours. KEEP CHAMBER MOIST.
3. Rinse conjugate off the slides with PB-sucrose.
   Wash in PB-sucrose: 3 changes of 10 minutes each.
4. Incubate in 2% glutaraldehyde-0.1 M sodium phosphate buffer pH 7.2 for 30 minutes.
5. Wash in PB-sucrose: 3 $\times$ 10 minutes.
6. Incubate in DAB solution for 30 minutes.
7. Incubate in DAB-H_2O_2 solution for 2 minutes.
8. Wash in PB-sucrose: 3 $\times$ 10 minutes.
9. Apply a drop of 2% OsO_4-0.1 M sodium phosphate buffer pH 7.2 as in step 2 above and incubate in a moist chamber for one hour.
10. Wash in PB-sucrose: 3 $\times$ 10 minutes.
11. Dehydrate in graded ethanols to 100%.
12. Embed by inverting a gelatin capsule filled with fresh Epon-Araldite over each section.
13. Polymerize the Epon by incubating at 37° overnight and at 60° 24–48 hours. The block and its tissue section can then be removed from the slide by *very* briefly heating the slide over a Bunsen burner flame.

    Note the longer incubation time and the use of HRP-labeled Fab′ fragments, obtained by pepsin digestion of IgG [29], which are required to achieve adequate antibody penetration for intracellular staining at the EM level.

Results: The DAB reaction product reduces and chelates OSO_4, leaving an electron-dense, insoluble precipitate at the site of enzyme action.

METHOD OF HRP-LABELED IgG PREPARATION [34]

1. Dissolve 5 mg of HRP (Sigma HRP Type VI, RZ approximately 3.0) in 1.0 ml of freshly-made 0.3 M sodium bicarbonate.
2. To the above solution, add 25 μl 0.32% p-formaldehyde (0.2 ml 8% + 4.2 ml H_2O). Mix gently for ½ hour at room temperature.
3. Add 1.0 ml of 0.04 M $NaIO_4$ in distilled water. Mix gently for 30 minutes at

room temperature. The color of the solution will become slightly green-yellow.

4. Add 1.0 ml of 0.16 M ethylene glycol in distilled water. Mix gently for one hour at room temperature.
5. Dialyze the solution against one liter of 0.01 M sodium carbonate buffer, pH 9.5 at 4°C overnight.
6. Add 10 mg of IgG (or 5 mg Fab′) in approximately 1 ml carbonate buffer solution. Mix gently for two hours at room temperature.
7. Add 5 mg $NaBH_4$.* Leave at 4°C two hours.
8. Dialyze at 4°C against PBS. A small amount of precipitate may form and should be removed by centrifugation.
9. Apply the sample to an 85 × 1.5 cm column of Sephadex G-100 or G-200 equilibrated in PBS. Read the absorbance of the fractions at 280 and 403 nm. Pool the fractions of HRP-labeled IgG, the first peak.

With the above procedure, approximately 70% of the HRP should couple with IgG and more than 90% of the IgG should be labeled with HRP. No significant losses of either IgG immunologic or HRP activities occur.

Reagents

PLP Fixative (Periodate-lysine-paraformaldehyde) [31]

Stock A = 0.1 M lysine – 0.05 M sodium phosphate, pH 7.4

1. Dissolve 1.827 gm L-lysine HCl (Sigma) in 50 ml H_2O. (0.2 M lysine HCl)
2. Adjust pH to 7.4 with 0.1 M Na_2HPO_4.
3. Make up to 100 ml with 0.1 M sodium phosphate buffer, pH 7.4. Osmolality should be approximately 300 mOs.
4. Store at 4°C for a maximum of ten days.

Stock B = 8% paraformaldehyde

1. Mix 8 gm paraformaldehyde (Matheson, Coleman, Bell) in 100 ml H_2O.
2. Heat to 60°C with stirring.
3. Slowly and 1–3 drops 1 N NaOH until clear.
4. Filter.
5. Store at 4°C.

Just before use, combine three parts A with one part B and add sodium m-periodate to 0.01 M (i.e., 21.4 mg $NaI0_4$/10 ml). Final composition is therefore 0.01M $NaI0_4$, 0.075 M lysine, 0.0375 M sodium phosphate buffer, and 2% paraformaldehyde. The final pH will be approximately 6.2 and osmolality around 700 mOs.

Albumin-coated Slides [33]

1. Clean glass microscope slides in 1% HCl-70% ETOH by immersion and agitation for several minutes.
2. Dip slides in 100% acetone and allow to air dry.
3. Combine one egg white and 1 ml concentrated (28%) NH_4OH with 500 ml distilled water. Stir ten minutes. Filter through a paper towel or six layers of gauze.
4. Place cleaned slides into freshly-made egg albumin solution for one minute. Drain off excess solution on a paper towel and place slides in racks to dry.

* To stabilize the Schiff base formed by the reactions of HRP-aldehyde with IgG.

5. Dry at 60°C for two hours to overnight.
   Up to 500 slides can be coated from the above egg mixture and they can be kept indefinitely.

DAB-H_2O_2 (Adapted from Graham and Karnovsky [28])

1. Dissolve 25 mg 3,3′-diaminobenzidine-4HCl in 100 ml 0.05 M Tris-HCl buffer pH 7.6. Filter if not completely colorless.
2. Adjust pH to 7.6 with 0.05 M Tris-base or with HCl if necessary.
3. Add 0.1 ml of 5% H_2O_2 to 100 ml DAB-Tris solution.
   Use within one to two hours.

4-Cl-1-naphthol [32]

1. Dissolve 25 mg 4-Cl-1-naphthol (K & K Laboratories) in 0.5 ml ETOH and then add 100 ml 0.05 M Tris-HCl buffer pH 7.6. Stir at room temperature 30–60 minutes. Filter.
2. Adjust pH to 7.6 with 0.05 Tris-base or with HCl if necessary.
3. Add 0.1 ml of 5% H_2O_2 to 100 ml 4-Cl-1-naphthol-Tris solution.
   Use within one hour.

GENERAL APPROACHES TO THE REDUCTION OR ELIMINATION OF BACKGROUND STAINING

The following suggestions or methods may help solve background problems. Their usefulness in a particular system has to be determined empirically.

1. *Background due to nonspecific affinities*
   a. Tissue-related
      1. Preincubation with 10–100% normal serum of species of second antiserum (e.g., to minimize nonspecific adherence of Fc portion of primary antibody).
      2. Increase tissue washing to remove excess fixative and/or staining reagents. Residual aldehydes (from fixative) may be reduced with sodium borohydride ($NaBH_4$) – 0.1 mg/ml H_2O for 10 mins.
      3. Use freshly cut frozen sections.
      4. Use 100% paraffin, not Paraplast, for embedding since the latter appears to have affinity for peroxidase-labeled antibodies.
   b. Reagent-related
      1. Avoid hemolyzed serum (hemoglobin has peroxidase activity).
      2. Dilute antisera as much as possible.
      3. Use absorption methods (e.g., tissue powders, tissue culture medium).
      4. Enrich antibody titer by affinity chromatography.
      5. Use Fab fragments (or Fab_2 except for intracellular staining at EM level where Fab′ fragments should be used).
      6. Avoid heavily peroxidase-labeled antibodies.
2. *Methods for Destruction of Endogenous Peroxidase Activity (primarily in red and white blood cells)*
   a. LM (WBC and RBC): fixation in 100% ethanol to which HCl (0.075%) has been added.
   b. LM and EM (WBC only) of PLP-fixed tissue: add 0.005 M sodium azide (NaN3) to DAB-H_2O_2 solution.

c. LM and EM (WBC and RBC): use before applying primary antisera (only if tissue is well fixed by aldehyde fixation):
   1) 0.01 periodic acid (H_5IO_6) for 10 mins., followed by
   2) $NaBH_4$ (0.1 mg/ml H_2O) for 10 mins. (to reduce aldehydes generated by H_5IO_6).

28. Graham RC Jr, Karnovsky MJ: The early stages of absorption of injected horseradish peroxidase in the proximal tubules of mouse kidney: ultrastructural cytochemistry by a new technique. J Histochem Cytochem 14: 291, 1966.
29. Grey HE, Kunkel HG: H chain subgroups of myeloma proteins and normal 7S γ-globulin. J Exp Med 120: 253, 1964.
30. Humason, G.L.: *In:* Animal Tissue Techniques. San Francisco, W.H. Freeman, 1967, p 14.
31. McLean IW, Nakane PK: Periodate-lysine-paraformaldehyde fixative. A new fixative for immunoelectron microscopy. J Histochem Cytochem 22: 1077, 1974.
32. Nakane PK: Simultaneous localization of multiple tissue antigens utilizing the peroxidase-labeled antibody method: A study on pituitary glands of the rat. J Histochem Cytochem 16: 557, 1968.
33. Nakane PK: Localization of hormones with the peroxidase-labeled antibody method. *In:* Colowick SP, Kaplan NO (Eds): Methods in Enzymology, Vol. VII, part B. New York, Academic Press, 1975, p 133.
34. Nakane PK, Kawaoi A: Perioxidase-labeled antibody. A new method of conjugation. J Histochem Cytochem 22: 1084, 1974.
35. Nakane PK, Pierce GB Jr: Enzyme-labeled antibodies: preparation and application for the localization of antigens. J. Histochem. Cytochem. 4: 929, 1966.
36. Pearse AGE: Histochemistry, Theoretical and Applied, Vol. II. Boston, Little Brown, 1972.
37. Stefanini M, DeMartino C, Zamboni L: Fixation of ejaculated spermatozoa for electron microscopy. Nature 216: 173, 1967.
38. Sternberger LA, Cuculis JJ: Method for enzymatic intensification of the immunocytochemical reaction without use of labeled antibodies. J Histochem Cytochem 17: 190, 1969.

LABELING OF LYMPHOCYTE SURFACE IMMUNOGLOBULIN USING IMMUNOELECTRONMICROSCOPIC MARKERS

ULTRASTRUCTURAL LABELING FOR TRANSMISSION ELECTRON MICROSCOPY

A step-by-step procedure for labeling lymphocyte surface immunoglobulin by immunoelectron microscopic technics will be presented. Two types of marker molecules, ferritin and hemocyanin, have been used with success in our laboratory. These markers are easily obtainable and can be coupled to antibody or immunoglobulin by covalent (chemical coupling) or noncovalent (antibody-antigen coupling) methods. The glutaraldehyde conjugation technic is relatively simple, but the degree of conjugation is difficult to control. However, a method for coupling IgG to both ferritin and hemocyanin using glutaraldehyde has been developed. Another method, complexing Ig with ferritin or hemocyanin by making soluble immune complexes with anti-ferritin or anti-hemocyanin, has also proved valuable. Both of these markers can be purified from uncomplexed globulins by ultracentrifugation and/or gel filtration chromatography. The markers can be tested by immunoelectrophoresis or inhibition of passive hemagglutination. These may also be used in a mixed anti globulin labeling technic (modified Coombs reaction) for detecting surface Ig. The type of surface Ig detected can be either endogenous (due to Fc receptor) or antibody-directed to cell surface antigens. A method for cell fixation, dehydration and embedding for electron microscopy is also described.

I. Preparation of Soluble Immune Complexes: Ferritin-Anti-Ferritin [39]

1. Test antiserum for anti-ferritin antibody by a quantitative precipitin assay with ferritin. Determine the precipitation equivalence point. Either a whole serum or the IgG fraction can be used.
2. Mix 40–50 X equivalent point of antigen (Ferritin) with the antiserum. For example, the eq. pt. might be 20 mg Fe per ml of serum. Therefore, 80–100 mg Fe should be mixed with 0.1 ml of serum.
3. Incubate the mixture for 30 min. at 37°C, then overnight at 4°C.
4. The complexes are then ready to be purified from the uncomplexed serum proteins by gel chromatography or ultracentrifugation. No attempt to remove uncomplexed or free antigen should be made (antigen excess is necessary).

   Note: The same procedure can be successfully used with other antigens, e.g., hemocyanin [49].

THEORY:

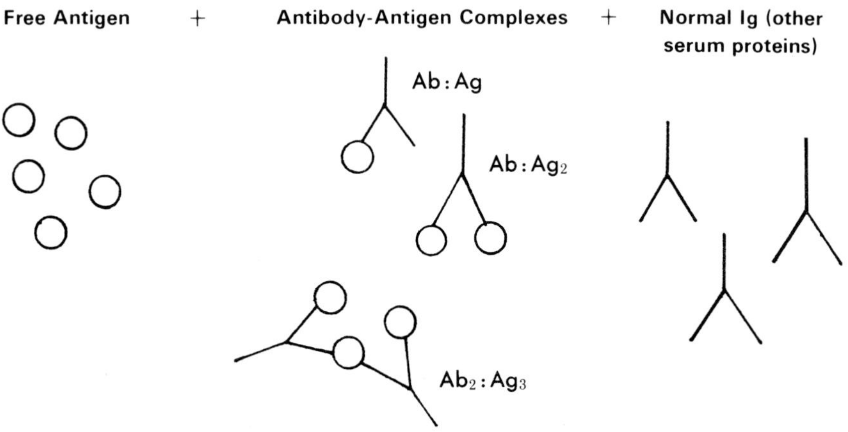

II. One Step Conjugation of Ferritin and IgG with Glutaraldehyde [40,47]

1. Ferritin (6X crystallized E.M. grade) should be used. Determine the concentration of ferritin by determining the optical density at 280 nm and 440 nm. Using the extinction coefficients $E^{1\%}_{280nm} = 100$ and $E^{1\%}_{440nm} = 13.5$ determine the conc. in mg/ml.
2. Determine the concentration of IgG to be conjugated by reading the OD at 280 nm and using $E^{1\%}_{280} = 14.2$.
3. A molecular ratio of *4 IgG: 1 Fe: 600 Glutaraldehyde molecules* has been found to give a satisfactory coupling.
4. Make up a solution containing 50 mg Fe and 10 mg IgG with a final volume of 1.0 ml. or less, using phosphate buffer, pH 7.0.
5. With constant stirring add 0.3 ml of 0.5% glutaraldehyde. Stir vigorously without bubbles or foaming for 1 hr at room temperature.
6. Place the solution in an ice bath and add 0.3 ml. of 0.1 M NH_4Cl or $(NH_4)_2$ CO3 for 15 min. to quench any unreacted glutaraldehyde.
7. Dialyze against 0.1 M $(NH_4)_2CO_3$ for 2–3 hr at 4°C. Dialyze against phosphate buffered saline, pH 7.0, overnight at 4°C.
8. Centrifuge at 10,000 × g for 20 min and recover the supernatant, discarding the pellet, if any.
9. Purify the conjugate from unconjugated IgG by ultracentrifugation or gel filtration chromatography. For *Ultracentrifugation,* spin at 100,000 × g for 3 hrs (31.5 K, SW50L rotor). Discard the supernatant and allow the pellet to redissolve in PBS (1–2 ml) overnight at 4°C. Repeat twice. Add 0.1% NaN_3 as preservative and store at 4°C. Centrifuge at 10,000 × g prior to use to remove any precipitate. For *Gel Filtration,* pour a 1 × 80 cm 6% Agarose (BIORAD 200–400 Mesh exclusion vol. 5 × 10^6 mw) column and wash with PBS. Apply conjugate and collect 1–2 ml fractions. Read OD at both 280 nm and 440 nm. The first peak will be aggregated Fe. The shoulder of the second peak is Fe-IgG, and the trailing end is free Fe and free IgG. Pool the Fe-IgG tubes and concentrate.

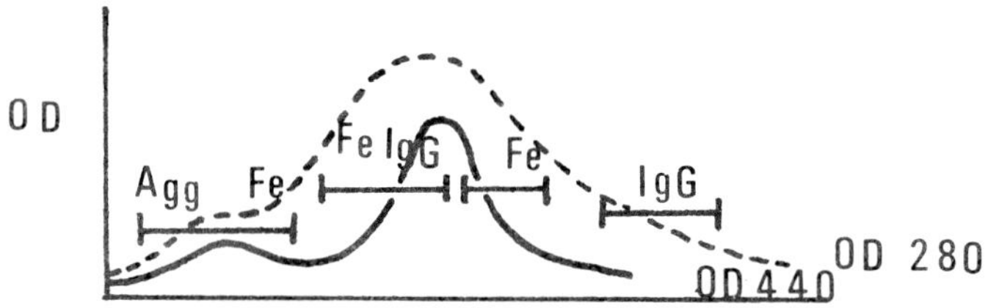

III. One Step Glutaraldehyde Conjugation of IgG and Hemocyanin [40]

1. Prepare hemocyanin by bleeding live marine whelks (*Busycon*). Centrifuge the hemolymph at 10,000 × g for 20 min to remove debris. Isolate hemocyanin (WH) by pelleting at 10,000 × g for 2 hrs at 4°C (Beckman L2, Rotor SW50L, 31,500 RPM, 5 ml nitrate tubes). Resuspend the pellet in 1% NaCl for 24–48 hrs (may take longer) at 4°C. Repeat centrifugation twice as before and finally resuspend the pellet in 3% NaCl containing 0.1% sodium azide. Determine the concentration of protein by reading the OD at 280 nm (E1% = 15.6) or at 350 nm (El% = 4.2).

2. Dialyze a small amount of WH against 0.1 M PO_4 buffer pH 6.8, overnight at 4°C. Centrifuge at 10,000 × g to remove any precipitate.
3. Make up a mixture of hemocyanin and IgG of about a 1:9 molecular ratio. This is about 10 mg WH and 2.5 mg IgG. Dissolve in 1–2 ml of 0.1 M PO_4 buffer.
4. Stir in small vial on magnetic stirrer without bubbles or foaming. Add 0.1 volume (e.g. 0.1–0.2 ml) of 0.5% glutaraldehyde. Continue stirring for 1 hr at room temperature.
5. Add 0.1 volume of NH_4Cl (0.1 M) or $(NH_4)_2CO_3$ (0.1 M) to quench excess glutaraldehyde, and place on ice for 15–30 mins.
6. Dialyze against 0.1 M $(NH_4)_2CO_3$ for 2–3 hours at 4°C; change dialysis to phosphate buffered saline for overnight at 4°C.
7. Centrifuge at 100,000 × g for 10–15 mins to remove any large aggregates or precipitated material. Discard pellet.
8. Centrifuge the conjugate at 100,000 × g for 2 hrs as in (1). Repeat 2×, each time resuspending the pellet in PBS.
9. Finally, resuspend the pellet in PBS containing 0.1% NaN_3 and store at 4°C.
10. Prior to use, centrifuge at 10,000 × g to remove any precipitate.

IV. Methods for Purification of Soluble Immune Complexes and Conjugates

Ultracentrifugation:

1. Centrifuge the mixture at 10,000 × g for 20 mins at 4°C to remove any large precipitates (10,000 RPM, Beckman L2, SW50 rotor). Recover the supernatant and discard pellet.
2. Centrifuge at 100,000 × g for 2 hrs (hemocyanin immune complexes) or 3 hrs (ferritin immune complexes) at 4°C (Beckman L2, SW50 rotor, 31,500 RPM). Fill the 5 ml nitrate tubes with the complex and add PBS pH 7.0.
3. Discard the supernatant and redissolve the pellet in PBS overnight at 4°C. It may take 48 hrs to redissolve the pellet completely.
4. Repeat centrifugation in step (2), then repeat step (3).
5. Centrifuge the dissolved complex at 10,000 × g to remove any insoluble material (see step 1).
6. Store the complex at 4°C in the presence of 0.1% sodium azide as a bacteriostatic preservative. Centrifuge as in step (1) prior to use.

Gel Filtration Chromatography:

1. Immune complexes purified by gel filtration chromatography should be prepared with 7S or IgG antibody.
2. Apply the complex to a Sephadex G200 column in phosphate buffered saline (PBS). The column should be precalibrated. The void volume and 7S peak should be known.
3. Collect 3–5 ml fractions, read the OD's at 280 nm, and plot the protein peaks. Concentrate the fractions with molecular sizes above the 7S peak (i.e., the void volume) to 2–5 ml by negative pressure dialysis. Dialyze against the column buffer.
4. Repeat step (2), then step (3).
5. Store the complex at 4°C in the presence of 0.1% sodium azide as a bacteriostatic preservative. Centrifuge at 10,000 × g for 20 mins at 4°C prior to use.

V. Testing Immunoglobulin in Conjugates or Immune Complexes by Immunoelectrophoresis [3]

Materials: Rabbit anti-ferritin serum (Anti-Fe). Anti-rabbit Ig produced in sheep or goat (GARG). IgG-Fe conjugate or soluble immune complex (Fe-anti-Fe).

Aim: To determine if free uncomplexed Ig is contaminating the IgG marker complex or conjugate. Uncomplexed markers will not cause any problems in labeling technics and therefore attempts to remove these are not made.

Analysis: If the IgG and marker (either Fe or H) are coupled, the electrophoretic mobility of the complex will be somewhere between free IgG (cathodal migrating) and that of the marker (Fe is anodal migrating; H is a slow anodal migrating). If there is no coupling, both proteins will be free to migrate to their respective positions, as seen in a simple mixture of the two proteins.

VI.

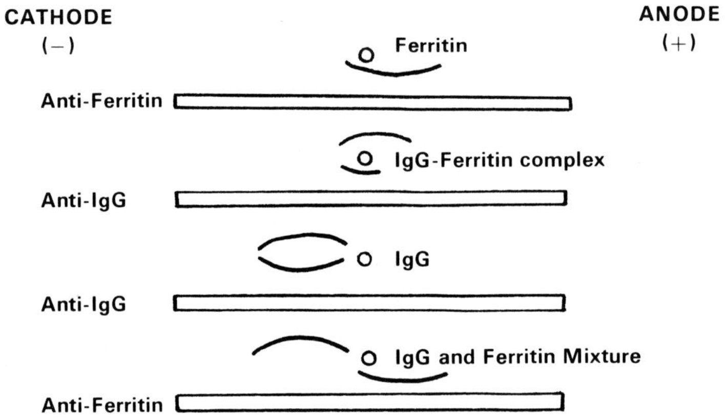

Spectrophotometric Determination of Fe and IgG concentrations

Determination of ferritin by spectrophotometry is unreliable. Ferritin has three major absorbancy bands: 260–290, 470 and 800–1000 nm [43]. This absorbancy overpowers any contribution due to protein (e.g., tyrosine residues). Since each ferritin preparation will have different iron:protein ratios, new extinction coefficients must be calculated, based on protein determination such as micro-Kjeldahl nitrogen analysis (factor 9.7). In our laboratory, the following E values have been calculated, using E.M. grade Ferritin from Polysciences Co.:

$$E^{1\,\text{per cent}}_{280\text{nm};\,1\text{cm}} = 100.8 \qquad E^{1\text{mg/ml}}_{280\text{nm};\,1\text{cm}} = 10.08$$

$$E^{1\,\text{per cent}}_{440\text{nm};\,1\text{cm}} = 13.5 \qquad E^{1\text{mg/ml}}_{280\text{nm};\,1\text{cm}} = 1.35$$

To determine the amount of IgG in a Fe-IgG conjugate, the absorbancy due to ferritin must be subtracted from the 280 absorbancy. At 440 nm, only ferritin absorbs, and the concentration can be determined by dividing the absorbancy by the extinction coefficient. The amount of ferritin absorbancy at 280 nm is then subtracted from the total absorbancy at 280; the remainder is due to IgG. However, this technic is unreliable.

The following formula is used:

$$\frac{A_{280} - (A_{440} \times 7.3)}{1.4} = \text{mg/ml of IgG}$$

Spectrophotometric Determinations of Ig and WH concentrations

Determination of the amount of IgG or WH can be accomplished by using the following absorbancy extinction coefficients (pathlength = 1 cm):

Hemocyanin: $E^{1\,\text{per cent}}_{280\text{nm};\,1\text{cm}} = 15.6$ *OR* $E^{1\text{mg/ml}}_{280\text{nm};\,1\text{cm}} = 1.56$

$E^{1\,\text{per cent}}_{345\text{nm};\,1\text{cm}} = 4.3$ $E^{1\text{mg/ml}}_{345\text{nm};\,1\text{cm}} = 0.43$

Rabbit IgG: $E^{1\,\text{per cent}}_{280\text{nm};\,1\text{cm}} = 14.2$ $E^{1\text{mg/ml}}_{280\text{nm};\,1\text{cm}} = 1.42$

$E_{345\text{nm}}$ not significant

To calculate the concentration of WH or IgG, divide the absorbancy (optical density) by the extinction coefficient; e.g., hemocyanin only

$$\frac{A_{280\text{nm}}}{E^{1\text{mg/ml}}_{280\text{nm}}} = \text{conc. in mg/ml}$$

To calculate the concentrations of each in a conjugate is a little more difficult since both proteins contribute to the absorbancy at 280. However, only hemocyanin absorbs at 345 nm and its concentration can be calculated and subtracted from the absorbancy at 280 nm; the remainder of the absorbancy at 280 nm is due to IgG in the conjugate; e.g.,

$$\frac{A_{280} - (A_{345} \times 3.39)}{1.4} = \text{conc. of IgG in mg/ml}$$

Note: Absorbancy ratio 280/345 for hemocyanin is 3.39.

VII. Purification of Human and/or Rabbit Blood Lymphocytes by Sedimentation and Ficoll-Hypaque Fractionation [42,48]

1. Defibrinate blood by stirring with wood stick or glass rod.
2. For rabbit blood, add 1/3 to 1/2 volume gelatin (3% pig skin gel in phosphate buffered saline heated for 20 min, then cooled to 37°C) and allow to settle for 30–60 min at 37°C.
3. For gelatin sedimented preparations, recover lymphocyte enriched supernatant and wash 2× in warm Eagle's medium. Resuspend in 2–5 ml of medium. For Dextran sedimentations, recover lymphocyte enriched supernatant and wash 1 × in cold medium. Resuspend in 2–5 ml.
4. Make up Ficoll-Hypaque solution: Ficoll (9%) 6.35 ml; Hypaque (40%) 1.8 ml; distilled water 0.85 ml; mix thoroughly.
5. Carefully pipet the cell suspension onto the Ficoll-Hypaque solution in a heavy duty 40 cc centrifuge tube.
6. Centrifuge at 2800 rpm for 20 min at 4°C. Turn up the speed slowly to avoid mixing.

7. Four fractions result: The upper layer consisting of serum and Eagle's medium; a white opalescent layer consisting of lymphocytes; a clear layer of granulocytes; and a pellet of erythrocytes.

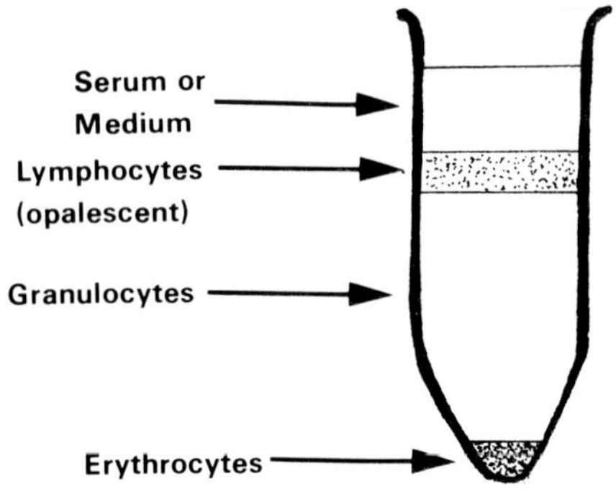

8. For pure lymphocytes (> 95%), aspirate off the top layer. Pipet off the lymphocyte layer, and a portion of the granulocyte layer. For an impure (~ 90%) lymphocyte preparation, take all the cell layers except the rbc pellet.
9. Wash 2× in phosphate buffered saline with 0.2% bovine serum albumin.

VIII. Mixed Antiglobulin Labeling for Lymphocytes [39,44,45]

1. Purified lymphocytes are washed in phosphate buffered saline, pH 7.0, containing 0.2% bovine serum albumin (300 × g or 800 rpm for 10 min).
2. Lymphocytes ($2\text{–}3 \times 10^6$) in 0.2 ml are placed in 10 × 75 mm culture tubes.
3. The sensitizing antiserum is added. Usually 0.1–0.05 ml of anti-allotype antisera is needed. Antibody excess is required since free antibody ligands are needed for binding the labeling complex. The cells are incubated in the presence of antiserum for 30 min. Sodium azide (0.1%) may be added to inhibit cell metabolism or the cells may be incubated at 0°C (ice bath) to prevent receptor redistribution. Decomplemented serum should be used to prevent possible cell lysis.
4. The cells are washed thrice at 0°C with ice-cold PBSA as in (1)
5. The cells are resuspended in 0.2 ml of PBSA.
6. The immune complex label is added. The optimal concentration of the label (50–100 μl) is determined by trial. Labeling for this step is usually carried out at 0°C for 30 min.
7. The cells are washed three times as before.
8. Additional procedures may be implemented (e.g., culture at 37°C) or the cells can be immediately fixed in ice-cold 1% glutaraldehyde in 0.1 M cacodylate buffer.

Theory:

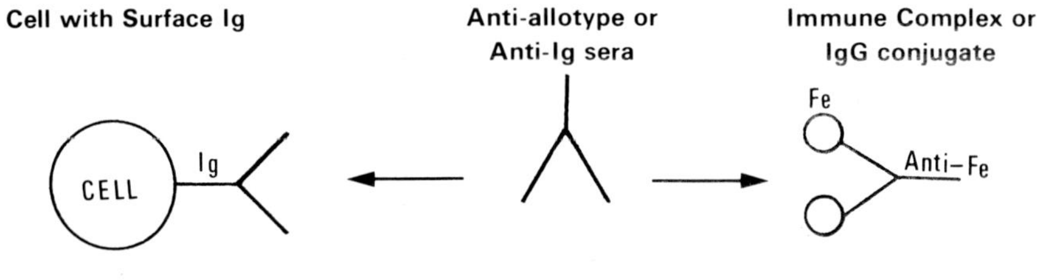

IX. Lymphocyte Fixation, Dehydration and Resin Embedding for Electron Microscopy [44,46]

Materials:

1. 0.1 M cacodylate buffer.
2. Activated Epon 812 resin: mixture of 6 ml resin A and 14 ml resin B with 0.2 ml of DMP-30 activator. Should be mixed fresh and kept in dessicator. Resin A and B can be premixed and stored at −20°C.
   Resin A: 5 ml Epon and 8 ml DDSA. Resin B: 8 ml Epon and 7 ml NMA

Method:

1. Slowly add 5 ml of 2% glutaraldehyde in 0.1 M cacodylate buffer, pH 7.2 to a cell suspension in 12 × 75 mm tubes. Keep at 4°C for 10 min, then centrifuge at 1200 rpm (300 × g) for 10 min. This will prevent the sticking and fixation of the cells to the sides of the tube; it will also prevent the cells from fixing as a tight pellet. Draw off the supernatant, resuspend the cells carefully with a cleaned and annealed pasteur pipet in 0.2 ml and transfer the cells to a pointed BEEM capsule. If necessary, fill the capsules with fresh fixative and centrifuge as before. Let cells sit for an additional 30–40 min at 4°C. Total fixation time should be at least 1 hr.
2. If the cells are to be left overnight, remove the supernatant after centrifugation and wash once in 0.15 M cacodylate buffer. Resuspend in buffer and leave at 4°C overnight.
3. For post-fixation, add 1% OsO_4 in 0.1 M cacodylate buffer. This must be done in a fume hood because osmium is very volatile and dangerous to eyes and mucus membranes. Cap the capsules and keep at 4°C for 60 min. Optional: Spin the cells down after 30 min of fixation and add fresh osmium for the remaining 30 min.
4. Begin dehydration in acetone. Cells must be centrifuged in a warm and dry centrifuge (clinical table top) after each acetone change. Dehydrate for 5–10 min with 35%, 50%, 70%, 80%, and 95% acetone and twice for 5–10 min with 100%. The cells may be centrifuged while undergoing dehydration.
5. Add mixture of activated Epon resin and 100% Acetone (1:1 ratio). Leave in dessicator for 1–2 hrs uncapped to allow the acetone to evaporate.
6. Spin down the cells (1500 rpm, 10 min) and remove resin with aspirator. Add fresh activated resin and label. Let sit overnight, uncapped, in the dessicator to allow any excess acetone to escape.
7. To harden blocks, incubate at 75°C for 4–6 hrs.

39. An T, Miyai K, Sell S: Electron microscopic localization of rabbit immunoglobin allotype b4 on blood lymphocytes by an indirect ferritin immune complex labeling technique. J Immunol 108: 1271–1277, 1972.
40. Avrameas S: Coupling of enzymes to proteins with glutaraldehyde. Use of the conjugates for the detection of antigens and antibodies. Immunochemistry 6: 43–52, 1969.
41. Borek F, Silverstein AM: Characterization and purification of ferritin-antibody globulin conjugates. J Immunol 87: 555–561, 1961.
42. Boyum A: Separation of leucocytes from blood and bone marrow. Scand J Clin Lab Invest 21: Suppl 97, pp 77–89, 1968.
43. Granick S: Ferritin: Its properties and significance for iron metabolism. Chem Rev 38: 379–403, 1946.

44. Linthicum DS, Sell S: Surface immunoglobulin on rabbit lymphoid cells. I. Ultrastructural distribution and endocytosis of b4 allotypic determinants on peripheral blood lymphocytes. Cell Immunol 12: 443–458, 1974.
45. Linthicum DS, Sell S: Surface immunoglobulin on rabbit lymphoid cells. II. Ultrastructural distribution of b4 allotypic determinants and morphology of spleen lymphocytes. Cell Immunol 12: 459–471, 1974.
46. Luft JH: Improvements in epoxy resin embedding methods. J Biophys Biochem Cytol 9: 409–414, 1961.
47. Nicolson G, Singer S: Ferritin-conjugated plant agglutinins as specific saccharide stains for electron microscopy: Application to saccharides bound to cell membranes. Proc Natl Acad Sci USA 68: 942–946, 1971.
48. Perper RT, Zee TW, Mickelson M: Purification of lymphocytes and platelets by gradient centrifugation. J Lab Clin Med 72: 842, 1968.
49. Weigle WO: Immunochemical properties of hemocyanin. Immunochemistry 1: 295–302, 1964.

AUTORADIOGRAPHY

^{125}I-labeled antibodies can be used for autoradiographic immunohistology at both the light microscopic and electron microscopic levels [50–52]. Surface markers can be detected with iodinated IgG molecules and intracellular antigens can be detected with iodinated Fab′ fragments.

Light Microscope Level

1. *Preparation of Slides:*

A. *Smears.* Slides should be cleaned so that no oil film remains. Emulsing of the smears is easier if the smear can be limited to approximately half of the slide, with an identifying label at the opposite end. If desired, smears may be fixed in methyl alcohol and air dried. Unless blood smears are fixed in some way, the cells tend to hemolyze and come off the slide. Several samples of each of the specimens are usually made to allow slides to be developed after varying times of exposure. Just prior to the actual emulsing procedure the slides are hydrated by placing them in distilled water for a few minutes and then drained so that the smears are moist but without much excess water remaining. This assures a more uniform flow of the emulsion over the slides.

B. *Paraffin Sections.* Tissues are embedded and sectioned in the usual manner and sections mounted on slides so as not to occupy more than half of the slide. An identifying label is marked at the opposite end of the slide. Before the emulsion is applied, the embedding paraffin must be removed. This should be timed to be done just before emulsing so that the sections do not dry out, as otherwise they will crack and peel from the slide.

2. *Preparation of Emulsion:*

This procedure is for the use of Kodak Nuclear Track Emulsion NTB-2 in the gel form.

The recommended safelight is a Wratten #1 filter turned toward the wall giving a more diffuse light.

Fill a 50 ml coplin jar to half level with distilled water. Mark top level with a black felt pen. In darkroom use a Wratten filter #1. Add enough NTB-2 Kodak gel to bring the water level up to a marked area. Place coplin jar in a light-tight container (with about 1/2 inch water level for better heat conductivity) and allow the gelatin to melt in a preheated water bath (42°C) for 20–30 minutes (not longer—heat increases background grains). After gelatin has melted, stir with a clean blank slide 2 or 3 times (*very gently*) to insure mixing and the slide should remove excess bubbles or foam.

3. *Storage of Emulsed Slides:*

As soon as slides have been emulsed, they should be stored in light-proof bakelite boxes. Drierite wrapped in lens tissue and held with a piece of Scotch tape should be placed behind the last slide in each box, being careful not to endanger the emulsion. Also, pieces of applicator sticks taped to the bottom of the boxes will lessen the danger of emulsion on the sides of the slides adhering to the boxes. Seal the boxes by wrapping them around the cover seam with masking tape. Be sure to mark clearly which way is up, so that they may be stored with the slides flat and emulsion side up. The theory involved in keeping the slides flat is that the emulsion never attains a completely solid state and will tend to flow if the slides are stored on their sides.

4. *Emulsion of Slides:*

Have slides arranged in whatever system they are to be stored. Be sure that all labels are at the same end and all slides are facing in the same direction to eliminate any danger of emulsing the wrong side of the slide. Have storage boxes ready with Drierite packs and a supply of paper towel pads on hand. Dip slides one at a time into the dissolved emulsion with a slow steady motion so as not to stir the emulsion and thus form bubbles, drain slide against top of jar for about 20 seconds, quickly wipe back of slide with pad to remove emulsion. Stand slide upright against a stable surface. Place slide in bakelite box with emulsion side up and allow emulsion to slightly gel. Do not expose slide to the safelight or to any possible dust contamination any longer than necessary. When finished, discard the emulsion. Never use emulsion that has been previously used or even heated.

5. *Development of Slides:*

Autoradiography slides should always be developed in fresh solutions, with care being taken that all solutions be maintained at the same temperature at least through fixation. If slides have been stored in the cold, they should be allowed to come to room temperature before developing. The usually recommended development temperatures vary between 64 and 68°F. The colder temperatures help to keep the grain size small. Although some agitation is needed in insure even development, an excessive amount serves to speed the rate of development and increase grain size. A Wratten #1A filter is used.

NTB-2 emulsion developing schedule as follows:

1. *D-19 developer:* 3 minutes (agitate slowly but constantly for the first 20 seconds, then for a few seconds only once a minute afterward).
2. *Wash briefly:* 10 seconds.
3. *Acid fix*—25% Na thiosulfate: 5 minutes (agitate as for developer).
4. *Water:* Wash 20 minutes in running water.

If a large number of slides is to be developed at one time, two staining baskets of slides may be developed in the same aliquot of D-19 without risk of contamination, provided the second one is developed almost immediately after the first. As long as the fixer is active and clean in appearance, it may be reused.

After washing, smears may be allowed to air dry and be stained as desired at a later time. Tissue sections, however, should not be allowed to dry before staining and coverslipping as they will crack and peel off of the slide. Since it is also possible for the emulsion to loosen and float off with prolonged standing in water, advance arrangements should always be made to insure their proper handling. Tissues are usually stained with hematoxylin only, as eosin tends to mask the grains.

Electron Microscope Level (thin sections)

A. *Preparation of Sections:*

Cut gold sections (700–800 Å) with as large as face as feasible and mount on grids having a carbon-coated collodion substrate. Care should be taken that the sections are mounted as flat as possible. Sufficient sections should be cut so that there are at least three or four grids for each sample, because of the many technical difficulties involved.

B. *Preparation of Grids for Emulsion:*

Make or purchase plastic peg boards (with lids) with 6 to 24 pegs per board for convenient handling. Cut strips from band-aids about 1/2 cm wide. Holding one end with forceps, sticky side up, secure *edge* of grid with tissue side down to *edge* of cut strip. Stick strip to the peg so that grids are centered on the top of the peg.

C. *Preparation of Emulsion:*

The Ilford Nuclear Research Emulsion L-4 in gel form is the emulsion used in this procedure, primarily because of its small grain size of 0.12 micron as compared to the 0.27 micron of Kodak NTB-2. The L-4 in gel form is in dehydrated spaghetti-like ropes that must be melted and diluted to the proper concentration to form monolayer films.

Put 25 drops of 0.05% Na lauryl sulfate in a 10 ml graduated cylinder. Add distilled water until 6.0 ml level. Use Wratten #1 safelight filter. Add L-4 Ilford until the water in the cylinder reaches the 10 ml level. Place cylinder in light-tight container with about 25 ml of water. Allow the emulsion to melt in a 42°C water bath for 20 minutes. When melted, cover top of cylinder with parafilm and invert emulsion 2 or 3 times *very gently* and pour into a 10 ml beaker held in a safe-light film can. (Add 3 ml of water to can first for good heat conductivity.) Chill the emulsion in an ice bath for 10 minutes, then bring to room temperature by holding can cupped in hands for approximately 10 minutes.

Before the actual emulsing, check darkroom for all necessary equipment: small forceps, masking tape, a 2 cm platinum loop with a hemostat as a handle, bakelite boxes (with drierite packaged in lens paper and scotch-taped), slides numbered with corresponding grid numbers. The Wratten #1A safelight filter may now be used.

D. *Emulsion of Sections:*

Holding the loop parallel to the emulsion, dip it into the beaker and slowly withdraw it. Hold the loop up to the safelight and closely examine the film stretched across it. If the film appears very thin and mostly even throughout, place the loop over the peg, making sure the best area is over the grids. The film will break free from the wire and adhere to the grids. If the film formed is quite wet and runs in streams the emulsion is not thick enough. Carefully wipe the loop with a piece of lens tissue and try again. If the loop comes out dry and with no film cast, probably the emulsion is too thick and needs to be warmed a little. If many bubbles are formed, it is difficult to remove the loop without either breaking the film or picking up some of the bubbles too. Unless the proper quality film is put over the grids there isn't any point in doing the entire procedure—so take all the time and patience required to make good films. If one aliquot doesn't respond to juggling of temperatures, discard it and take another one. Do not leave the freshly coated grids exposed to the safelight or to any possible dust contamination any longer than necessary.

E. *Storage of Emulsed Sections:*

Remove tape from peg with forceps and stick it on prelabeled glass slide. Check corresponding numbers. Put slides, grid side up, in bakelite, safelight box containing drierite. Seal boxes with double wrapping of masking tape. Store in 40°C refrigerator.

F. *Development of Sections:*

Use the same precautions of fresh solutions and constant temperatures as with the NTB-2 emulsion. A Wratten #1A safelight filter is used.

Developing L4 Emulsion:

1. *D-19 developer:* 3 minutes (agitate very slowly for the first 20 seconds then for a few seconds only once a minute afterward).
2. *Wash briefly:* 10 seconds.
3. *Acid fix*—25% Na thiosylfate: 3 minutes (agitate as for developer).
4. *Water:* Wash 20 minutes in running water.

If a large number of slides is to be developed at one time, two or three staining baskets of slides may be developed in the same aliquot of D-19 without risk of contamination, provided they are developed at the same time. Solutions are never kept and reused.

After washing, each grid is carefully lifted from the band-aid strip and blotted on lens tissue. They are stained with 2% Uranyl Acetate for 20 min and .01% Lead Citrate for 3 min. Since each grid is handled numerous times, there is a high breakage rate of the substrate film, so it should be handed as gently as possible and picked up only at the extreme edges.

50. Greenwood EC, Hunter WH, Glover JS: The preparation of I^{131} labelled human growth hormone of high specific radiochemistry. Biochem J 89: 114, 1963.
51. Marchalonis JJ: An enzyme method for the trace iodination of immunoglobulins and other proteins. Biochem J 113: 299, 1969.
52. Rogers AW: Techniques of Autoradiography. Amsterdam, Elsevier, 1967.

Molecular Cloning

Dean H. Hamer, Charles A. Thomas, Jr.

INTRODUCTION

The cloning of a segment of DNA is accomplished by connecting it to a suitable plasmid or viral vehicle and introducing the recombinant DNA molecule into an appropriate bacterial on animal host cell where it is replicated. Since the descendents of this host cell, or virus particle, will contain identical copies of the recombinant DNA molecule, this scheme provides an ideal way of purifying and producing large quantities of a given segment of DNA. If the culture conditions are such that cells derived by replication remain together in a single tube or colony, then individual clones of the recombinant DNA molecule can be isolated. The only remaining problem is to identify these clones that contain the recombinant DNA molecule of interest. This entire operation is called the "molecular cloning of DNA."

In the laboratory exercise presented here, some of the basic operations are performed: (1) The cleavage of certain simple DNAs with a restriction endonuclease and the analysis of the DNA segments produced; (2) The rejoining of restriction segments by annealing and treatment with polynucleotide ligase; (3) The transformation of *E. coli* with the recombinant DNA; (4) The isolation of a recombinant plasmid DNA from a culture derived from a single cell.

To demonstrate these technics, we will construct a recombinant plasmid containing the *E. coli* nonsense suppressor gene *su*$^+$*III*. The vehicle used in this experiment is pSF2124, an 11,000 base pair derivative of ColEl. This plasmid, which replicates under relaxed control, specifies immunity to colicin El (ImmEl$^+$), production of colicin El (Colicin El$^+$), and resistance to ampicillin (Ap^R). It has a single site for restriction endonuclease *Eco*RI, and insertion of new DNA at this site inactivates colicin production but no other plasmid functions. The suppressor gene is obtained from the transducing phage ϕ80p*su*$^+$*III*, which carries a single copy of the tRNA^{Tyr} *su*$^+$*III* sequence located within a 6000 base pair *Eco*RI fragment.

EXPERIMENTS

Experiment 1. *Eco*RI cleavage of pSF2124 and ϕ80p*su*$^+$*III* DNAs

Introduction

*Eco*RI is a type II restriction endonuclease isolated from *E. coli* strain RY13. The only requirement for activity is Mg^{++}. Under standard conditions (100 mM Tris, 50 mM NaCl, 10 mM $MgCl_2$, pH 7.4) the enzyme cleaves at the palindromic sequence

```
   ↓
X—G—A—A—T—T—C—X
X—C—T—T—A—A—G—X
             ↑
```

Lowering the ionic strength and raising the pH of the reaction mixture (25 mM Tris, 2 mM $MgCl_2$, pH 8.5) relaxes the specificity of the enzyme such that it will cleave at the sequence

```
 ↓
X—A—A—T—T—X
X—T—T—A—A—X
         ↑
```

The activity responsible for cleavage at this site is referred to as *Eco*RI*.

DNA fragments produced by *Eco*RI have symetric, cohesive termini, and can anneal to one another at low temperature in head-to-tail, head-to-head, and tail-to-tail orientation. The resulting hydrogen-bonded structures can be covalently joined by treatment with polynucleotide ligase. This provides a simple method for the biochemical construction of recombinant DNA molecules.

Materials

The *Eco*RI endonuclease was purified by the method of Greene et al. [17], and has an activity of approximately 800 units/ml (one unit is defined at the amount of enzyme required to completely digest 1 μg of λ DNA in 2 hours at 37°). *Eco*RI can also be purchased from Miles or from New England Biolabs; however, this is quite expensive, especially since the suppliers invariably overstate the activity of their preparations. Plasmid pSF2124 DNA is prepared as described in Experiment 5, while ϕ80p*su*$^+$*III* DNA is isolated by phenol extraction of purified phage particles. Agarose is from Sigma. Gel tubes are made by cutting down Corning 5 ml serologic pipets and smoothing the ends with sandpaper. Power supplies and gel tanks are purchased from any of a number of sources, such as Buchner or BioRad. The short wave UV light is from UV products.

Method

All buffers, glassware, and pipets are autoclaved. The DNA solutions are sterilized by filtration through 0.45μ Millipore filters.

Set up the 25 μl reactions in small glass tubes, seal with Parafilm, and incubate at 37° for 2 hours. Stop the reactions by adding 2.5 μl of 0.13 M EDTA (pH 8.0), 50%

glycerol containing a trace of bromophenol blue. Electrophorese a 10 μl sample of each reaction.

To prepare gels, cover the bottoms of the glass tubes with Parafilm. Dissolve 1 gm of agarose in 100 ml of SEB buffer (40 mM Tris, 5 mM Na acetate, 1 mM EDTA, pH 7.8) by boiling or autoclaving. Cool the agarose to 50° and pour into tubes, leaving about 1 cm at the top. Let the agarose solidify for 1 hour. Secure a small square of nylon mesh on the top of each tube with a rubber grommet. Then remove the tube from the rack, invert it, remove the Parafilm, and let the gel slide down. This means that the DNA sample will be loaded onto the flat surface which was originally at the bottom of the gel.

Place the gels in the tank and fill the bottom and top reservoirs with SEB. Load the reaction mixtures onto the gels using a screw device and micropipet. Run at 1 milliamp per gel, which should require a current of about 20 volts. Stop the run when the blue tracking dye is at the bottom of the gel. This should require about 16 hours for 12 cm gels.

Stain the gels in ethidium bromide (0.5 μg/ml in H_2O or SEB) for about 20 minutes. Visualize under UV light using a barrier filter. Photograph with a Polaroid Model 195 camera equipped with a close-up lens containing a Wratten #22 filter at f-5.6 for 10–20 seconds using B&W type 107 film.

Hints: Ethidium bromide is a powerful mutagen. Handle with care. Don't look at UV light without goggles.

The mobility and sharpness of bands is dependent upon the ionic strength and volume of the sample loaded. Ideally, the ionic strength of the sample should be less than that of the electrophoresis buffer, and the volume less than 50 μl.

EXPERIMENT 1

Rx.	pSF2124 DNA (67 μg/ml)	φ80p*su*$^+$*III* DNA (250 μg/ml)	H_2O	1.0 M Tris (pH 7.4)	0.1 M $MgCl_2$	0.5 M NaCl	*Eco* RI
1	10	0	7.5	2.5	2.5	2.5	0
2	10	0	7.0	2.5	2.5	2.5	0.5
3	10	0	6.5	2.5	2.5	2.5	1
4	10	0	3.5	2.5	2.5	2.5	4
5	10	0	0	2.5	2.5	2.5	7.5
6	0	10	7.5	2.5	2.5	2.5	0
7	0	10	7.0	2.5	2.5	2.5	0.5
8	0	10	6.5	2.5	2.5	2.5	1
9	0	10	3.5	2.5	2.5	2.5	4
10	0	10	0	2.5	2.5	2.5	7.5

Note: All volumes are in μl.

Experiment 2. Ligation of pSF2124/*Eco*RI to ϕ80p*su*$^{+}$*III*/*Eco*RI

Introduction

DNA fragments produced by *Eco*RI can be covalently closed by treatment with polynucleotide ligase. This enzyme establishes the 3′—O—P—O—5′ phosphodiester linkage of the polynucleotide chain by esterification of the 5′ phosphoryl group to the 3′ hydroxyl group of DNA chains that have been properly aligned. A complete discussion of the two types of ligase so far known is found in the review article by Richardson [33].

DNA molecules with two cohesive termini can undergo both intramolecular joining, to yield circles, and intermolecular joining to yield linear concatemers. The equilibrium ratio of circles to linears depends upon the length and concentration of the molecules. A short molecule will cyclize more readily than a long molecule. This is because a short molecule has a smaller end-to-end distance, and hence greater concentration of one end in the neighborhood of the other, than does a long molecule. High DNA concentrations favor intermolecular joining. This, of course, is because the probability of finding the end of one molecule in the neighborhood of another molecule is proportional to DNA concentration. A complete mathematical treatment of this problem is found in the article by Wang and Davidson [37].

In order to form pSF2124/*Eco*RI – ϕ80p*su*$^{+}$*III*/*Eco*RI recombinant molecules, one wants to maximize intermolecular joining, and to minimize intramolecular joining of the vehicle. This will be accomplished by using a high DNA concentration and a high ratio of phage to plasmid DNA.

Materials

Polynucleotide ligase from T4-infected *E. coli* was prepared by the method of Weiss et al. [38] and has an activity of 84 units/ml (a unit is defined by Weiss et al.). Ligase can also be purchased from Miles. ATP is dissolved in 10 mM Tris (pH 7.0) and stored at –20°.

Method

1. Scale-up the *Eco*RI cleavage reactions of Experiment 1 by a factor of 4 for pSF2124 and a factor of 8 for ϕ80p*su*$^{+}$*III*. Use the minimum concentration of endonuclease which was found to give complete cleavage.
2. Following incubation for 2 hours at 37°, heat the reaction mixtures for 5 minutes at 65° to inactivate the *Eco*RI.
3. Combine the pSF2124/*Eco*RI (100 μl, 2.7 μg) and ϕ80p*su*$^{+}$*III*/*Eco*RI (200 μl, 20 μg) digests. Withdraw 20 μl of the mixture, add 2 μl of EDTA-glycerol-bromophenol blue, and save at 0.°.
4. Hold the remainder of the mixture at 0° for 30 minutes. Then add 15 μl of 1.32 mM ATP and 5 μl of ligase. Incubate at 12.5° for 12 to 16 hours.

5. Add an additional 15 μl of ATP and 5 μl of ligase and incubate at 12.5° for 12 to 16 hours.

6. Withdraw 20 μl of the reaction mixture and add 2 μl of EDTA-glycerol-bromophenol blue. Analyze the control and ligated samples by 1% agarose gel electrophoresis as in Experiment 1.

7. Dialyze the remainder of the ligase reaction mixture versus 10 mM Tris, 0.13 mM EDTA, pH 7.4, for 24 hours.

Hint: Ligase is very unstable at room temperature and somewhat unstable at 0.°. Store it in 50% glycerol at −20°C.

Experiment 3. Transformation of *E. coli* and identification of *su⁺III* recombinant plasmids

Introduction

Treatment of *E. coli* with Ca^{++} renders them competent for the uptake of DNA. This is the standard method for the introduction, and subsequent propagation and cloning, of recombinant DNA molecules.

Little data is available on the mechanism and optimum conditions for transformation. In our hands, only about 1 out of 10^3 cells is competent. This can be shown either by transforming with saturating levels of DNA, or by transforming with two types of DNA and determining the frequencies of single and double transformants. The efficiency of transformation is about 10^4 transformants per μg of plasmid DNA.

Nagaishi et al. [30] have studied the effect of the genetic background of the host on transformation efficiency. They found that the presence of F (integrated or autonomous) and the *recA* mutation (commonly used to minimize recombination and rearrangement of cloned sequences) decrease the transformation efficiency. In contrast, mutations in the genes for endonuclease I and the *recBC* nuclease increase transformation efficiency for chromosomal markers. To clone foreign DNA fragments, it is necessary to use a strain lacking a host specificity restriction system. Such strains are as readily transformed as their wild type counterparts.

After treating cells with drug-resistant plasmid DNA, it is necessary to grow the cells for at least 15 minutes under nonselective conditions before plating for drug-resistant transformants. Presumably, this is because a certain amount of time is required for the synthesis and functioning of the molecules (e.g., membrane proteins) which protect the cells from the drug. The number of independent transformants is calculated as $N = R_t(T_0/T_t)$, where N is the number of independent transformants, R_t is number of transformants at time t, and T_0 and T_t are the number of cells at time 0 and time t, respectively. Of course, this calculation assumes that all the cells grow at the same rate under nonselective conditions. This assumption could be avoided by a single-burst experiment.

In this experiment, the pSF2124/*Eco*RI – ϕ80p*su*$^+$*III*/*Eco*RI ligation reaction mixture will be used to transform *E. coli* strain CA274, which is HfrC lac^-_{amber} trp^-_{amber} su^-. Three types of ampicillin-resistant transformants should be obtained. Those which harbor rejoined pSF2124 will be $ImmEl^+$ $ColicinEl^+$ Lac^- Trp^-. Those which harbor recombinants between pSF2124 and phage DNA fragments other than the *su*$^+$*III* fragment will be $ImmEl^+$ Colicin El^- Lac^- Trp^-. Those which harbor recombinants between pSF2124 and the *su*$^+$*III* phage fragment will be $ImmEl^+$ Colicin El^- Lac^+ Trp^+, due to the fact that both host amber mutations are *su*$^+$*III*-sensitive.

Materials

E. coli strains CA274, K12, K12(Col E1), and K12 (Col E2) can be obtained from the Yale stock center. Media are from Difco. Ampicillin is Omnipen N from Searle, and is used at 50 μg/ml.

Method

Use aseptic technic throughout. Materials can be sterilized by autoclaving, flaming, exposure to UV light, or immersion in alcohol, chloroform, Wescodyne, or 5% chlorox. Follow the "Standard Microbiological Practices" described at the conclusion of this Experiment.

A. *Transformation:*

1. Pick a colony of CA274 HfrC lac^-_{amber} trp^-_{amber} into 2 ml of Penassay broth and grow overnight with shaking at 37°.
2. Dilute 0.5 ml of the overnight (about 2×10^9/ml) into 10 ml of Penassay broth in a screw-cap centrifuge tube and grow with shaking at 37° for 1 hour (about 5×10^8/ml).
3. Spin down cells in Sorvall at 5000 rpm for 5 minutes. Discard supernatant.
4. Resuspend cells in 5.0 ml of iced 0.03 M $CaCl_2$. Hold at 0° for 20 minutes.
5. Spin down cells, discard supernant, and resuspend in 1.0 ml of iced 0.03 M $CaCl_2$.
6. Mix cells, $CaCl_2$, DNA, and H_2O as specified in the protocol. Hold at 0° for 1 hour.
7. Place the mixtures at 42° for 2 minutes. Add 3.0 ml of Penassay broth. Titer for total cells as specified in the protocol.
8. Incubate with shaking at 37° for 30 minutes. Then titer for total cells and transformants as specified in the protocol.
9. Incubate the plates at 37° overnight. Count the colonies and determine the number of transformants which were obtained.

EXPERIMENT 3. Transformation Mixtures

Rx.	pSF2124/*Eco*RI – ϕ80*psu*$^+$*III*/*Eco*RI ligation reaction mixture	10 mM Tris, 0.13 mM EDTA, pH 7.4	0.03 M$CaCl_2$	0.3 M$CaCl_2$	Cells
1	100	0	200	10	0
2	0	100	—	10	200
3	100	0	0	10	200

Note: All volumes are in μl.

EXPERIMENT 3. Dilution Schedule

Rx.	t = 0 NA	t = 30 minutes NA	t = 30 minutes NAAp
1		1×10^0	1×10^0
2			1×10^0
3	1×10^{-4}	1×10^{-4}	4×10^0
	1×10^{-5}	1×10^{-5}	4×10^{-1}
	1×10^{-6}	1×10^{-6}	4×10^{-2}

Notes: Prepare dilutions by serial transfers of 0.2 ml into 1.8 ml of 0.1% peptone in H_2O. Plate 0.2 ml. NA is nutrient agar. NAAp is nutrient agar containing 50 μg/ml ampicillin.

B. *Genetic tests:*

1. Replica pick (using sterile toothpicks) 200 of the ampicillin-resistant transformants onto minimal lactose agar, then nutrient agar, then nutrient agar containing ampicillin. Each plate should contain 16 colonies, arranged in a 4 × 4 grid. Colonies of CA274, K12, and K12 (ColE1) should be included as controls in the final series of plates. Incubate at 37° overnight for the nutrient plates and 2 days for the minimal plates.

2. The clones are tested for colicin production as follows. Place a sheet of filter paper in the top of each of the nutrient agar plates. Add a few drops of $CHCl_3$ to the filter papers and incubate the plates upside down and closed for 15 minutes to kill the cells. Replace the tops with new ones and incubate the plates upside down and half open for 10 minutes to remove $CHCl_3$. Then overlay the plates with about 2×10^7 cells of strain K12 in 3 ml of top agar. Incubate overnight at 37°. A clear halo in the overlay lawn will be observed around the Colicin E1$^+$ clones.

3. Clones are tested for immunity to Colicin E1 as follows. Prepare indicator plates by picking a colony of K12, K12(ColE1) and K12(ColE2) into nutrient agar. Incubate overnight at 37°, and treat with $CHCl_3$ just before use. Pick all of the Colicin E1$^-$ Lac$^+$ Trp$^+$, all of the Colicin E1$^-$ Lac$^-$ Trp$^-$, and 10 of the Colicin E1$^+$ Lac$^-$ Trp$^-$ clones into 2 ml of Penassay broth containing ampicillin and incubate overnight at 37°. Add a drop of each overnight to 3 ml of top agar and use this to overlay an indicator plate. ImmE1$^+$ clones will give a halo around the K12(ColE2) colony, but not the K12 or K12(ColE1) colonies.

4. Calculate the frequency of each type of transformant.

Hints: $CaCl_2$-treated cells can be kept at 0° for up to 24 hours with no loss in transformability.

Some component (ATP?) of the mixture used for EcoRI cleavage and ligation inhibits transformation up to 10-fold. This component is completely removed by overnight dialysis.

Frequently a flocculent precipitate will appear when the transformation mixture is diluted with Penassay broth. This does not have any noticeable effect on the transformation efficiency or cell viability.

Cells grown in Penassay broth or L broth are just as competent as cells grown in minimal medium.

Standard Microbiological Practices

The construction of pSF2124-ϕ80p*su*$^+$*III* recombinants is an "EK1-P1" experiment, according to the current N.I.H. guidelines. Appropriate containment and decontamination procedures are described below.

"The control of biohazards at the P1 level is provided by standard microbiological practices, of which the following are examples: (i) Laboratory door should be kept closed while experiments are in progress. (ii) Work surfaces should be decontaminated daily and following spills of recombinant DNA materials. (iii) Liquid wastes containing recombinant DNA materials should be decontaminated before disposal. (iv) Solid wastes contaminated with recombinant DNA materials should be decontaminated or packaged in a durable leak-proof container before removal from the laboratory. (v) Although pipetting by mouth is permitted, it is preferable that mechanical pipetting devices be used. When pipetting by mouth, cotton-plugged pipettes shall be employed. (vi) Eating, drinking, smoking, and storage of food in the working aiea should be discouraged. (vii) Facilities to wash hands should be available. (viii) An insect and rodent control program should be provided. (ix) The use of laboratory gowns, coats, or uniforms is discretionary with the laboratory supervisor." (From "Guidelines for Research Involving Recombinant DNA Molecules," National Institutes of Health, June 1976).

Experiment 4. Rapid purification and gel electrophoretic analysis of plasmid DNA

Introduction

Bacterial plasmids are small, circular, autonomously replicating DNA molecules. In attempting to isolate plasmid DNA, one usually uses one or more of the following properties: difference in base composition from chromosomal DNA, transferability, small size relative to the chromosome, and circularity. Various methods based on these properties are reviewed by Freifelder [14].

Plasmids can replicate under either stringent or relaxed control. Stringent plasmids are present in about one copy per cell in logarithmically growing cells. Their replication is tightly coupled to the replication of the host chromosome, and requires continuous protein synthesis but not DNA polymerase I. In contrast, relaxed plasmids are present at about 20 copies per cell under normal growth conditions. Their replication requires polymerase I, but not protein synthesis. Treatment with the antibiotic chloramphenicol, which inhibits protein synthesis, stops the synthesis of chromosomal DNA but not of relaxed plasmid DNA. Up to 50% of the total DNA in such cells is plasmid DNA. Obviously, this is a tremendous advantage for preparative purposes.

In this experiment, crude plasmid DNA preparations will be made from various strains isolated in Experiment 3 which harbor relaxed plasmids. These preparations will be analyzed by gel electrophoresis with and without EcoRI cleavage.

Materials

Brij 58 is obtained from Ruger Chemical Co. (Irvington-on-Hudson, N.Y.). Phenol is distilled under nitrogen, stored over H_2O at $-20°$, and thawed and equilibrated just before use. Lysozyme is from Sigma.

Method

1. Pick various transformant clones obtained in Experiment 3 into 2 ml of Penassay broth containing 50 μg/ml ampicillin. Grow overnight with shaking at 37°.
2. Inoculate 5 ml of Penassay broth containing ampicillin with 0.25 ml of each overnight. Grow with shaking at 37° for 1 hour (about 5×10^8/ml).
3. Add 50 μl of chloramphenicol (25 mg/ml). Continue incubation at 37° overnight.
4. Ice cells. Collect at 5000 rpm for 5 minutes in the Sorvall. Discard the supernatant. Rinse the pellet with 2.5 ml of iced 10 mM Tris, 1 mM EDTA, pH 8. Collect cells at 5000 rpm for 5 minutes.
5. Resuspend in 0.25 ml of iced 25% sucrose (in 50 mM Tris, pH 8).
6. Add 0.05 ml of lysozyme (10 mg/ml in 250 mM Tris, pH 8, freshly prepared). Incubate at 0° for 5 minutes.
7. Add 0.1 ml of 0.25 M EDTA, pH 8. Incubate at 0° for 5 minutes.
8. Transfer the suspension to a plastic centrifuge tube. Add 0.4 ml of Brij mix (1% Brij 58, 0.4% Na deoxycholate, 63 mM EDTA, 50 mM Tris, pH 8) and mix gently. Incubate at 0° for 5 minutes, then at 37° for 5 minutes.
9. Centrifuge in Sorvall at 17,000 rpm for 1 hour. Transfer the supernatant to a small glass tube.
10. Add 1 volume of phenol (equilibrated with 0.2 M Tris, pH 8, then with 10 mM Tris, 1 mM EDTA, pH 8). Rock for 5 minutes.

11. Centrifuge in Sorvall at 5000 rpm for 5 minutes to separate phases.
12. Remove the lower organic phase. Extract the remaining aqueous phase with 1 volume of $CHCl_3$. Centrifuge to separate phases.
13. Remove the upper aqueous phase and precipitate the DNA by adding 2 volumes of cold, absolute EtOH. Leave at −20° for at least 2 hours.
14. Collect the DNA by centrifugation at 10,000 rpm for 30 minutes in the Sorvall. Resuspend the pellet in 0.5 ml of 10 mM Tris, 0.13 mM EDTA, pH 7.4. Dialyze extensively versus the same buffer.
15. Analyze 50 μl aliquots of these preparations by 1% agarose gel electrophoresis as in Experiment 1. The covalently closed circular plasmid species should be well separated from the chromosomal DNA (near the top of the gel) and rRNA (near the bottom of the gel).
16. Cleave 50 to 100 μl aliquots with *Eco*RI and analyze by 1% agarose gel electrophoresis as in Experiment 1. The discrete fragments of plasmid DNA should be readily apparent over the broad distribution of fragments originating from the chromosomal DNA.

Hints: Phenol burns bare skin! Wash burns with alcohol.

$CHCl_3$ fumes are toxic.

For long plasmids (greater than 20,000 bp for 0.8% agarose gels), the plasmid DNA will not be separated from chromosomal DNA. To circumvent this problem, boil (usually for about 1 minute) the supernatant before gel electrophoresis. This denatures the linear chromosomal DNA, but not the covalently closed circular plasmid DNA.

The crude plasmid DNA preparations obtained by this method can be used for transformation.

PROBLEMS

Problem 1:

Suppose we had three different DNA preparations, each with a number average length (l_m) of 5,000 bp, prepared from *E. coli* (genome = 4×10^6 bp), *Drosophila* (2×10^8bp), man (4×10^9bp), and *Necturus* (1×10^{11} bp). (a) How many independent transformants must be prepared in order for there to be a 90% chance of having the desired segment in at least one clone? We assume that joining of the vehicle molecule and DNA fragment occurs randomly. (b) Presuming that one can only screen 10^4 clones for a desired segment, how would you go about finding a single copy human gene?

Problem 2:

If the genomic DNA is *randomly* broken to produce a random distribution of lengths ($Nt/No = p^2(1-p)^{t-1}$, where $p = {}^{1/l}m$, and *t* the length of the DNA containing t bp), is it more likely to obtain clones with large or small insets? How does this depend on the magnitude of l_m?

Problem 3:

Consider three possible mechanisms for the *Eco*RI reaction:

1. The enzyme makes a double-stranded break at the recognition site.
2. The enzyme makes a single-stranded break at the recognition site; nicked and intact sites are equally susceptible to the enzyme.
3. The enzyme makes a single-stranded break at the recognition site; a nicked site is much more susceptible than an intact site.

Given a covalently closed plasmid with one *Eco*RI site, how would you distinguish between these three mechanisms?

Problem 4:

You have discovered a new restriction endonuclease, endo R·X, which produces fragments with cohesive termini. You want to know if the

cleavage site is a perfect palindrome (e.g., G↓A A T T C / C T T A A↑G) or an inverted

repeat with an internal asymmetric nucleotide (e.g., G↓T A C T A C / C A T G A T↑G).

Fortunately, you have some ligase and some *Eco*RI in the freezer. Also, a plasmid with the following structure:

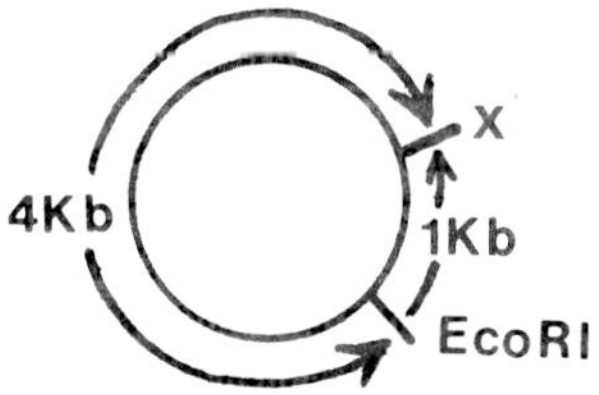

Describe two experiments to distinguish between the alternatives. (Hint: You need an EM and a gel electrophoresis setup.)

Problem 5:

Describe two *independent* methods to determine what fraction of a population of cells is competent for the uptake and subsequent expression of plasmid DNA.

Problem 6:

You mix 0.2 ml of competent *E. coli* with 1.8 μg of pSC101 DNA in 0.1 ml of buffered $CaCl_2$. Following incubation at 0° then at 42°, you add 3.0

ml of Penassay broth and immediately titer on nutrient agar plates. Two hours later, you titer on both nutrient agar and nutrient agar plus tetracycline. The results are:

at time 0, 0.2 ml of a 10^{-5} dilution = 46 colonies on NA

at 2 hours 0.2 ml of a 10^{-6} dilution = 112 colonies on NA

0.1 ml of a 10^{-2} dilution = 37 colonies on NA + tetracycline.

How many transformants/μg of DNA have you obtained? What is the efficiency (transformants/molecule)? (pSC101 is 9,000 bp.)

Unfortunately, you later discover that the presence of pSC101 has a significant effect on growth rate in the absence of tetracycline. Describe an experiment which allows you to calculate the number of transformants regardless of this fact.

Problem 7: The origin of DNA replication is, presumably, a specific nucleotide sequence. Starting with pSC101 (a 9000 bp tetracycline resistance factor which replicates under stringent control and which has one *Eco*R1 site in a nonessential portion of the genome) and ColE1 (a 6000 bp colicinogenic factor which replicates under relaxed control and which has one *Eco*R1 site in the gene for colicin production), how would you go about isolating mutations in the pSC101 origin of replication? What genetic properties would be expected of such mutants? (Hint: The first step utilizes methods you have learned in this course.)

Problem 8:

Plasmid pColE1*lac*$^+$ is a ColE1 derivative that carries the *E. coli lac* operon and also specifies immunity to colicin E1. This plasmid, which has a single SalI site, replicates under relaxed control. Plasmid pColTc is a ColE2 derivative that specifies resistance to tetracycline and immunity to colicin E2. It also has a single SalI site and replicates under relaxed control. A mixture of pColE1*lac*$^+$ and pColE2Tc DNAs is cleaved with *Sal*I, treated with ligase, then introduced into a *lac* deletion strain. Transformants are selected on nutrient agar containing tetracycline, then tested for their ability to ferment lactose and for immunity to colicins E1 and E2. The results are as follows:

Line	Tc	Lac	ImmE1	ImmE2	colony-forming units/ml
1	R	+	+	+	8.4×10^2
2	R	+	−	+	0
3	R	−	+	+	0
4	R	+	+	−	9.6×10^2
5	R	−	−	+	1.2×10^4
6	R	+	−	−	0
7	R	−	+	−	6.2×10^2
8	R	−	−		0

Explain. (Hints: ColE1 and ColE2 are compatible. The biochemical manipulations do not create any unexpected structures, and on pressure no bizarre reconstructional variants are produced.)

REFERENCES

1. Berg P, Gayda R, Avri H et al: Cloning of *Escherichia coli* DNA that controls cell division and capsular polysaccharide synthesis. Proc Natl Acad Sci USA 73: 697–701, 1976.
2. Brown WM, Watson RM, Vinograd J et al: The structure and fidelity of replication of mouse mitochondrial-pSC101 EcoRI recombinant plasmids grown in *E. coli* K12. Cell 7: 517–530, 1976.
3. Cabello F, Timmis K, Cohen S: Replication control in a composite plasmid constructed by *in vitro* linkage of two distinct replicons. Nature 259: 285–290, 1976.
4. Cameron J, Panasenko S, Lehman I, Davis R: *in vitro* construction of bacteriophage λ carrying segments of the *Escherichia coli* chromosome: Selection of hybrids containing the gene for DNA ligase. Proc Natl Acad Sci USA 72: 3416–3420, 1975.
5. Carroll D, Brown D: Adjacent repeating units of *Xenopus laevis* 5S DNA can be heterogenous in length. Cell 7: 477–486, 1976.
6. Chang A, Cohen S: Genome construction between bacterial species *in vitro*. Replication and expression of Staphylococcus plasmid genes in *Escherichia coli.* Proc Natl Acad Sci USA 71: 1030–1034, 1974.
7. Chang A, Lansman R, Clayton D, Cohen S: Studies of mouse mitochondrial DNA in *Escherichia coli:* Structure and function of the eukaryotic-procaryotic chimeric plasmids. Cell 6: 231–244, 1975.
8. Clarke L, Carbon J: Biochemical construction and selection of hybrid plasmids containing specific segments of the *Escherichia coli* genome. Proc Natl Acad Sci USA 72: 4361–4365, 1975.
9. Clewell DB, Helinski DR: Supercoiled circular DNA-protein complex in *Escherichia coli:* Purification and induced conversion to an open circular DNA form. Proc Natl Acad Sci USA 62: 1159, 1969.
10. Cohen S, Chang A: Recicularization and autonomous replication of a sheared R-factor DNA segment in *Escherichia coli* transformants. Proc Natl Acad Sci USA 70: 1293–1297, 1973.
11. Cohen S, Chang A, Boyer H, Helling R: Construction of biologically functional bacterial plasmids *in vitro.* Proc Natl Acad Sci USA 70: 3240–3244, 1973.
12. Cohen S, Chang A, Hsu L: Nonchromosomal antibiotic resistance in bacteria: Genetic transformation of *Escherichia coli* by R-factor DNA. Proc Natl Acad Sci USA 69: 2110, 1972.
13. Dugaiczyk A, Boyer H, Goodman H: Ligation of EcoRI endonuclease-generated DNA fragments into linear and circular structures. J Mol Biol 96: 171–184, 1975.
14. Friefelder D: Isolation of extrachromosomal DNA from bacteria. Meth Enzymol 21: 153, 1971.
15. Ganem D, Nussbaum A, Davoli D, Fareed G: Propagation of a segment of bacteriophage λ-DNA in monkey cells after covalent linkage to a defective simian virus 40 genome. Cell 7: 349–359, 1976.
16. Glover D, White R, Finnegan D, Hogness D: Characterization of six cloned DNAs from *Drosophila melanogaster,* including one that contains the genes for rRNA. Cell 5: 149–157, 1975.
17. Greene P, Betlach M, Boyer H: The EcoRI restriction endonuclease. *In:* Wickner R (Ed): Methods in Molecular Biology, 7. New York, Marcel Dekker, 1974, p 88.

18. Grunstein M, Hogness D: Colony hybridization: A method for the isolation of cloned DNAs that contain a specific gene. Proc Natl Acad Sci USA 72: 3961–3965, 1975.
19. Hedgpeth J, Goodman H, Boyer H: DNA nucleotide sequence restricted by the RI endonuclease. Proc Natl Acad Sci USA 69: 3448, 1972.
20. Hershfield V, Boyer H, Yanofsky C et al: Plasmid ColE1 as a molecular vehicle for cloning and amplification of DNA. Proc Natl Acad Sci USA 71: 3455–3459, 1974.
21. Jackson D, Symons R, Berg P: Biochemical method for inserting new genetic information into DNA of simian virus 40: Circular SV40 DNA molecules containing lambda phage genes and the galactose operon of *Escherichia coli.* Proc Natl Acad Sci USA 69: 2904–2909, 1972.
22. Kato K, Goncalves M, Houts G, Bollu F: Deoxynucleotide-polymerizing enzymes of calf thymus gland. II. Properties of the terminal deoxyribonucleotidyltransferase. J Biol Chem 242: 2780–2789, 1967.
23. Kedes L, Chang A, Houseman D, Cohen S: Isolation of histone genes from unfractionated sea urchin DNA by subculture cloning in *E. Coli.* Nature 255: 533–538, 1975.
24. Little J: An exonuclease induced by bacteriophage λ. II. Nature of the enzymatic reaction. J Biol Chem 242: 679–686, 1967.
25. Lobban P, Kaiser A: Enzymatic end-to-end joining of DNA molecules. J Mol Biol 78: 453–471, 1973.
26. Mandel M, Higa A: Calcium-dependent bacteriophage DNA infection. J Mol Biol 53: 154, 1970.
27. Mertz J, Davis R: Cleavage of DNA by R_1 restriction endonuclease generates cohesive ends. Proc Natl Acad Sci USA 69: 3370–3374, 1972.
28. Morrow J, Cohen S, Chang A et al: Replication and transcription of eukaryotic DNA in *Escherichia coli.* Proc Natl Acad Sci USA 71: 1743–1747, 1974.
29. Murray N, Murray K: Manipulation of restriction targets in phage λ to form receptor chromosomes for DNA fragments. Nature 251: 476–481, 1974.
30. Nagaishi H, Skurray R, Clark A: The influence of genetic background or transformation frequencies of pCR11 and pSC101. Nucleic Acid Scientific Memoranda NAR9.
31. Nussbaum A, Davoli D, Ganem D, Fareed G: Construction and propagation of a defective simian virus 40 genome bearing on operator from bacteriophage λ. Proc Natl Acad Sci USA 73: 1068–1072, 1976.
32. Polisky B, Greene P, Garfin D et al: Specificity of substrate recognition by the EcoRI endonuclease. Proc Natl Acad Sci USA 72: 3310, 1975.
33. Richardson CC: Enzymes in DNA metabolism. Ann Rev Bichem 38: 798, 1969.
34. Sgaramella V, Khorana H: Studies on polynucleotides. CXVI. A further study of the T_4 ligase-catalyzed joining of DNA at base-paired ends. J Mol Biol 72: 493–502, 1972.
35. Sgaramella V, van de Sande J, Khorana H: Studies on polynucleotides. C. A novel joining reaction catalyzed by the T_4-polynucleotide ligase. Proc Natl Acad Sci USA 67: 1468–1475, 1970.
36. Thomas M, Cameron J, Davis R: Viable molecular hybrids of bacteriophage lambda and eukaryotic DNA. Proc Natl Acad Sci USA 71: 4579–4583, 1974.
37. Wang J, Davidson N: On the probability of ring closure of lambda DNA. J Mol Biol 19: 469–482, 1966.
38. Weiss B, Jacquemin-Sablon A, Live T et al: Enzymatic breakage and joining of deoxyribonucleic acid. VI. Further purification and properties of polynucleotide ligase from *Escherichia coli* infected with bacteriophage T_4. J Biol Chem 243: 4543–4555, 1968.
39. Wensink P, Finnegan D, Donelson J, Hogness D: A system for mapping DNA sequences in the chromosomes of *Drosophila melanogaster.* Cell 3: 315–325, 1974.

The Human Environment
a new publication

C. A. Thomas, Jr.

The origin of what I am about to tell you (to the best of my recollection) had its origins at the Given Institute last year. It was a session, not unlike this one, dealing with Science and Society—on Science and the Law—or something of that nature. I can't recall most of what was said, but the room contained a number of important personages: Killian was in a back row, perhaps the chairman of the board of CBS; closer at hand was Ted Puck; Judge Bazelon was present, and Don King (in his usual fashion) was benignly moderating the proceedings from the wings.

Judge Bazelon made some interesting remarks. He said (in effect): "I don't care about the *issues*. As a matter of fact, I don't understand most of the issues that I deal with. My job, and the job of the court, is to assure the integrity of the *process* by which decisions are made, and I interpret this to mean that all relevant facts and points of view are represented and debated—otherwise any decision will be faulty."

A few days later, I left Aspen to return to Harvard, where, as you remember, we have a number of highly visible (and often very accomplished) scientists who are, or profess to be, deeply concerned about the scientific issues and their impact on society. To me, their presentations to the press and other media have not been very helpful to the public, the political process or to Biological Science. Their arguments have been selective in the sense that they have been made to urge the adoption of a policy that they have selected in advance. Certainly, all the evidence and a variety of points of view are not being fully presented nor can they be in the public arena. Finally, and most importantly, these scientists, speaking for the media or in articles to the various general scientific journals seem strangely liberated from the normal restraints that prevail in written or oral scientific discourse. Whatever the reason for this, the net effect is one of *irresponsibility*, which further erodes the public esteem of scientists and of science in general.

The third component of the background for the proposal to be described shortly was the NIH Advisory Committee on Recombinant DNA Molecules, on which I served until last month. This group had the task of devising a policy (the Guidelines) as well as trying to obtain and understand the relevant facts relating to the possible biohazards of certain recombinant DNAs. All this took place in the presence of numerous reporters and others who represented various public interest groups. The proceedings were not happy ones: the scientific basis upon which the hazards were estimated were *imaginary* at best. Hand-waving arguments took the place of experiments and policy was devised to mollify the most vocal objectors and little consideration was given to the long-term effects of what we now can see will be a potent bureaucratic inhibition on the future development of science.

I began to think not of the issues, whether they be XYY, recombinant DNA, environmental mutagens, nuclear waste disposal, or any one of the hundreds of others that might be named, but of the *process*. What kind of *process* could one envisage that would allow, and indeed encourage, scientists to contribute constructively to public debate? Could one devise a game that could be played only on responsible terms, yet be open to all, the result of which would improve the ability of Americans to make free choices as well as enhance the political process by which we make public policy?

What might be some of the features of this process?

1) It must take place in private, away from the glare of the media, thereby focusing the attention of the participants on the subject at hand and not upon its public effect.

2) The scientific contributions must be subject to the normal constraints of any responsible scientific communication. After all, science is about verifiable knowledge and we have in existence a fully developed procedure to test the validity of scientific knowledge.

3) During the entire process, no comment on statement should be made regarding policy—either personal or public policy. Policy statements, even by indirection, have a way of preventing a full exploration of the evidence. Moreover, if they were allowed, the participants would be quickly accused of being self-serving and thereby vitiating the utility of the whole process. There is an even more compelling reason: our society is rooted in a conception of man, his rights and his liberties, that is probably much more important to the American future than anything else. Our objective is not to improve the health of future Americans, but to encourage them to develop a humane civilization. It is the character of future Americans that we are truly concerned with. In one form or another, we believe that people learn best by making decisions for themselves. Personal policy is an individual affair: decisions made and implemented by individual people. Public policy, in the purest sense, is made by the pluralistic political and economic processes that are already in place and functioning—perhaps imperfectly. Therefore, any handful, or even a few hundred, scientists should never advocate a policy based on their presumed special knowledge of a certain arcane science.

4) The process must have some *utility* in order to survive. The end result should improve the ability of Americans to make personal and public decisions that will affect their lives. This last feature suggests that the results of the process should be some kind of written document that would be freely available to all. The only problem is that scientific evidence is not freely available to all, because it requires years of special study to have a scientific competence in even a narrow area.

5) The document must be understandable to the reading public. How can one marry the requirements of *scientific integrity* with *understandability*?

A PROPOSAL—*The Human Environment*

The Harvard School of Public Health is seriously considering publishing a series of volumes entitled, "The Human Environment" (unless someone can come up with a better name). Each volume will be devoted to a single agent, toxin or practice that affects the health and well-being of Americans. Thus, for example, there might be volumes devoted to Red Dye #2, cyclamates, diethyl pyrocarbonate, nitrosamines, hexachloraphene, XYY, recombinant DNA, organic chlorides in the upper atmosphere, etc. Each of the volumes will consist of two parts, the first part consisting of scientific papers contributed by any and everyone who has primary scientific evidence to offer, the second of interpretations or translations of the foregoing scientific evidence by two or more invited interpreters. The scientific contributors will receive preprints of all other scientific contributions, and will be encouraged to comment on the validity of the various contributions, and some or all of this commentary can appear in their contributions. The number of contributors will be controlled by the definition of the topic—not by refusing a contribution. We desire *all* the evidence available to be exposed and recorded.

The interpreters will be allowed, by editorial policy, only to comment on the evidence submitted in written form that will appear in the first part of the volume. Should they desire to comment on facts not previously introduced into the evidentiary portion, then a new contribution must be sought, preprints circulated, etc., so that this firm policy may be maintained. Neither contributors nor interpreters may advocate any policy. The entire process will take place in private.

The result will be a published volume that should be an authoritative compilation of the evidence on isolated issues as of a given date. These volumes may be updated from time to time as new knowledge accumulates. These volumes should be useful to regulatory agencies, and to the various branches of the Federal government. They should be useful to the media and to ordinary people who have practical decisions to make (Should I allow my children to eat bacon? Should I abort the XYY fetus I carry? Should I vote for the relocation of a nuclear power facility?). Lastly, they should have a beneficial effect on the development of science in these areas. Just as the Cold Spring Harbor volumes have a way of focusing attention on incomplete evidence or erroneous interpretations, these volumes would do the same. In addition, they would serve as a highly visible volume in which to publish primary scientific papers which otherwise could be buried in little known reports. In this way, one might hope that the volumes might encourage and improve the quality of the applied science in these areas.

The Organization: Basically, the volumes are produced by the *editors* who recruit the contributors and select the interpreters. They apply, but do not make, editorial policy. This is made by the *editorial policy committee* which is directly sponsored by the Dean. Every year a third group called the *monitors* review the operation of the entire game and make recommendations to the *editorial policy committee.* At this point, we are beginning a small working seminar of five or so individuals who will be trying to select a suitable topic for the first volume. We are also trying to recruit our first editor, and attempting to obtain funding (which I would prefer to obtain from a variety of private sources).

CONCLUSION

The objective of this exercise is *not* to prevent cancer or improve health, although this would be a pleasing side effect, but rather to develop a process by which the scientific community can participate more constructively and responsibly in public affairs.

It is important to recognize the limitations of what we are trying to do. We should not be attempting to influence policy decisions, which, to repeat, are more properly to be made by others. It is my personal belief that scientists as a group have no more right to influence policy than any other group in America. What we can do, and should do, is to implement a more rational basis for policy decisions.

At the outset we must select issues that are not too controversial, because we are acutely aware of the fact that the format that has been tentatively adopted may not be the optimal one. We must learn by doing. If there is some merit to this plan, we can proceed to tackle more hotly debated problems—but there will be plenty of time for that.

Finally, there are more issues that need consideration than can ever be handled by one editorial group. The kind of publication that is outlined here could be sponsored by dozens of universities. But this is in the future. At this moment your ideas and support will be most welcome.